Lecture Notes in Computer Science

Commenced Publication in 1973
Founding and Former Series Editors:
Gerhard Goos, Juris Hartmanis, and Jan van Leeuwen

T0076397

Editorial Board

John S. Baras   Jonathan Katz
Eitan Altman (Eds.)

Decision and Game Theory for Security

Second International Conference, GameSec 2011
College Park, Maryland, USA, November 14-15, 2011
Proceedings

 Springer

Volume Editors

John S. Baras
University of Maryland
College Park, MD 20742, USA
E-mail: baras@umd.edu

Jonathan Katz
University of Maryland
College Park, MD 20742, USA
E-mail: jkatz@cs.umd.edu

Eitan Altman
INRIA
335 Chemin des Combes, 84140 Montfavet, France
E-mail: eitan.altman@sophia.inria.fr

ISSN 0302-9743   e-ISSN 1611-3349
ISBN 978-3-642-25279-2   e-ISBN 978-3-642-25280-8
DOI 10.1007/978-3-642-25280-8
Springer Heidelberg Dordrecht London New York

Library of Congress Control Number: 2011940677

CR Subject Classification (1998): C.2.0, J.1, D.4.6, K.4.4, K.6.5, H.2-3

LNCS Sublibrary: SL 4 – Security and Cryptology

Typesetting: Camera-ready by author, data conversion by Scientific Publishing Services, Chennai, India

Printed on acid-free paper

Springer is part of Springer Science+Business Media (www.springer.com)

Preface

Securing complex and networked systems has become increasingly important as these systems play an indispensable role in all aspects of modern life. Security, trust, authentication, and privacy of communications, data, and computing are critical for many applications and infrastructures, and their analysis and establishment pose novel and difficult challenges. These challenges are further exacerbated by the heterogeneity of communication networks, and by their distributed and asynchronous operation. Human, social, and economic factors play an important role in security and performance of such networked systems, and pose additional challenges that require innovative methodologies and at the same time challenge the foundations of conventional methods in computer science, mathematics, economics, and sociology. The investigation of security, trust, and privacy in such systems involves inference and decision making at multiple levels and time scales, given the limited and time-varying resources available to both malicious attackers and administrators defending these complex networked systems. Decision and game theory — in a broad sense — provides a rich and increasingly expanding arsenal of methods, approaches, and algorithms with which to address the novel resource allocation, inference, and decision-making problems arising in security, trust, and privacy of networked systems.

GameSec 2011, the Second Conference on Decision and Game Theory for Security, took place on the campus of the University of Maryland, College Park, during November 14–15, 2011, under the sponsorships of the Maryland Cybersecurity Center (MC^2) and other technical sponsors. GameSec brings together researchers who aim to establish a theoretical foundation for making resource-allocation decisions that balance available capabilities and perceived security risks in a principled manner. The conference focuses on analytical models based on game, information, communication, optimization, decision, and control theories that are applied to diverse security topics. At the same time, the connections between theoretical models and real-world security problems are emphasized to establish the important feedback loop between theory and practice. Given the scarcity of venues for researchers who try to develop a deeper theoretical understanding of the underlying incentive and resource allocation issues in security, GameSec aims to fill an important void and to serve as a distinguished forum.

This edited volume contains the summaries of the two plenary keynote addresses, and the 16 contributed full papers, presented at GameSec 2011. These 18 articles are categorized into the following seven sessions:

- "Plenary Keynotes" contains summaries of the two plenary keynote addresses, which present inspiring, visionary, and innovative ideas in game theory and its interplay with social and economic considerations within the context of security and trust in complex networked systems.

– "Attacks, Adversaries, and Game Theory" has two articles discussing game-theoretic approaches to intrusion-detection systems and the role of adversaries' risk profiles.
– "Wireless Adhoc and Sensor Networks" contains three articles, which investigate attacks and defense in infrastructureless wireless communication and sensor networks.
– "Network Games" has three articles focusing on analytical investigations of games related to security problems in networks.
– "Security Insurance" contains two articles on the new field of economic insurance considered as a component of the overall security infrastructure for complex networks and systems.
– "Security and Trust in Social Networks" has four articles investigating, analytically and experimentally, game-theoretic methods in the important area of social networks.
– "Security Investments" contains two articles investigating the value and effectiveness of investments for security mechanisms in the Internet.

Considering that inference and decision making for human–machine networked systems is still an emerging research area, we believe that this edited volume as well as the GameSec conferences will be of interest to both researchers and students who work in this challenging and multidisciplinary area.

November 2011   John Baras
                                                                   Jonathan Katz
                                                                    Eitan Altman

Organization

Steering Committee

Tansu Alpcan	Technical University of Berlin and T-Labs., Germany
Nick Bambos	Stanford University, USA
Tamer Başar	University of Illinois at Urbana-Champaign, USA
Anthony Ephremides	University of Maryland, College Park, USA
Jean-Pierre Hubaux	EPFL, Switzerland

Program Committee

General Chair	John Baras	University of Maryland, USA
TPC Co-chairs	Jonathan Katz	University of Maryland, USA
	Eitan Altman	INRIA, France
Publicity Chair	Sennur Ulukus	University of Maryland, USA
Publication Chair	Gang Qu	University of Maryland, USA
Finance Chair	Ion Matei	National Institute of Standards and Technology and University of Maryland, USA
Local Chair	Shah-An Yang	University of Maryland, USA
Secretary	Kimberly Edwards	University of Maryland, USA

Sponsoring Institutions

Gold Sponsors

Maryland Cybersecurity Center (MC^2)
Maryland Hybrid Networks Center (HyNet)
Lockheed Martin Chair in Systems Engineering

Silver Sponsors

Institute for Systems Research (ISR)

Technical Co-sponsors

IEEE Control System Society (IEEE CSS)
International Society of Dynamic Games (ISDG)
In cooperation with the ACM Special Interest Group on Security, Audit, and Control (SIGSAC)

Technical Program Committee

Tansu Alpcan	TU Berlin and T-Labs, Germany
Venkat Anantharam	University of California Berkeley, USA
Sonja Buchegger	KTH Stockholm, Sweden
Levente Buttyán	Budapest University of Technology and Economics, Hungary
Srdjan Capkun	ETH Zurich, Switzerland
Alvaro Cardenas	Fujitsu Labs of America, USA
Song Chong	KAIST, Republic of Korea
T. Charles Clancy	Virginia Tech, USA
Laura Cottatellucci	Eurecom, France
Merouane Debbah	SUPELEC, France
Andrey Garnaev	St. Petersburg State University, Russia
Jens Grossklags	Pennsylvania State University, USA
Joseph Halpern	Cornell University, USA
Sushil Jajodia	George Mason University, USA
Tao Jiang	Intelligent Automation Inc., USA
Arman Khouzani	University of Pennsylvania, USA
Iordanis Koutsopoulos	University of Thessaly, Greece
Richard La	University of Maryland, USA
Armand Makowski	University of Maryland, USA
Fabio Martignon	University of Bergamo, Italy
Pietro Michiardi	EURECOM, France
Ariel Orda	Technion, Israel
Manoj Panda	Indian Institute of Science, India
Radha Poovendran	University of Washington, USA
Balakrishna Prabhu	LAAS CNRS, France
Gang Qu	University of Maryland, USA
Alonso Silva	INRIA, France
Rajesh Sundaresan	Indian Institute of Science, India
Georgios Theodorakopoulos	EPFL, Switzerland
Wade Trappe	Rutgers University, USA
Tunca Tunay	Stanford University, USA
Kavitha Voleti Veeraruna	INRIA, France
Jean Walrand	University of California Berkeley, USA
Nan Zhang	George Washington University, USA

Table of Contents

Security Insurance

Security and Trust in Social Networks

Security Investments

Beyond Nash Equilibrium:
Solution Concepts for the 21st Century

Joseph Y. Halpern

Computer Science Department
Cornell University
Ithaca, NY 14853, USA
halpern@cs.cornell.edu

An often useful way to think of security is as a game between an adversary and the "good" participants in the protocol. Game theorists try to understand games in terms of *solution concepts*; essentially, this is a rule for predicting how the game will be played. The most commonly used solution concept in game theory is *Nash equilibrium*. Intuitively, a Nash equilibrium is a *strategy profile* (a collection of strategies, one for each player in the game) such that no player can do better by deviating. The intuition behind Nash equilibrium is that it represent a possible steady state of play. It is a fixed point where each player holds correct beliefs about what other players are doing, and plays a best response to those beliefs. Part of what makes Nash equilibrium so attractive is that in games where each player has only finitely many possible deterministic strategies, and we allow mixed (i.e., randomized) strategies, there is guaranteed to be a Nash equilibrium [11] (this was, in fact, the key result of Nash's thesis).

For quite a few games, thinking in terms of Nash equilibrium gives insight into what people do (there is a reason that game theory is taught in business schools!). However, as is' well known, Nash equilibrium suffers from numerous problems. For example, the Nash equilibrium in games such as repeated prisoner's dilemma is to always defect. It is hard to make a case that rational players "should" play the Nash equilibrium in this game when "irrational" players who cooperate for a while do much better! Moreover, in a game that is only played once, why should a Nash equilibrium arise when there are multiple Nash equilibria? Players have no way of knowing which one will be played. And even in games where there is a unique Nash equilibrium (like repeated prisoner's dilemma), how do players obtain correct beliefs about what other players are doing if the game is played only once? (See [10] for a discussion of some of these problems.)

I will argue that Nash equilibrium is particularly problematic when it comes to security. Not surprisingly, there has been a great deal of work in the economics community on developing alternative solution concepts. Various alternatives to and refinements of Nash equilibrium have been introduced, including, among many others, *rationalizability, sequential equilibrium, (trembling hand) perfect equilibrium, proper equilibrium,* and *iterated deletion of weakly dominated strategies.* (These notions are discussed in standard game theory texts, such as [4,12].) Despite some successes, none of these alternative solution concepts address the following four problems with Nash equilibrium, all quite relevant to security.

J.S. Baras, J. Katz, and E. Altman (Eds.): GameSec 2011, LNCS 7037, pp. 1–3, 2011.

– Although both computer science and distributed computing are concerned with multiple agents interacting, the focus in the game theory literature has been on the strategic concerns of agents—rational players choosing strategies that are best responses to strategies chosen by other player, the focus in distributed computing has been on problems such as fault tolerance and asynchrony, leading to, for example, work on Byzantine agreement [3,13]. Nash equilibrium does not deal with "faulty" or "unexpected" behavior, "irrational" players, or colluding agents. When dealing with security concerns, all these issues are critical.
– Nash equilibrium does not take computational concerns into account. We need solution concepts that can deal with resource-bounded players, concerns that are at the heart of cryptography.
– Nash equilibrium presumes that players have common knowledge of the structure of the game, including all the possible moves that can be made in every situation and all the players in game. But in security applications, sometimes the largest problems come from actions that were totally unanticipated, and not on anyone's radar screen beforhand (other than the attacker's!).
– Nash equilibrium presumes that players know what other players are doing (and are making a best response to it). But how do they gain this knowledge in a one-shot game, particularly if there are multiple equilibria?

The full version of this paper [5] has an overview of all the relevant issues and further references; more details can be found in [1,2,7,8,6,9].

References

1. Abraham, I., Dolev, D., Gonen, R., Halpern, J.Y.: Distributed computing meets game theory: robust mechanisms for rational secret sharing and multiparty computation. In: Proc. 25th ACM Symposium on Principles of Distributed Computing, pp. 53–62 (2006)
2. Abraham, I., Dolev, D., Halpern, J.Y.: Lower Bounds on Implementing Robust and Resilient Mediators. In: Canetti, R. (ed.) TCC 2008. LNCS, vol. 4948, pp. 302–319. Springer, Heidelberg (2008)
3. Fischer, M.J., Lynch, N.A., Paterson, M.S.: Impossibility of distributed consensus with one faulty processor. Journal of the ACM 32(2), 374–382 (1985)
4. Fudenberg, D., Tirole, J.: Game Theory. MIT Press, Cambridge (1991)
5. Halpern, J.Y.: Beyond Nash Equilibrium: solution concepts for the 21st century. Lectures in Game Theory for Computer Scientists, pp. 264–289. Cambridge University Press, Cambridge (2011); an earlier version of the paper can be found in the Proceedings of Twenty-Seventh Annual ACM Symposium on Principles of Distributed Computing, pp. 1–10 (2008) and the Proceedings of the Eleventh International Conference on Principles of Knowledge Representation and Reasoning (KR 2008), pp. 6–14 (2008)
6. Halpern, J.Y., Pass, R.: Iterated regret minimization: A more realistic solution concept. In: Proc. Twentieth International Joint Conference on Artificial Intelligence (IJCAI 2007), pp. 153–158 (2007); a longer version of the paper, with the title Iterated regret minimization: a new solution concept, will appear in Games and Economic Behavior

7. Halpern, J.Y., Pass, R.: Game theory with costly computation. In: Proc. First Symposium on Innovations in Computer Science (2010)

8. Halpern, J.Y., Pass, R.: I don't want to think about it now: Decision theory with costly computation. In: Principles of Knowledge Representation and Reasoning: Proc. Twelfth International Conference, KR 2010 (2010)

9. Halpern, J.Y., Rêgo, L.C.: Extensive games with possibly unaware players. In: Proc. Fifth International Joint Conference on Autonomous Agents and Multiagent Systems, pp. 744–751 (2006), full version available at **arxiv.org/abs/0704.2014**

10. Kreps, D.M.: Game Theory and Economic Modeling. Oxford University Press, Oxford (1990)

11. Nash, J.: Equilibrium points in n-person games. Proc. National Academy of Sciences 36, 48–49 (1950)

12. Osborne, M.J., Rubinstein, A.: A Course in Game Theory. MIT Press, Cambridge (1994)

13. Pease, M., Shostak, R., Lamport, L.: Reaching agreement in the presence of faults. Journal of the ACM 27(2), 228–234 (1980)

Network Security Games:
Combining Game Theory, Behavioral Economics,
and Network Measurements

Nicolas Christin

Carnegie Mellon University
Information Networking Institute and CyLab
Pittsburgh, PA 15123
nicolasc@cmu.edu

Computer and information networks are a prime example of an environment where negative externalities abound, particularly when it comes to implementing security defenses. A typical example is that of denial-of-service prevention: ingress filtering, where attack traffic gets discarded by routers close to the perpetrators, is in principle an excellent remedy, as it prevents harmful traffic not only from reaching the victims, but also from burdening the network situated between attacker and target. However, with ingress filtering, the entities (at the ingress) that have to invest in additional filtering are not the ones (at the egress) who mostly benefit from the investment, and, may not have any incentive to participate in the scheme. As this example illustrates, it is important to understand the incentives of the different participants to a network, so that we can design schemes or intervention mechanisms to re-align them with a desirable outcome.

Game theory offers a solid bedrock for formally assessing the incentives of non-cooperative participants. In this talk, I will start by discussing a framework for network security games [4,5] that we devised to help model how rational, individual, end-users would respond to security threats in large-scale networks. We decouple security decisions between self-insurance (which does not present any externalities) and self-protection (which does present externalities). Assuming fully rational players, acting with perfect information, and with the ability to perfectly execute their security decisions, we can derive results showing how much of a negative impact externalities can have on security decision-making. I will also introduce extensions of this work which deal with more limited information cases [6].

However, humans are not acting perfectly rationally when it comes to security decision-making. Prospect theory tells us that humans tend to be risk-averse when it comes to gains; and risk-seeking when it comes to losses [7]. In other words, people tend to "gamble" more than they should when it comes to security risks. I will further show, through an experiment related to our framework [3] that in addition to these biases, users have very limited "computational" ability; in particular, they seem unable to strategize over more than one decision variable at a time. I will present complementary experimental results [1] that suggest that Peltzman effects [11] also apply in computer security. Much like drivers wearing seat belts or helmets tend to drive faster, people tend to behave more insecurely online when they believe they have adopted secure precautions, such as installing an anti-virus scanner. As a result, I will postulate that game-theoretic modeling either needs to be complemented by behavioral analysis (for

J.S. Baras, J. Katz, and E. Altman (Eds.): GameSec 2011, LNCS 7037, pp. 4–6, 2011.

individual users) or is better suited to describing institutional users (e.g., corporations, governments, ISPs...).

In the second part of this presentation, I will make the case that to provide improved resilience to attacks, we must be simultaneously mindful of the capabilities of the attackers, as well as their own economic incentives. Indeed, since the early- to mid-2000's, attackers have become mostly profit-driven [9]. By primarily conditioning their actions on their best financial interest, attackers are more and more behaving rationally in the economic sense of the term, and are considerably more predictable than attackers driven by less mundane ideals. Trying to disrupt the economic incentives that drive attackers to commit their forfeits appears to be a defensive strategy worth investigating, as a complement to the technical approaches that have been proposed.

I will contend that modeling attacker behavior is easier than modeling defender behavior. First, attackers show much stronger economic rationality than defenders: the success of the attack directly conditions their profits, while for defenders, security precautions are often viewed as sunk costs. Second, attackers' actions are often publicly observable: attacks such as phishing, malware distribution or search-engine manipulation leave a visible footprint. I will present a couple of recent measurement studies we conducted [2, 8, 10] in an effort to acquire more information on attacker behavior, and will show that a priori disparate attacks all present *concentration points*. Specifically, very often, the number of actual perpetrators behind entire class of attacks (e.g., search engine manipulation) are small. This in turn helps us inform security games where we want to model attackers as players, rather than exogenous entities.

Finally, I will conclude by presenting a roadmap for future research integrating network measurements and formal, game theoretic, modeling.

References

1. Christin, N., Egelman, S., Vidas, T., Grossklags, J.: It's all about the Benjamins: Incentivizing users to ignore security advice. In: Proceedings of IFCA Financial Cryptography 2011, Saint Lucia (March 2011)
2. Christin, N., Yanagihara, S., Kamataki, K.: Dissecting one click frauds. In: Proc. ACM CCS 2010, Chicago, IL (October 2010)
3. Grossklags, J., Christin, N., Chuang, J.: Predicted and observed behavior in the weakest-link security game. In: Proceedings of the 2008 USENIX Workshop on Usability, Privacy and Security (UPSEC 2008), San Francisco, CA (April 2008)
4. Grossklags, J., Christin, N., Chuang, J.: Secure or insure? A game-theoretic analysis of information security games. In: Proceedings of the 2008 World Wide Web Conference (WWW 2008), Beijing, China, pp. 209–218 (April 2008)
5. Grossklags, J., Christin, N., Chuang, J.: Security and insurance management in networks with heterogeneous agents. In: Proceedings of the 9th ACM Conference on Electronic Commerce (EC 2008), Chicago, IL, pp. 160–169 (July 2008)
6. Johnson, B., Grossklags, J., Christin, N., Chuang, J.: Are Security Experts Useful? Bayesian Nash Equilibria for Network Security Games with Limited Information. In: Gritzalis, D., Preneel, B., Theoharidou, M. (eds.) ESORICS 2010. LNCS, vol. 6345, pp. 588–606. Springer, Heidelberg (2010)
7. Kahneman, D., Tversky, A.: Prospect theory: An analysis of decision under risk. Econometrica XLVII, 263–291 (1979)

8. Leontiadis, N., Moore, T., Christin, N.: Measuring and analyzing search-redirection attacks in the illicit online prescription drug trade. In: Proceedings of USENIX Security 2011, San Francisco, CA (August 2011)
9. Moore, T., Clayton, R., Anderson, R.: The economics of online crime. Journal of Economic Perspectives 23(3), 3–20 (2009)
10. Moore, T., Leontiadis, N., Christin, N.: Fashion crimes: Trending-term exploitation on the web. In: Proceedings of ACM CCS 2011, Chicago, IL (October 2011)
11. Peltzman, S.: The effects of automobile safety regulation. Journal of Political Economy 83(4), 677–726 (1975)

Indices of Power in Optimal IDS Default Configuration: Theory and Examples

Quanyan Zhu and Tamer Başar[*]

Coordinated Science Laboratory and
Department of Electrical and Computer Engineering,
University of Illinois at Urbana Champaign,
1308 W. Main St., Urbana, IL, USA, 61801
{zhu31,basar1}@illinois.edu

Abstract. Intrusion Detection Systems (IDSs) are becoming essential to protecting modern information infrastructures. The effectiveness of an IDS is directly related to the computational resources at its disposal. However, it is difficult to guarantee especially with an increasing demand of network capacity and rapid proliferation of attacks. On the other hand, modern intrusions often come as sequences of attacks to reach some pre-defined goals. It is therefore critical to identify the best default IDS configuration to attain the highest possible overall protection within a given resource budget. This paper proposes a game theory based solution to the problem of optimal signature-based IDS configuration under resource constraints. We apply the concepts of indices of power, namely, Shapley value and Banzhaf-Coleman index, from cooperative game theory to quantify the influence or contribution of libraries in an IDS with respect to given attack graphs. Such valuations take into consideration the knowledge on common attack graphs and experienced system attacks and are used to configure an IDS optimally at its default state by solving a knapsack optimization problem.

Keywords: Intrusion Detection Systems, IDS Configuration, Cooperative Games, Shapley Value, Banzhaf-Coleman Index.

1 Introduction

The issue of optimal IDS configuration and provisioning has always been difficult to deal with, mainly due to the overwhelming number of parameters to tune. IDSs are generally shipped with a number of attack detection libraries (also known as categories [13] or analyzers [12]) with a considerable set of configuration parameters. The current version of the Snort IDS [13], for example, has approximately 10,000 signature rules located in fifty categories. Each IDS also comes with a *default configuration* to use when no additional information or expertise is available. It is not trivial to determine the optimal configuration of an

---

[*] This work was supported in part by the U.S. Air Force Office of Scientific Research (AFOSR) under grant number AFOSR MURI FA9550-09-1-0249, and in part by Boeing Company through the Information Trust Institute of the University of Illinois.

J.S. Baras, J. Katz, and E. Altman (Eds.): GameSec 2011, LNCS 7037, pp. 7–21, 2011.

IDS because it is essential to understand the quantitative relationship between the wide range of analyzers and tuning parameters. This explains the reason why current IDSs are configured and tuned simply based on a trial-and-error approach. Although there have been recent approaches, such as in [15, 18, 20], to optimize IDS resource consumption, we still need to deal with resource constraints and make the best use of an IDS with available resource budgets. On the other hand, most of current computer attacks do not come in one shot but in several steps, by which attackers can acquire an increasing amount of knowledge and privileges to attack the target system. To describe such multi-stage behaviors, *attack graphs or trees* are commonly used as tools to model security vulnerabilities of a system and all possible sequences of exploits used by intruders.

In this paper, we develop a novel game theory based solution to the problem of optimal default signature-based IDS configuration under resource limitations. The solution considers the costs and functionalities of libraries and defender's knowledge on common attack graphs to configure an IDS optimally at its default state.

The contribution of this paper can be summarized as follows. We introduce the concept of *detectability* of an attack sequence with respect to a given set of IDS libraries and devise metrics to measure the detectability and the efficacy of detection. From a game theoretical perspective, we view a configuration as a coalition among libraries and apply the indices of power, namely, *Shapley value* and *Banzhaf-Coleman index*, to rank the overall importance of a library for the purpose of intrusion detection, which can be used in a knapsack problem for finding the optimal default configuration. In addition, we extend our results to general attack graphs based on multilinear extension and propose a scheme to approximate the indices of power when the number of libraries is large.

The rest of the paper is organized as follows. In the next section, we summarize some recent related work on IDS configuration and cooperative games. In Section 3, we define the important notion of detectability and establish a mathematical model for attackers and detectors. In Section 4, we formulate a cooperative game framework to evaluate the indices of power for a given attack sequence. In Section 5, we introduce multilinear extension as a general framework and an approximation technique to evaluate the indices of power. Finally, in Section 6, we conclude the paper.

2   Related Work

We find a recent growing literature on performance characterization of IDSs in the computer science community. Some of the related work is summarized as follows.

2.1   IDS Performance Evaluation

Gaffney et al. in [4] use a decision analysis that integrates and extends Receiver Operating Characteristics (ROCs) to provide an expected cost metric. They demonstrate that the optimal operation point of an IDS depends not only on the system's own ROC curve and quantities such as the expected rate of false positives, false negatives, and the cost of operation, but also on the degree of hostility of an environment in which the IDS is situated, such as the probability and the type of an intrusion. Hence, the performance evaluation of an IDS has to take into account both the defender's side and attacker's side.

In [23], a network security configuration problem is studied. A nonzero-sum stochastic game is formulated to capture the interactions among distributed intrusion detection systems in the network as well as their interactions against exogenous intruders. The authors have proposed the notion of security capacity as the largest achievable payoff to an agent at an equilibrium to yield performance limits on the network security, and a mathematical programming approach is used to characterize the equilibrium as well as the feasibility of a given security target.

Zhu and Başar in [22] use a zero-sum stochastic game to capture the dynamic behavior of the defender and the attacker. The transition between different system states depends on the actions taken by the attacker and the defender. The action of the defender at a given time instant is to choose a set of libraries as its configuration, whereas the action of the attacker is to choose an attack from a set of possible ones. The change of configuration from one instant to the next implies for the defender to either load new libraries or features to the configuration or unload part of the current ones. The actions taken by the attacker at different times constitute a sequence of attacks used by the attacker. An online Q-learning algorithm is used to learn the optimal defense response strategies for the defender based on the samples of outcomes from the game.

In this paper, we address the issues of optimal default configuration, which is complementary to the one addressed in [22]. We find an optimal configuration which can serve as an initial or starting profile for dynamic IDS configuration.

To identify important factors for the performance of an IDS is another crucial investigation. In [14], Schaelicke et al. observe several architectural and system parameters that contribute to the effectiveness of an IDS, such as operating system structure, main memory bandwidth and latency as well as the processor micro-architecture. Memory bandwidth and latency are identified as the most significant contributors to sustainable throughput. CPU power is important as well; however, it has been overlooked in the experiments due to the existence of other closely related architectural parameters, such as deep pipelining, level of parallelism, and caching.

In [2], the authors investigate the prediction of resource consumption based on traffic profile. An interesting result, which we assume to be available in this paper, is that both CPU and memory usage can be predicted with a model linear in the number of connections. Equally important is the confirmation that the factoring of IDS resource usage with per-analyzer and per-connection scaling is a reasonable assumption. The authors use this finding to build an analyzer selection and configuration tool that estimates resource consumption per analyzer to determine whether a given configuration is feasible or not. The constraint used is a target CPU load below which the load should remain for a predefined percentage of time. The actual selection of a feasible analyzer set is however left as a manual task for the IDS operator. In our work, we propose a more informed and automated way of IDS configuration decision that takes into account the resource utilization per IDS library as well as the expected intrusion context based on experienced attack sequences or graphs.

2.2 Attack Graphs

The generation of attack graphs has received considerable attention in the literature [6,7,9,17,16,19,21]. Sheyner et al. present in [16,17] a tool for automatically generating attack graphs and performing different kinds of formal vulnerability analysis on them. Attack graphs have also been used in intrusion containment. Foo et al. develop in [3] the ADEPTS intrusion containment system in the context of E-commerce environments. The system builds a graph of intrusion goals, localizes intrusions, and deploys responses at the appropriate services to allow the system to work with minimum overall performance degradation. The system takes into consideration the financial impact of an attack and derives response actions that go beyond the simple deactivation or isolation of the infected service/host by considering interaction effects among multiple components of the protected environment. Finally, in [10], attack graphs are used to derive optimal IDS placement in a network so as to minimize intrusion risk. The authors developed the TVA tool (Topological Vulnerability Analysis), which can be used to model a network and populate it with information regarding vulnerabilities. The tool is claimed to have the ability to avoid state-space explosion through attack graph reduction. In this paper, we assume that such knowledge of attack graphs is given or has been acquired previously through experience.

2.3 Game-Theoretical Methods

Game-theoretical methods appear to be an appropriate framework that connects the performance evaluation of an IDS with the attack sequences or graphs on the intruder's side. The concepts in cooperative game theory become natural to study the contribution of each IDS component to the attack sequence, especially when we view a configuration as a coalition among IDS libraries.

Cooperative game theory studies the outcome of a game when coalitions are allowed among multiple players. The concepts of the core and stable sets are regarded as solutions to $N-$person cooperative games. However, the lack of general existence theorem has led game theorists to look for other solution concepts. Currently, indices of power such as *Shapley value*, *Banzhaf-Coleman index* of power and their multilinear extensions have been widely used in a variety of literature involving resource allocation and estimation of power in a group of decision-making agents. In [11], examples are given on the application of indices of power in the analysis of presidential election games with a quantitative conclusion that voters in some states are assuming more power than voters in other states in the election. In [1], Shapley value is used to allocate profit in a multi-retailer and a single supplier cooperative game where players can form inventory-pooling coalition. In [5], an efficient measurement allocation in unattended ground sensor networks is suggested based on Shapley values. It is shown that by allocating measurements proportional to the Shapley value, the observability of localizing a target increases. A similar approach was also found to allocate unit start-up costs among electricity consumers, load and retailers.

It appears that there has been very little work on using game theoretical methods to study IDS configurations. Similar to the problems involving resource allocations and presidential elections, cooperative game theory lends itself naturally also to studying the relations among libraries in an adversarial environment.

3   Attacker and Detector Model

We let $\mathcal{L} = \{l_1, l_2, \cdots, l_N\}$ denote the set of a finite number of libraries and $\mathcal{L}^*$ denote the set of all the possible subsets of $\mathcal{L}$, with cardinality $|\mathcal{L}^*| = 2^N$. We let $F_i \in \mathcal{L}^*, i \in \{1, 2, \cdots, 2^N\}$ be a configuration set of libraries, which is a subset of $\mathcal{L}$. Each library has a cost associated with it, i.e., there is a mapping function $\mathcal{C} : \mathcal{L} \rightarrow \mathbb{R}_+$ that determines the cost of each library $c_i = \mathcal{C}(l_i)$. Assuming the cost of each library is independent of the others [2], we define the cost of a configuration F_i by $C_{F_i} = \mathcal{C}^*(F_i) = \sum_{x \in F_i} \mathcal{C}(x)$, where $\mathcal{C}^* : \mathcal{L}^* \rightarrow \mathbb{R}_+$ is a mapping function of configuration cost.

The attacker, on the other hand, has different types of attacks a_i. Let $a_i \in \mathcal{A}$ be a specific action of attack and $\mathcal{A} = \{a_1, a_2, \cdots, a_M\}$ be the set of possible attacks. We define a sequence of attacks S_i to be a tuple of elements of $\mathcal{A}$, and $\mathcal{A}^*$ be the set of all possible sequences of attacks. The order of the elements in S_i indicates a sequential strategy of an intrusion. Every attack $a_i \in \mathcal{A}$ incurs a damage d_i, given by the mapping function $\mathcal{D} : \mathcal{A} \rightarrow \mathbb{R}_+$, i.e., $d_i = \mathcal{D}(a_i), \forall a_i \in \mathcal{A}$. Assuming that the damage caused by a sequence of attacks does not depend on the order of the sequence and the damage by one attack is independent of other attacks, we define the damage caused by an attack sequence $S_i \in \mathcal{A}^*$ by $D_{S_i} = \mathcal{D}^*(S_i) = \sum_{x \in S_i} \mathcal{D}(x)$.

Each library l_i can only effectively detect certain attacks. We define the set $P_{l_i} \subset \mathcal{A}$ as its scope of detection. A library l_i is capable of detecting an attack a_i if and only if $a_i \in P_{l_i}$, otherwise the library l_i is sure to fail to detect. The definition of detectability of a library configuration follows from the scope of detection of each library.

Without losing generality, we can further assume that $\bigcap_{l_i \in \mathcal{L}} P_{l_i} = \emptyset$ because we can always define libraries to have functions that do not overlap with each other. This is particularly true in practice with signature-based libraries.

Definition 1. *An attack sequence S_i is detectable by a library configuration F_i if $S_i \subseteq T_i$, where $T_i := \cup_{l_k \in F_i} P_{l_k}$. An attack sequence S_i is undetectable by F_i if $S_i \subseteq \overline{T}_i$, where $\overline{T}_i := \mathcal{A} \setminus T_i$.*

Based on Definition 3, we can separate an attack sequence S_i into two separate subsequences S_i° and $S_i^\star$, where S_i° is undetectable and $S_i^\star$ is detectable. These

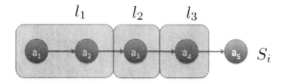

Fig. 1. A library configuration F_i that consists of libraries l_1, l_2, l_3 is used to detect an attack sequence $a_1 \rightarrow a_5$

two sequences satisfy the properties that they are mutually exclusive, i.e., $S_i^\circ \cup S_i^\star = S_i$ and $S_i^\circ \cap S_i^\star = \emptyset$.

An example is given in Fig. 1, where we have a sequence of five attacks a_1, a_2, a_3, a_4 and a_5. Detection library l_1 can be used to monitor a_1 and a_2 effectively whereas libraries l_2 and l_3 can only detect a_3 and a_4 respectively. However, a_5 is alien to the detection system and no library can be used to detect a_5. An IDS will rely on the successful detection of earlier known attack stages to prevent the last unknown one.

The definition of detectability assumes that each library can detect a certain signature or anomaly-based attack with success once it is loaded. However, due to many practical reasons such as delay and mutations of attacks, we can only successfully detect with some true positive (TP) rate. We use α_{ij}^P to denote the probability of successful detection of an attack $a_i \in \mathcal{A}$ using library l_j and by definition $\alpha_{ij}^P = 0$ for i not in P_{l_j}. The probability of undetected attacks when attacks occur, or the false negative (FN) rate, is thus given by $\alpha_{ij}^N = 1 - \alpha_{ij}$.

We also provide a metric that measures the detectability of an attack sequence S_i with respect to a configuration F_j, and the efficiency of detection for S_i.

Definition 2. *Let function $v : \mathcal{A}^* \rightarrow \mathbb{R}$ be a value function defined on attacker's set of sequences, satisfying*

$$v(A_1 \cup A_2) \leq v(A_1) + v(A_2), \tag{1}$$

where $A_1, A_2 \in \mathcal{A}^*$. Given a library component l_j, its coverage P_j, and an attack sequence S_i, we define detection effectiveness, $\eta_{ij} \in [0,1]$, as follows:

$$\eta_{ij} := \frac{v(S_i \cap P_j)}{v(S_i)}. \tag{2}$$

Remark 1. *One simple choice of v is the cardinality of a set, i.e., $v(S_i) = card(S_i)$. We can also have more complicated value functions; for example, we may have more weights on particular important attacks or final attacks.*

Given a configuration F which consists of a finite number of libraries, we use definition in (3) to define detectability of an IDS configuration with respect to an attack sequence S_i as follows.

$$\eta_i := \sum_{l_k \in F} \eta_{ik} = \frac{v(S_i \cap T)}{v(S_i)} = \frac{\sum_{l_k \in F} v(S_i \cap P_k)}{v(S_i)}. \tag{3}$$

Using the concepts of TP/FN, we can weight our definition of detectability in (3) by true positive rate α_{ij}^P. Thus we have the definition of weighted detectability as follows.

Definition 3. *Given a configuration F and attack sequence S_i, α^P-weighted detectability is defined as*

$$\eta_i^\alpha = \sum_{l_k \in F} \alpha_{ik}^P \eta_{ik}. \tag{4}$$

The notion of detectability shows the effectiveness of detection configuration F_j with respect to attack sequence S_i. We will later use detectability as a metric to optimize the performance of an IDS since higher detectability yields better detection results.

On the other hand, we can also describe efficiency of a detection using value function v.

Definition 4. *Given an attack sequence S_i and configuration F, we let ζ_i describe the efficiency of detection*

$$\zeta_i = \frac{v(S_i \cap T)}{v(T)} = \sum_{l_k \in F} \frac{v(S_i \cap P_k)}{v(T)}, \tag{5}$$

and let ζ_i^α denote the weighted detection efficiency given by

$$\zeta_i^\alpha = \sum_{l_k \in F} \alpha_{ik}^P \frac{v(S_i \cap P_k)}{v(T)}. \tag{6}$$

where T is a coverage of configuration F.

Proposition 5. *With (1) and the metrics η_i and ζ_i defined in (3) and (5) respectively, we have the following relation between these two metrics:*

$$\frac{1}{\eta_i} + \frac{1}{\zeta_i} \leq 1, \forall i \in \mathcal{A}^*. \tag{7}$$

Proof. Using the definitions in (3), (5) and (1), we arrive at

$$\frac{1}{\eta_i} + \frac{1}{\zeta_i} = \frac{v(S_i) + v(T)}{v(S_i \cap T)} \leq \frac{v(S_i \cup T)}{v(S_i \cap T)} \leq 1. \tag{8}$$

The inequality (7) provides a fundamental tradeoff relationship between detectability and efficiency.

4   Cooperative Game Model

In this section, we review essential concepts of indices of power and use them in the context of optimal default IDS configuration. We can view each possible configuration as a coalition of different libraries and hence each library can be associated with an index of power, signifying its contribution to the detection of intrusions.

We introduce the notion of ω-effective detection to quantity the goal of intrusion detection.

Definition 6. *We call a configuration F of an IDS $\omega-$effective for attack sequence S_i if the detectability does not fall below ω, that is $\eta_i \geq \omega$, and $(\omega, \alpha)-$effective if the weighted detectability does not fall below ω, that is $\eta_i^\alpha \geq \omega$.*

Parameter ω is a level of intrusion detection performance an IDS wants to achieve. We call a configuration goal achieving if it is (ω, α)-effective, and unsatisfactory otherwise.

4.1   Shapley Values and B-C Index

To describe an N-person cooperative game using game-theoretical language, we let $\mathcal{L}$ be the set of the players, and any subset of $\mathcal{L}$, or a configuration $F \in \mathcal{L}^*$, be a coalition. We let $f : \mathcal{L}^* \rightarrow \{0, 1\}$ be a characteristic function of the game having basic properties that

(P1) $f(\emptyset) = 0$;
(P2) $f(F_1 \cup F_2) \geq f(F_1) + f(F_2)$, $F_1, F_2 \in \mathcal{L}^*$ and $F_1 \cap F_2 = \emptyset$;
(P3) $f(\{i\}) = 0$, for all $i \in \mathcal{L}$.
(P4) $f(\mathcal{L}) = 1$.

By having the characteristic function f taking values 0 and 1 only, we have defined a *simple* game. The value 1 from the mapping f of a coalition or configuration means a winning or goal achieving library configuration, whereas the value 0 yields a non-winning or unsatisfactory library configuration.

A carrier of the cooperative game is a library which does not contribute to any configurations to attain the goal of detection. Mathematically, a carrier is a coalition, $F_c \in \mathcal{L}^*$, such that $f(F) = f(F \cap F_c)$, for all F. We can always remove from our library list those dummy libraries which do not contribute or disregard them in our cooperative game when they are found to be carriers.

The Shapley value of the i-th library l_i is given by ϕ_i, for all $l_i \in \mathcal{L}$.

$$\phi_i = \sum_{R \subset \mathcal{L}} \frac{(r-1)!(N-r)!}{N!} \left[f(R) - f(R - \{l_i\}) \right] \qquad (9)$$

The Shapley value ϕ_i given in (9) evaluates the contribution of each library toward achieving $\omega-$effective detection. Since the characteristic mapping f only takes value in 0 and 1, Shapley value can be further simplified into

$$\phi_i = \sum_{R' \subset \mathcal{L}} \frac{(r-1)!(N-r)!}{N!}, \qquad (10)$$

where, for a given l_i, R' is the winning coalition such that the configuration can achieve $\omega-$effectiveness with $l_i \in R'$, whereas $R' - \{l_i\}$ fails to achieve the goal. With a smaller scale problem, Shapley value is relatively easy to compute. However, in large problems, the evaluation of the weights can create computational overhead and the complexity increases exponentially with the size of the library. An easier index of power to compute is the Banzhaf-Coleman index of power, or B-C index, which depends on counting the number of swings, i.e., number of coalitions or configurations that wins when l_i is included but loses when is not.

Definition 7. *(B-C Index, [11]) The normalized Banzhaf-Coleman index $\beta_i, \forall l_i \in \mathcal{L}$ is given by*

$$\beta_i = \frac{\theta_i}{\sum_{j=1}^{N} \theta_j}, \tag{11}$$

where θ_i is the number of swings for l_i; a swing for $l_i \in \mathcal{L}$ is a set $R \subset \mathcal{L}$ such that R is a goal-achieving configuration if $l_i \in R$, and $R - \{l_i\}$ is not.

Shapley value and B-C index are closely related. They can both be evaluated by multilinear extension (see Section 5). The difference lies in the weighting coefficients used. In Shapley value, the weights are varied according to the coalition size, whereas in the B-C index, the weights are all equal.

4.2   An Example

Suppose we are given an attack sequence S_i as depicted in Fig. 1, where we have five attack actions ordered by $a_1 \rightarrow a_2 \rightarrow a_3 \rightarrow a_4 \rightarrow a_5$. There are 3 libraries and the sets $P_{l_i}, i = 1, 2, 3$ are given as follows: $P_{l_1} = \{l_1, l_2\}$, $P_{l_2} = \{l_3\}$, $P_{l_3} = \{l_4\}$. It is obvious that the sequence S_i can only be partially detected as a_5 is alien to the existing libraries of the IDS system. Suppose that each library has TP rates equal to 1 and v is the cardinality of the set. We can use Shapley value and B-C index to quantify the contribution to the detection of the sequence S_i. Let $\omega = 3/5$. The set of swings for l_1, l_2 and l_3 are $\{(l_1, l_2), (l_1, l_3), (l_1, l_2, l_3)\}, \{(l_1, l_2)\}$, and $\{(l_1, l_3)\}$, respectively. The Shapley values are thus given by $\phi_1 = \frac{1}{3}, \phi_2 = \frac{1}{6}$, and $\phi_3 = \frac{1}{6}$; and the B-C indices are thus $\beta_1 = \frac{3}{5}, \beta_2 = \frac{1}{5}$, and $\beta_3 = \frac{1}{5}$. To achieve $\omega = \frac{3}{5}$ level of detection, l_1 is most important and l_2 and l_3 are equally important. Therefore, in terms of the priority of loading libraries, l_1 should be placed first and then one should consider l_2 and l_3. Such evaluation via Shapley value and BC-index is useful for IDS system to assess the influence of each library and make decisions on which libraries to load initially when cost constraints are present.

5   Multiple Attack Sequences and Multinear Extension

In section 4, we introduced a cooperative game and proposed the concept of $\omega-$effectiveness to determine the winning or losing coalitions for a given known sequence. In this section, we extend this framework to deal with multiple cooperative games with respect to different sequences in an attack graph. We look at multilinear extensions in this section for two reasons: one is that it can be used to approximate the Shapley value when the library size grows; and the other reason is that it is a general framework that can yield B-C index.

5.1   Multilinear Extension (MLE)

A multilinear extension is a continuous function that can be used to evaluate Shapley value and B-C index as special cases. We let each library $l_i \in \mathcal{L}$ be associated with a continuous variable $x_i \in [0,1]$, and f_j be a characteristic function for detecting particular attack sequence $S_j \in \mathcal{A}^*$. We now introduce the multilinear extension for an attack sequence S_j, denoted by h_j, as follows:

Definition 8. *The multilinear extension of the cooperative game with characteristic function f_j is a function $h_j : [0,1]^N \to \mathbb{R}_{++}$ given by*

$$h_j(x_1, x_2, \cdots, x_N) = \sum_{R \subset \mathcal{L}} \left\{ \prod_{l_i \in R} x_i \prod_{l_i \in \mathcal{L} - R} (1 - x_i) \right\} f_j(R). \quad (12)$$

The function h_j can be used to evaluate Shapley's value by (13) below and B-C index by (14) below.

$$\phi_{ij} = \int_0^1 \left. \frac{\partial h_j(x_1, x_2, \cdots, x_N)}{\partial x_i} \right|_{x_1 = t, x_2 = t, \cdots, x_N = t} dt, \quad (13)$$

$$\beta_{ij} = \left. \frac{\partial h_j(x_1, x_2, \cdots, x_N)}{\partial x_i} \right|_{x_1 = \frac{1}{2}, x_2 = \frac{1}{2}, \cdots, x_N = \frac{1}{2}}, \quad (14)$$

where ϕ_{ij} and β_{ij} are the Shapley value and B-C index of library l_i for detecting sequence S_j, respectively. The set $\mathcal{M}$ is a subset of $\mathcal{A}^*$ that models a set of attacks known to detectors.

To aggregate the effect of a library of detecting a set of sequences $\mathcal{M} \subset \mathcal{A}^*$, we define an aggregated MLE $\bar{h}$ as a sum of MLEs over all the sequences, as follows

$$\bar{h} = \sum_{S_j \in \mathcal{M}} p_j h_j, \quad (15)$$

where p_j is a weight on h_j, indicating the frequency of occurrence of the attack sequence S_i. It is a normalized parameter that satisfies $p_j \geq 0$ and $\sum_{S_j \in \mathcal{M}} p_j = 1$.

Proposition 9. *The Shapley value ϕ_i for detecting multiple sequences is given by*

$$\phi_i = \sum_{S_j \in \mathcal{M}} p_j \phi_{ij}, \quad (16)$$

and B-C index β_i for detecting multiple sequences is given by

$$\beta_i = \sum_{S_j \in \mathcal{M}} p_j \beta_{ij} = \sum_{S_j \in \mathcal{M}} \left(\frac{p_j \theta_{ij}}{\sum_{k=1}^N \theta_{kj}} \right), \quad (17)$$

where θ_{ij} is the number swings for detecting sequence S_j.

Proof. The proposition can be proved using the linearity of MLE.

5.2 Multilinear Approximation

As is well known, the multilinear extension in (12) has a probabilistic interpretation and it can be used to approximate the indices of power when the number of libraries grows large. We can view the variable $x_i \in [0,1]$ as the probability of a library l_i in a random coalition $\mathcal{S} \subset \mathcal{L}$ such that $f_j(\mathcal{S}) = 1$ when $l_i \in \mathcal{S}$, and $f_j(\mathcal{S}) = 0$ otherwise. Since the event that a library is in a random coalition is independent from the event of other libraries in a coalition, we have that the probability of forming the random coalition $\mathcal{S}$ as a particular coalition R is given by $\mathbb{P}(\mathcal{S} = R) = \prod_{l_i \in R} x_i \prod_{l_i \in \mathcal{L} - R}(1 - x_i)$. The definition in (12) can be interpreted as the expectation of $f(\mathcal{S})$, i.e., $h_j(x_1, x_2, \cdots, x_N) = \mathbb{E}(f_j(\mathcal{S}))$.

Let Z_j be a random variable such that

$$Z_j = \begin{cases} \sum_{i \in \mathcal{M}} \eta_{ij} =: \eta_j^S, & \text{if } l_j \in \mathcal{S} \\ 0 & \text{if } l_j \in \mathcal{L} - \mathcal{S} \end{cases} \tag{18}$$

and let Y be another random variable defined by $Y = \sum_{l_j \in \mathcal{S}} \eta_j^S = \sum_{l_j \in \mathcal{S}, j \neq i} Z_j$. Since Z_j's are independent, Y has the mean and variance

$$\mu(Y) = \sum_{j \neq i, l_j \in S} \eta_j^S x_j, \tag{19}$$

$$\sigma^2(Y) = \sum_{j \neq i, l_j \in S} \eta_j^S x_j (1 - x_j). \tag{20}$$

Hence, $h_i(x_1, \cdots, x_N)$ is the probability that a coalition wins to be ω-effective with respect to a set of sequences $\mathcal{M}$ but loses if l_i is removed from the coalition. From the definition of ω-effectiveness, we can express h_i as

$$h_i(x_1, x_2, \cdots, x_N) = \mathbb{P}(\omega \leq Y \leq \omega + \eta_i^S) \tag{21}$$

When the size of the library grows, the random variable Y can be approximated by a normal random variable $\bar{Y}$, with mean and variance given in (19) and (20). Hence,

$$h_i(x_1, x_2, \cdots, x_N) = \mathbb{P}\left(\omega - \frac{1}{2} \leq \bar{Y} \leq \omega + \eta_i^S - \frac{1}{2}\right). \tag{22}$$

The Shapley value can thus be computed from (13) and (15) by $h_i(t, t, \cdots, t)$ with the random variable $\bar{Y}$ having the mean and variance $\mu(\bar{Y}) = t \sum_{j \neq i} \eta_j^S$ and $\sigma^2(\bar{Y}) = t(1 - t) \sum_{j \neq i} \eta_j^S$, respectively. The B-C value can be approximated by evaluating $h_i(t, t, \cdots, t)$ at $t = \frac{1}{2}$ using (22).

5.3 Optimal Default Configuration

The indices of power rank the importance of each library from high to low. We can make use of these indices to make a decision on which libraries to load when the system is subject to some cost constraint C_0. Toward that end, we arrive at an integer programming problem as follows:

$$\max_{\mathbf{z}} \quad \sum_{l_i \in \mathcal{L}} z_i \phi_i \tag{23}$$
$$\text{s.t.} \quad \sum_{l_i \in \mathcal{L}} z_i c_i \leq C_0$$
$$z_i \in \{0, 1\}, \forall l_i \in \mathcal{L}$$

We can use the B-C index as well in the objective function. The optimization problem (23) can be viewed as a knapsack problem [8]. The knapsack problem is well-known to be NP-complete. However, there is a pseudo-polynomial time algorithm using dynamic programming and a fully polynomial-time approximation scheme, which invokes the pseduo-polynomial time algorithm as a subroutine.

5.4   An Example

In this section, we continue with the example in Section 4.2, but with an extended attack tree depicted in Fig. 2. The libraries that can be used for detection are l_1, l_2, l_3 and l_4 whose coverages are $P_{l_1} = \{a_1, a_2\}, P_{l_2} = \{a_3, a_7\}, P_{l_3} = \{a_4, a_8\}$ and $P_{l_4} = \{a_6\}$, respectively. There are 4 known attack sequences in the attack tree. Let S_1 be the sequence $a_1 \rightarrow a_2 \rightarrow a_3 \rightarrow a_4 \rightarrow a_5$; S_2 denote $a_1 \rightarrow a_2 \rightarrow a_6$; S_3 denote $a_1 \rightarrow a_2 \rightarrow a_3 \rightarrow a_7$; and S_4 be $a_1 \rightarrow a_2 \rightarrow a_3 \rightarrow a_4 \rightarrow a_8$.

The Shapley values and B-C indices are summarized in the Table 1 and Table 2, respectively. Suppose each library is equally expensive with 1 unit per library. With the capacity constraint being 2 units, we can load library l_1 and l_2 to optimize the default library. This choice is intuitively plausible because l_1 and l_2 covers the major routes in the attack tree. l_4 does not contribute to the result of detection as much as other libraries, and when $\omega = \frac{3}{5}$, the impact of l_4 becomes negligible when l_1 is used. When the size of the tree grows, we need to evaluate indices of power in an automated fashion and use a polynomial-time algorithm to find the solution to the knapsack problem (23).

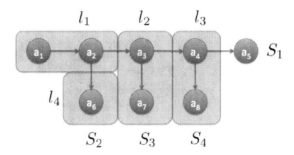

Fig. 2. Attack tree with attacks $a_j, j = 1, 2, \cdots, 8$, and libraries $l_i, i = 1, 2, 3, 4$, that are used to detect attacks

Table 1. Shapley value for the attack tree

Sequences	l_1	l_2	l_3	l_4
S_1	$\frac{1}{3}$	$\frac{1}{6}$	$\frac{1}{3}$	0
S_2	1	0	0	0
S_3	$\frac{1}{2}$	$\frac{1}{2}$	0	0
S_4	$\frac{1}{3}$	$\frac{1}{3}$	$\frac{1}{3}$	0

Table 2. B-C index for the attack tree

Sequences	l_1	l_2	l_3	l_4
S_1	$\frac{3}{5}$	$\frac{1}{5}$	$\frac{1}{5}$	0
S_2	1	0	0	0
S_3	$\frac{1}{2}$	$\frac{1}{2}$	0	0
S_4	$\frac{1}{3}$	$\frac{1}{3}$	$\frac{1}{3}$	0

6   Conclusion

In this paper, we have adopted a cooperative game approach to study the influence of each library when forming a configuration or a coalition to detect intrusions according to some known attack graphs. We have used the game approach to connect the detector and the attacker, and developed novel notions of detectability and efficacy of detection. The paper has described the applications of Shapley value and B-C index in a combinatorial knapsack optimization problem, which gives rise to an optimal configuration under the resource and cost constraints. The multilinear extension offers a technique to generalize the two indices discussed in the paper and, in addition, offers an approach to estimate these values when the number of libraries grows.

Acknowledgement. The first author acknowledges the fruitful discussions with members of the security research group at the University of Waterloo, in particular, Dr. I. Aib and Prof. R. Boutaba.

References

1. Bartholdi, J., Kemahlioglu-Ziya, E.: Using Shapley value to allocate savings in a supply chain. Supply Chain Optimization 98, 169–208 (2006)
2. Dreger, H., Feldmann, A., Paxson, V., Sommer, R.: Predicting the Resource Consumption of Network Intrusion Detection Systems. In: Lippmann, R., Kirda, E., Trachtenberg, A. (eds.) RAID 2008. LNCS, vol. 5230, pp. 135–154. Springer, Heidelberg (2008)

3. Foo, B., Wu, Y.S., Mao, Y.C., Bagchi, S., Spafford, E.: ADEPTS: adaptive intrusion response using attack graphs in an e-commerce environment. In: Proc. of International Conference on Dependable Systems and Networks (DSN), June 28-July 1, pp. 508–517 (2005)
4. Gaffney Jr., J.E., Ulvila, J.: Evaluation of intrusion detectors: a decision theory approach. In: Proc. of the IEEE Symposium on Security and Privacy (S&P), pp. 50–61 (2001)
5. Ghassemi, F., Krishnamurthy, V.: A cooperative game-theoretic measurement allocation algorithm for localization in unattended ground sensor networks. In: Proc. of International Conference on Information Fusion (2008)
6. Jajodia, S., Noel, S., O'Berry, B.: Topological analysis of network attack vulnerability. In: Proc. of the 2nd ASIAN ACM Symposium on Information, Computer and Communications Security, p. 2 (2007)
7. Lippmann, R.P., Ingols, K.: An annotated review of past papers on attack graphs. Tech. rep. MIT (March 31, 2005)
8. Martello, S., Paolo, T.: Knapsack Problems: Algorithms and Computer Implementations. John Wiley and Sons (1990)
9. Mehta, V., Bartzis, C., Zhu, H., Clarke, E., Wing, J.: Ranking Attack Graphs. In: Zamboni, D., Krügel, C. (eds.) RAID 2006. LNCS, vol. 4219, pp. 127–144. Springer, Heidelberg (2006)
10. Noel, S., Jajodia, S.: Optimal IDS sensor placement and alert prioritization using attack graphs. J. Netw. Syst. Manage. 16(3), 259–275 (2008)
11. Owen, G.: Game Theory, 3rd edn. Academic Press (1995)
12. Paxson, V.: Bro: a system for detecting network intruders in real-time. In: Proc. of the 7th Conference on USENIX Security Symposium (1998)
13. Roesch, M.: Snort–lightweight intrusion detection for networks. In: Proc. of the 13th Large Systems Administration Conference, LISA 1999 (1999)
14. Schaelicke, L., Slabach, T., Moore, B., Freeland, C.: Characterizing the Performance of Network Intrusion Detection Sensors. In: Vigna, G., Krügel, C., Jonsson, E. (eds.) RAID 2003. LNCS, vol. 2820, pp. 155–172. Springer, Heidelberg (2003)
15. Schear, N., Albrecht, D.R., Borisov, N.: High-speed Matching of Vulnerability Signatures. In: Lippmann, R., Kirda, E., Trachtenberg, A. (eds.) RAID 2008. LNCS, vol. 5230, pp. 155–174. Springer, Heidelberg (2008)
16. Sheyner, O., Haines, J., Jha, S., Lippmann, R., Wing, J.M.: Automated generation and analysis of attack graphs. In: Proc. of IEEE Symposium on Security and Privacy, pp. 273–284 (2002)
17. Sheyner, O.: Tools for Generating and Analyzing Attack Graphs. In: de Boer, F.S., Bonsangue, M.M., Graf, S., de Roever, W.-P. (eds.) FMCO 2003. LNCS, vol. 3188, pp. 344–371. Springer, Heidelberg (2004)
18. Sinha, S., Jahanian, F., Patel, J.M.: WIND: Workload-Aware Intrusion Detection. In: Zamboni, D., Krügel, C. (eds.) RAID 2006. LNCS, vol. 4219, pp. 290–310. Springer, Heidelberg (2006)
19. Swiler, L.P., Phillips, C., Ellis, D., Chakerian, S.: Computer-attack graph generation tool. In: Proc. of DARPA Information Survivability Conference & Exposition II, DISCEX 2001, vol. 2 (2001)
20. Vasiliadis, G., Antonatos, S., Polychronakis, M., Markatos, E.P., Ioannidis, S.: Gnort: High Performance Network Intrusion Detection Using Graphics Processors. In: Lippmann, R., Kirda, E., Trachtenberg, A. (eds.) RAID 2008. LNCS, vol. 5230, pp. 116–134. Springer, Heidelberg (2008)

21. Wang, L., Noel, S., Jajodia, S.: Minimum-cost network hardening using attack graphs. Comput. Commun. 29(18), 3812–3824 (2006)
22. Zhu, Q., Başar, T.: Dynamic policy-based IDS configuration. In: Proc. of the 48th IEEE Conference on Decision and Control (CDC), held jointly with the 2009 28th Chinese Control Conference (CCC), pp. 8600–8605 (December 2009)
23. Zhu, Q., Tembine, H., Başar, T.: Network security configurations: A nonzero-sum stochastic game approach. In: Proc. of American Control Conference (ACC), June 30-July 2, pp. 1059–1064 (2010)

Exploiting Adversary's Risk Profiles in Imperfect Information Security Games

Gabriel Fortunato Stocco and George Cybenko

Thayer School of Engineering at Dartmouth
{gfs,gvc}@dartmouth.edu

Abstract. At present much of the research which proposes to provide solutions to Imperfect Information Non-Cooperative games provides superficial analysis which then requires a priori knowledge of the game to be played. We propose that High Card, a simple Multiplayer Imperfect Information Adversarial game, provides a more robust model for such games, and further, that these games may model situations of real world security and international interest. We have formulated two such real world models, and have created a modeling bot, which when facing adversaries with equal or better performing risk profiles, achieves a 7-fold increase in win performance.

Keywords: High Card, Imperfect Information, Adversarial Game Theory.

1   Introduction

The field of research for imperfect information non-cooperative games can be said to comprise a number of areas of such research. To begin, we discuss two such areas - Poker research, and normal form simultaneous move games, in order to lay the groundwork for where our research fits into the field. Poker research, such as in Billings[3][9], Papp[16], focuses on creating AI or bot programs which play a game of poker - generally Texas Hold'em - based on some expert knowledge of the game of poker. These bots will sometimes perform in-depth opponent modeling, but generally their opponent models are rather simple in nature. However, even in the strongest case, this research focuses solely on the game of poker, and does not suggest that the results can be used as a model for other games or real world scenarios, such as security or international interest applications.

The second pertinent area of research is research that involves studies of specific normal form synchronous move games, such as Lye et al.[10] and Jiang et al.[11][15], which propose to solve a broader problem for which these games serve as a model. Unfortunately, these normal form games results are only theoretical as they require that we know a priori the payoffs for the game. The games are 'solved' to find the Nash Equilibrium. Unfortunately, imperfect information games are complex, and real world players don't play according to "optimal" strategies as they are far to difficult to compute, even, in many cases, in a theoretical sense. In addition, these results do not provide any additional knowledge about the nature of non-cooperative game theory as a field.

J.S. Baras, J. Katz, and E. Altman (Eds.): GameSec 2011, LNCS 7037, pp. 22–33, 2011.

We propose that High Card, a multi-player adaptation of von Neumann's betting game, similar to a single betting round in Poker, can serve as a model for security applications. High Card does not require that we assume a priori knowledge of payoffs, and further, it allows for multiplayer play. We will show that we have created a bot that creates opponent models based on past opponent play in order to estimate the secret information that adversaries have, and exploit those opponents to win additional resources.

2   High Card

Poker is a rather complex game to model. Much of the complexity of the game adds significant requirement for expert knowledge, but does not provide a benefit for use in modeling real world scenarios. As such, it is not uncommon to study a simplified version of the poker game, which preserves the basic elements of the game. Borel and von Neumann each simplify poker to a two-player zero-sum game, where instead of a hand of cards, each player is dealt a hand $X \in [1, S]$.[1] High Card is essentially a multi-player adaptation of such simplified poker models. In particular, the simplifications we have made to the poker game are a subset of the simplifications made for the von Neumann poker model, namely that this is a multiplayer, unlimited bet sequence game rather than a two-player limited bet sequence game.[2] These simplifications drastically reduce the complexity. For example, in a game of Texas Hold'em with n opponents there are $H = \prod_{k=1}^{n} \binom{52-2k}{2} \div k!$ possible starting hands. For 6 opponents, this works out to over 1 quadrillion combinations of starting hands. In High Card, by contrast, for the same 6 opponent game there are only 13 billion starting hand combinations. In addition, Texas Hold'em has additional added complexities by the nature of its multiple betting rounds, whereas High Card only has one.

2.1   Game Parameters

Upon starting the game, a player is selected at random to be the "Dealer" and each player is issued a set, equal amount of chips. The player to the dealer's right is the "Big Blind", which also refers to the size of the ante the player in the Big Blind seat makes. The player to the right of the Big Blind is the "Small Blind", and antes an amount of chips less than or equal to the amount that the Big Blind wagers, traditionally half of the Big Blind wager. Any chips wagered are immediately added to the "Pot", which the winning player receives in its entirety.

2.2   A Round of Play

At the beginning of each round the Blinds post their antes simultaneously. Betting then begins with the Dealer and proceeds to the left. Play continues until only M players remain. M is determined in advance.

Player take actions in turn, left around the table. When it is their turn a player may:

- Fold: abandon any claim to the Pot.
- Call: wager an amount of chips such that their total wager is equal to the largest wager made thusfar.
- Raise: wager an amount of chips larger than the largest wager so far by at least the Big Blind.
- Check: call a current additional wager of zero chips.

As a simple example, we have a game with a Big Blind of size 20, and a Small Blind of size 10, and each player starting with 50 chips. Players A, B and C sit around a table, B to A's left, and C to B's left. Player A puts in the Small Blind, Player B puts in the Big Blind. Player C chooses to Call the current wager of size 20, as set by B's Big Blind. Player A then chooses to Raise the wager to 50, requiring that A put in 40 additional chips. Player B decides to Call Player A's wager of 50, putting in 30 additional chips. Player C folds. There are now 120 chips in the Pot, and whichever of Player A or B has more chips will receive all 120 chips. The other player will be left with 0, and C will remain with 30.

2.3   Sidepots

If a player wagers an amount of chips larger than the amount of chips possessed by any other player who has not folded in the current round, a sidepot is created. If there are a number of players with chips smaller than the amount of chips wagered by a player, then more than one sidepot may need to be created. Each player is automatically a party to any sidepots which they are able to participate in. Once a sidepot is created, any raises will go into that sidepot, or if a new sidepot is required, into the new sidepot.

For example, Player A has 70 chips, Player B 50 and Player C 20. Player A wagers 70 chips. This puts 20 chips - because C only has 20 chips - into the Main Pot which all players are a party. 30 chips go into Side Pot 1 - because B only has 50 chips, 20 of which will go into the Main Pot should B call - to which Players A and B are Parties. The remaining 20 chips into Side Pot 2 which only Player A is a party to, because no other player has enough chips to participate in Side Pot 2. While it would not be realistic that A would wager chips into a Pot that only A could win, it does not effect the results in any way, as A is guaranteed to receive those chips back. Players must meet the bet in any sidepot to which they are a party in order to not fold, and Players may only win chips from Pots to which they are a party. Assume that then from the previous example, both B and C call Player A's wager. If Player A has a 30, Player B a 40 and Player C a 50, Player C will win 60 (20 + 20 + 20) chips from the Main Pot, Player B will win 60 (30 + 30) chips from Side Pot 1 and Player A will win (by default) his own 20 chips from Side Pot 2.

2.4   Round Resolution

When each player who has not folded, or caused the last raise or ante, has called or checked in succession since the last raise or ante, the betting is over. At that point the player with the best card wins the entirety of any pots to which they are a party. At the end of each round the Dealer, Big Blind and Small Blind positions pass one player to the left. Note, unlike in Poker where there are multiple rounds of betting in each hand, in High Card each hand only has one round of betting.

2.5   Rules in a Nutshell

At the start of each hand two players are forced to make an ante, and each player is dealt a card. Betting proceeds until all players have matched the same bet or folded out of the hand. The player with the highest card wins all the chips wagered. Which players are forced to make the ante changes, and play continues with a new hand until a predetermined number of players remain.

3   Examples

In this section in order to demonstrate the wider potential use of the model simulation, we propose a number of possible situations that could feasibly be modeled by High Card.

3.1   Diplomacy

We find that there is existing literature in which Poker is used as a metaphor for diplomatic relations. In particular, in Smith et al., no-limit poker in particular is used as a metaphor for North Korean - United States relations [18] and in Freeman it is used as a metaphor for Cold War nuclear disarmament discussions [19].

Such comparisons are apt because diplomatic relations constantly involve quarrels between countries often with outside actors intervening occasionally escalating to very high stakes. Focusing on these quarrels in particular, we propose that these relations can be modeled using a game of High Card. In particular, during such quarrels, it may be the case that a number of countries which possess damaging information about the other countries will be willing to risk some of their own credibility or resources in order to attempt to extract resources or credibility from the other countries. The end goal of course is that by doing so they will increase their own power in future negotiations.

We propose there exists a model for diplomatic relations in which, at each time period t two diplomatic adversaries have a small disagreement which forces them to risk some credibility. We assume that other actors may become involved in these diplomatic relations, each in turn deciding if they want to wager some credibility. During each actor's turn they may take one of three actions - matching the amount of credibility wagered by an opponent, withdrawing from the confrontation thus leaving behind any credibility that they have wagered, or raising the stakes and so by wagering additional credibility.

At the end of wagering the agent with the best secret information is able to gain diplomatic leverage over the other agents and by doing so use that information to take the credibility that was wagered on the quarrel.

A particular example of a diplomatic situation occurring right now in which such a model seems potentially useful is the United States attack on Osama bin Ladin's compound in Pakistan. This action has caused a diplomatic quarrel between the two countries, with other countries intervening, and each country holding secret information about the other. It is suspected that the United States holds damaging information about Pakistan related to Osama bin Ladin's un-detected life in the suburbs [4], and Pakistan has arrested five people who they claim are CIA informants [5]. At this point we are seeing the countries throwing in their wagers, but no one yet knows what cards they each truly hold.

Even better, the 1960 U-2 incident between the United States and the Soviet Union can be seen as a two-player betting game in action. In this example, the United States triggers the diplomatic incident, sending a spy plane into Soviet airspace, the Soviets then shoot the plane down. The United States, bets that it can win the confrontation, and puts additional credibility on the line, stating that the plane is a NASA weather plane which mistakenly drifted into Soviet airspace. Unbeknownst to the United States of course, the Soviets had captured the plane mostly intact and the pilot alive. At this point they reveal their secret information, and claim the credibility that the U.S. had staked upon the incident, embarrassing President Eisenhower at multiparty talks with Great Britain and France.[8]

3.2   Computer Security

In a model of Computer Security as a game of High Card each play is akin to a country competing against other countries to protect their own resources, and attempting to obtain the opponent resources. Each hand is akin to the 'big blind' being the defender, and the 'small blind' launching an initial cyber attack against their systems. We assume then that there is public information about these attacks, or that the attacks are frequent enough so as to be constant, and thus the players always able to participate. This is a reasonable assumption based on multiple statements from government officials, as well as actual security incidents that cyber attacks are a real, constant, ongoing threat[12][13].

The game proceeds as a normal game of High Card with each player who is not the Big Blind being presented the opportunity to participate in the attack. Whoever has the most sophisticated attack (best secret information) is then able to obtain resources. If they are defending, the resources they gain can be con-sidered to be techniques that the attackers can no longer use. For the attackers, the resources are information about attack techniques the other attackers have used, as well as whatever resources (data, operational security) the defender was attempting to protect.

4   Simulation

We created a simulation of a High Card game using bot players whose actions were determined by various utility functions. At the beginning of each round the players are allocated an equal number of chips, here 1000, and seated at random positions around a table. The Big Blind was set to 20, and the Small Blind 10. We ran simulations with M set to either 1 or 2, for 6 player games. Each round uses a shuffled complete 52 card deck drawn without replacement.

4.1 Probability of Winning

In High Card the probability of a player winning a game of equals the probability that every other player at the table was dealt a card of a lower value than the card that the player was dealt.

$$P_{win} = \frac{\binom{S-C}{0}\binom{C-1}{N}}{\binom{S}{N}} = \frac{\binom{C-1}{N}}{\binom{S}{N}}$$

Given the total number of cards remaining, S, the rank of your card, C, and the number of other players in the game, N, we propose that it is a reasonable assumption that the players who have folded at the time a player is taking an action had cards of a rank lower than the rank of that player's card. Call the number of players who have folded F, thus:

$$P_{win} = \frac{\binom{C-1-F}{N-F}}{\binom{S-F}{N-F}}$$

A plot of the probability of winning given the number of other players in the game, and the card you have assuming F=0 can be seen in Fig. 1.

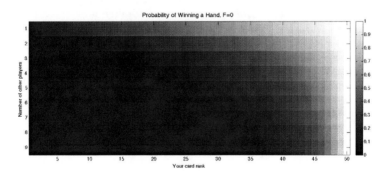

Fig. 1. A heatmap for probability of winning, given your card rank and the number of players in the game. As shown, as N increases at a fixed C, the probability of winning decreases, but your potential profit could still be increasing.

4.2 Utility Functions

We define $U_i(P, B, W)$ as the expected utility of player i betting on pot P with size of the bet to be made B, and the estimated probability to win W.

These bots use a number of different utility functions including:

- linear: $U(P, B, W) = W * P - (1 - W) * B$
- superlinear: $U(P, B, W) = W * \frac{P^{1+\rho}}{1+\rho} - (1 - W) * \frac{B^{1+\rho}}{1+\rho}$
- sublinear: $U(P, B, W) = W * \frac{P^{1-\rho}}{1-\rho} - (1 - W) * \frac{B^{1-\rho}}{1-\rho}$
- prospect[14]: $U(P, B, W) = \frac{(W*P)^{1-\rho_a}}{1-\rho_a} - \frac{((1-W)*B)^{1-\rho_b}}{1-\rho_b}$
- cumulative prospect[17]: $U(P, B, W) = \frac{(f(W)*P)^{1-\rho_a}}{1-\rho_a} - \frac{(f(1-W)*B)^{1-\rho_b}}{1-\rho_b}$

Linear. This utility function is the most obvious as there is no change value of a chip you expect to lose or a chip you expect to gain. In short, with this utility function you would be willing to wager exactly 50 chips for a 50% shot at 100 chips.

Superlinear. This utility function will overvalue large payoffs and bets. Thus you should be willing to bet more than 50 chips for a 50% shot at 100 chips.

Sublinear. This utility function undervalues large payoffs and bets. Thus you would only be willing to bet less than 50 chips for a 50% shot at 100 chips.

Prospect Utility. Prospect Utility is a social science theory which proposes that human actors will value possible gains differently than potential losses. In particular they will be very adverse to losing large amounts of money, whereas gains of large amounts of money are not that different than small amounts of money.

Selection. We determined that the Prospect Utility bot was the best choice for a number of reasons. In particular, the prospect utility bot was desirable due to its background as a model for actual human behavior as it relates to decision making under risk[14]. This is desirable in order to demonstrate that our results can be applied to real world games, in particular for quickly creating models of opponent behavior based on samples of real world data, estimating their risk curves, simulating possible outcomes, and in the process refining your own strategy.

In testing even weak versions of the prospect utility bot defeated the linear bot, as well as the super and sublinear bots for many different ρ values. The results of simulating 5 bots types against one another can be seen in Table 1.

Table 1. Win rates of the 5 Different bot Types 1-game Iterations with 5 players per game. The ProspectUtility bot was by far the strongest.

BotName	ρ_a	ρ_b	$WinRate_{M=1}$
PBot	0.8	0.1	0.69749
ExpBot	0.2		0.01709
LogBot	0.4		0.02312
LinearBot			0.01910
CumulativePBot	0.8	0.1	0.24322

Prospect Theory Utility Tuning. After evaluating the alternative algorithms and selecting the prospect utility bot, we tested different versions of the bot using various parameters varying from 0.1 to 0.9 for each ρ_a and ρ_b. To prevent restricting the field of bots to only a single utility function we determined that the top 10 bots had roughly $0.6 \leq \rho_a \leq 0.9$ and $0.1 \leq \rho_b \leq 0.3$ and at most a 20% deviation in win percentages. We mitigated any restriction to too narrow a field of bots by drawing the field of bot candidates uniformly on ρ_a and ρ_b over those ranges, in units of 0.1. Fig. 2 shows an example of a prospect utility curve. It appears that being strongly risk averse in specific is a strong strategy. That is undervaluing possible gains and overvaluing possible losses, rather than just under or overvaluing gains or losses using the same function as the super or sublinear utility functions do, provides a much stronger strategy in this game.

In particular, the no-limit nature leads the linear or superlinear bots to bet too aggressively, thus allowing them to push around the more conservative players for small amounts, ultimately causing them to be eliminated when a more conservative player gets the card they were waiting for. Since the game is multiplayer and the players have finite resources winning the overall game is much more important than winning the hand. We find that this too is applicable to

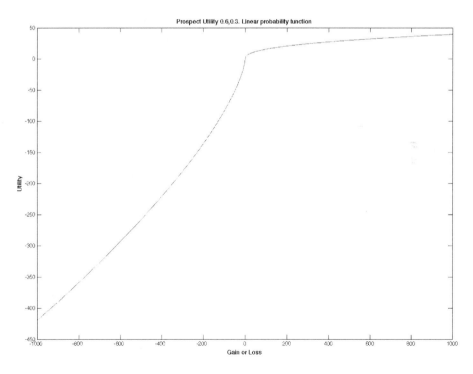

Fig. 2. The utility curve for potential gains (greater than zero) and losses (less than zero) used by a prospect utility bot with parameters 0.6,0.3. As you can see the prospect utility bots are highly risk-adverse.

real world scenarios. Actors in the games we have proposed would be loathe to wager all of their resources and risk being unable to participate in future rounds unless their success was all but guaranteed or they were on the verge of being eliminated anyway.

4.3   Non-modeling Bots

The bots that do not model behavior have their actions determined by a utility function generated when the bots are added to the game at the start of the round . These bots choose an action by maximizing the size of bet which returns positive utility based on their utility function:

$$\max_B(U(P, B, W) > 0)$$

We choose to do this rather than simply maximizing the utility of the players because simple maximization of utility would imply that the bots always wish to make the smallest bet possible which would allow them to win the pot. However, since there are other players in the game this is not a situation like a lottery, in which the player is presented with the option to spend an amount, B, to be guaranteed the chance, W, to win the pot, P. Instead, they must play against other players. Each bot, by maximizing the bet they will make are able to force out other players who may be more risk averse, or, if other players choose to match the larger bet, they are able to take more resources from those players should they win. Since the expected utility in such situations is still positive, it should be preferable in such a situation to actually make the largest bet for which the player expects a positive utility, rather than the smallest. The desire of the bots to make these larger bets is tempered by the fact that the bots are strongly risk averse, and they will rarely make huge bets. In this case there will often be actual back and forth play, whereas when they are always trying to minimize their bet, the game itself is not very interactive, with most bots very rarely participating if they are at all risk averse. A strategy similar to this is used by professional poker players called value-sizing [6][7].

4.4   Modeling Bot

In order to compete against the fixed strategy prospect utility bots we created a bot that would modify its approximation of the chances of winning a hand of High Card based on observed past behaviors of the players which it is playing against. This bot began each round with no information stored. As each hand was played, and a players hand was revealed at the end of a round (due to competing for winning the round), the bot stored the information of that player's play, indexed on the parameters at the time the action was taken, e.g. the pot size, required bet, number of players in the game, and the number of players who folded, and the card that the player had when they played their action in response to the action parameters.

Our bot played with the suboptimal, but highly conservative Prospect Utility strategy with parameters 0.6, 0.3. However, unlike the other bots who evaluate their chance of winning based only on the hypergeometric estimate of their chance of winning based only the card they possess; our modeling bot estimated its chance to win based on the historical play data it had for players who had previously played the same action which they had played this round.

The probability that the modeling bot (MB) will win is, as shown in Fig. 3, the product of the estimated probability that MB will beat each opponent, assuming that its probability of beating itself is 100%. If MB sees that there is an opponent who it has data for who only has shown higher cards in the same situations, it has no chance of winning. If MB has some information, MB estimates its probability as being a function of the range of those cards, as if they were uniformly distributed. If MB has no information, MB uses an approximation to the hypergeometric distribution for the remaining players. For example, assuming that the bot has a card ranked 40 and that there are two players for whom MB had no information for, MB would then use the equation in Fig. 4. It can treat these quantities as equal because for the size of the sample set we are drawing from and the number of items selected, selecting without replacement doesn't strongly effect the probabilities.

4.5   Results

Our simulation results consist of 2 sets of 1000 1-game tournaments. In these tournaments the modeling bot only has the information about hands played in the current game when modeling opponents. As depicted in the results in Table 2, the modeling bot achieves roughly a 7 fold increase in performance for

$$W = \prod_{i=0}^{N} P(C_{mod} > C_i)$$

$$P(C_{mod} > C_i) = \begin{cases} \dfrac{\binom{C_{mod}-1}{1}}{\binom{S}{1}} & \text{if } max_i = \emptyset \\ 1 & \text{if } C_{mod} = C_i \text{ or } C_{mod} > max_i \\ \dfrac{C_{mod}-min_i}{max_i-min_i-1} & \text{if } C_{mod} < max_i \text{ and } C_{mod} > min_i \\ 0 & \text{if } C_{mod} < min_i \end{cases}$$

Fig. 3. Probability Estimation Function for the Modeling Bot

$$\left(\frac{\binom{40}{1}}{\binom{51}{1}}\right)^2 = 0.6151 \approx \frac{\binom{40}{2}}{\binom{51}{2}} = 0.6118$$

Fig. 4. The hypergeometric can be approximated by calculating the hypergeometric for selecting one item and raising it to the power of the number of items you wish to select

Table 2. Win rates of the 12 Different bot Types and our ModelBot, 1-game Iterations with 6 players per game

BotName	ρ_a	ρ_b	$WinRate_{M=2}$	$WinRate_{M=1}$
ModelBot	0.6	0.3	0.39139	0.12510
PBot	0.6	0.1	0.22746	0.14930
PBot	0.6	0.2	0.13444	0.04859
PBot	0.6	0.3	0.05305	0.01813
PBot	0.7	0.1	0.52542	0.35881
PBot	0.7	0.2	0.30512	0.16637
PBot	0.7	0.3	0.16531	0.04928
PBot	0.8	0.1	0.63265	0.44704
PBot	0.8	0.2	0.43141	0.25092
PBot	0.8	0.3	0.20155	0.07869
PBot	0.9	0.1	0.34362	0.13656
PBot	0.9	0.2	0.17043	0.03340
PBot	0.9	0.3	0.03937	0.00511

both M=1 and M=2. The average number of hands played per round was 289 with M=2 and 615 for M=1.

5   Conclusions

Compared to the baseline Prospect Utility bot with $\rho_a = 0.6, \rho_b = 0.3$ our modeling bot saw an improvement in play ranging from a factor of 6.9 to 7.5, while using a modeling scheme without heuristic inference, or attempting to wholesale recreate the risk curves of opponents. These results are promising given that the bot had minimal observational inputs, yet still performed quite well.

Additionally, we have proposed that there exists a set of games, which hold practical interest for modeling real world scenarios. Further, we propose that no-limit High Card presents a simple model that can be used to simulate these highly complex scenarios, allowing for preloading of adversary information, in order to simulate possible outcomes from previous observations.

Acknowledgements. This research was partially supported by Air Force Research Laboratory contracts FA8750-10-1-0045, FA8750-09-1-0174 and AFOSR contract FA9550-07-1-0421, The opinions expressed in this article belong solely to the article's authors and do not reflect any opinion, policy statement, recommendation or position, expressed or implied, of the U.S. Department of Defense.

References

1. Ferguson, C., Ferguson, T.: On the Borel and von Neumann Poker Models. Game Theory and Applications 9, 17–32 (2003)

2. von Neumann, J., Morgenstern, O.: The Theory of Games and Economic Behavior. Princeton University Press (1944)
3. Billings, D.: Algorithms and Assessment in Computer Poker. Doctoral Dissertation, University of Alberta (2006)
4. A Forced Marriage Plague by Ever-Deepening Distrust. Spiegel Online International (May 7, 2011)
5. Detaining CIA moles: Haqqani defends decision. The Express Tribune (June 21, 2011)
6. Wikipedia, http://en.wikipedia.org/wiki/Value_poker
7. Flop Turn River, http://poker-strategy.flopturnriver.com/
   5-Biggest-Leaks-Of-Losing-NL-Player-3.php
8. Ferraro, V.: U-2 Incident (1960),
   http://www.mtholyoke.edu/acad/intrel/u2.htm
9. Billings, D., Papp, D., Schaeffer, J., Szafron, D.: Opponent Modeling in Poker. In: Proceedings of The Association for the Advancement of Artificial Intelligence (1998)
10. Jiang, W., et al.: Optimal Network Security Strengthening Using Attack-Defense Game Model. In: Sixth International Conference on Information Technology: New Generations, ITNG 2009, April 27-29, pp. 475–480 (2009)
11. Jiang, W., et al.: A Game Theoretic Method for Decision and Analysis of the Optimal Active Defense Strategy. In: 2007 International Conference on Computational Intelligence and Security, December 15-19, pp. 819–823 (2007)
12. UK Government Under Constant Cyber Attack,
    http://techland.time.com/2011/05/16/
    uk-government-under-constant-cyber-attack/
13. Cyber attack shows constant threat to key intel,
    http://www.cbsnews.com/stories/2011/06/01/national/main20067895.shtml
14. Kahneman, D., Tversky, A.: Prospect Theory: An Analysis of Decision under Risk. Econometrica 47(2) (1979)
15. Lye, K.-w., Wing, J.M.: Game strategies in network security. International Journal of Information Security 4(1), 71–86 (2005)
16. Papp, D.: Dealing With Imperfect Information in Poker. Master's Thesis, University of Alberta (1998)
17. Tversky, A., Kahneman, D.: Advances in prospect theory: Cumulative representation of uncertainty. Journal of Risk and Uncertainty 5, 297–323 (1992)
18. Smith, W.R., Lai, D.: United States vs. North Korea in No-Limit Poker: Alligator Blood or Dead Money? Korean Journal of Defense Analysis 17, 111–126 (2005)
19. Freeman, S.: The Highest Stakes Poker Game Ever Played: Ronald Reagan, Mikhail Gorbachev, and the Reykjavik Summit of 1986. Thesis, Vanderbilt University (2010)

An Anti-jamming Strategy for Channel Access in Cognitive Radio Networks

Shabnam Sodagari and T. Charles Clancy

Department of Electrical and Computer Engineering,
Virginia Tech, Arlington, VA 22203, USA
{shabnamsodagari,tcc}@vt.edu

Abstract. We address an anti-jamming strategy of channel access for secondary user in a cogntive radio network when some idle channels of the primary user are being jammed in each time slot. Given the secondary does not know what idle bands are under attack, using our method it tries to choose the best possible channel in each time slot to avoid the jammer. We show this problem can be formulated as a multi-armed bandit process and compare the results of different approaches for channel selection including ϵ-greedy, ϵ-first, and random. Simulatons verify that our method results in selecting channels with an average of almost 50% improved signal to noise ratio (SNR) over randomly selected channels.

Keywords: Cognitive radio, multi-armed bandit, security, jamming.

1 Introduction

With the ever increasing role of wireless data communications in various aspects of society, measurements have been conducted to prove that due to the bursty nature of wireless traffic, static spectrum licensing does not exploit the full capacity in time and frequency. Opportunistic spectrum access is a proposed solution. The cognitive radio (CR) communication paradigm, as opposed to the traditional one, is designed to access the unused spectral/temporal/spatial resources more efficiently. Unlike traditional communication systems, CRs have flexible operating parameters, e.g., modulation, power, and frequency, and are able to adapt themselves to the radio environment [1].

In cognitive radio networks (CRNs), as shown in Figure 1, secondary users sense the primary (licensed) spectrum usage, utilize idle resources, and release them when the primary returns. Learning capability is an integral part of CR.

The denial of service caused by radio jamming can be achieved in several ways. Some attackers try to increase interference and degrade signal to noise (SNR) of a user by directing signals to its vicinity [13]. Other attackers constantly send packets and never let the wireless spectrum be released [6].

Like any communication network, CRNs are vulnerable to jamming attacks, the goal of which is to prevent legitimate users from utilizing spectrum opportunities. However, here, attackers can also act adaptively to time-varying radio environment. Stochastic game modeling is a suitable tool for dynamic security

J.S. Baras, J. Katz, and E. Altman (Eds.): GameSec 2011, LNCS 7037, pp. 34–43, 2011.

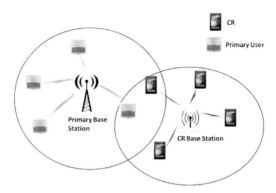

Fig. 1. Secondary CRs (unlicensed users) dynamically accessing unoccupied primary (licensed) bands

mechanisms [10]. In this Markov decision process model, states are defined as spectrum availability, channel quality, and the status of jammed channels observed at the current time slot. Our goal here is to demonstrate at a smaller scale that the multi-armed bandit strategy can also be adopted in this regard.

Various anti-jamming approaches have been studied for wireless networks. These can be divided into three main categories:

- **Network layer anti-jamming:** Network coding or adding network-level redundancy is a defense strategy against snooping and eavesdropping attacks. In network coding the nodes combine multiple packets and then forward them to other nodes in a network to increase the nework throughput. Particularly, random linear coding based schemes provide robust anti-jamming capability in CRNs [4] by achieving required throughput with small redundancy. Another example is called spatial retreat [14]. In this escaping method when mobile nodes are interfered with, they should move to a safe operating frequency. In other words, they should decide about the location they should move to in a coordinated manner.
- **Link layer anti-jamming:** Channel hopping is a familiar example of this strategy. Protecting IEEE 802.11 networks against jamming using channel hopping is discussed in [6]. This method first evaluates what the best possible channel scanning and jamming strategy against channel hopping can be and then decides how to best tune to the hopping strategy. A combination of software-based Markovian modeled reactive channel hopping and error-correction coding [5] can be used to maximize network throughput in spite of jamming. An anti-jamming protocol for IEEE 802.15.4-based hardware and its effectiveness against interrupt jamming, activity jamming, scan jamming, and pulse jamming has been investigated in [12]. Narrow-band RF interference to IEEE 802.11 networks, even with weak power, can degrade the performance of such networks to a great extent. In this regard, rapid channel hopping is a possible strategy [3].

- **Physical layer anti-jamming:** Beamforming, directional antennas, and spread spectrum fall into this category. For example, a combination of sectored antennas and mobility can maintain the connectivity of multihop *ad hoc* wireless networks in spite of jamming attacks [7].

2   Problem Modeling and Solution

Figure 2 shows the structure of the problem. In each time slot t the primary user releases some channels. The secondary user tries to avoid interference to the primary signals by spectrum sensing. The jammer, on the other hand, can perform spectrum sensing and since it has limited power, it jams j channels at random from the total channels released to the secondary. The random jamming strategy helps the jammer against a smart secondary user. We assume the secondary user can select one channel in each time slot and does not want to select a channel under attack. We present the best strategy for the secondary to maximize its throughput over time.

We assume the jammer does not jam the channels while the primary user is active and only targets the secondary user's access. This can hold true in several cases, e.g., when the primary can recognize and punish the jammers, when attacker is distant from the primary [10], or when the jammer has no incentive to disrupt primary access.

To come up with the solution, we note that this problem can be forumalted as a multi-armed bandit process, in which given unknown rewards of multiple levers, the player (here the secondary user) tries to pull the most rewarding lever at each time slot with the goal of maximizing its total reward over the time horizon. Here, the levers are associated with available sub-channels and the reward is proportional to the SNR of the secondary user over the chosen sub-channel.

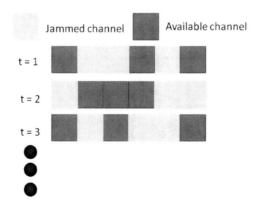

Fig. 2. Secondary user selects best strategy over time to avoid jammed channels among released primary bands

Obviously, if the secondary user mistakenly selects a jammed channel, its SNR is severely degraded and the reward decreases. If the jammer has a low power, the capacity may still be greater than zero, but without loss of generality, we consider the worst case and define the secondary user's reward over jammed sub-channel equal to zero. Otherwise it gets a reward proportional to the chosen channel throughput, proportional to its SNR over that channel.

The jammer can adopt a random frequency hopping strategy, in which it jams different sub-channels from each trial to the next. Also, we note that the SNR of non-jammed sub-channels varies from each time horizon to the next. We model this setting as a *restless* multi-armed bandit process [2]. In restless multi-armed bandit processes the state of non-pulled levers can evolve and change according to some hidden patterns. This means regardless of whether or not the CR user selects a sub-channel c at each round, it cannot tell about the reward of that sub-channel index, or its state, in the next round. In other words, the state of c changes over time.

Assume there are a total of C sub-channels and each sub-channel c can be in one of K states $p_c = \{p_c^1, p_c^2, \ldots, p_c^K\}$. These states correspond to SNR levels. $p_c^1 = 0$, which denotes sub-channel c is under attack. There is a set of time-variant transition probabilities for each state p_c^k, where $k \in \{1, \ldots, K\}$, denoted by $q^c(k, j, t)$. Obviously, $\sum_{j \neq k} q^c(k, j, t) \leq 1$. Suppose the sub-channel is in state p_c^k and is selected next after $t \geq 1$ rounds. It gives reward r_c^k proportional to p_c^k and transitions to one of the states p_c^k with probability $q^c(k, j, t)$. In the most general case, the transition probabilities for different sub-channels are independent. The goal of the CR user is to find a sub-channel selection strategy at each t, as in Figure 2, such that infinite time-horizon reward that is expressed in terms of selected channel's SNR is maximized. The solution lies in solving Whittle's linear program [11], as in equations (1) to (5). Here, $x_{p_c^k t}^c$ and $y_{p_c^k t}^c$ denote the probabilities, in the optimal policy, that sub-channel c in state $p_c^k \in p_c$ is selected or not selected by CR user t timesteps after it was last accessed. To solve this linear program, the CR user has to have an estimate of state transition probabilities.

On the other hand, if the jammer's strategy is fixed, the problems turns into a stochastic multi-armed bandit process, which leans itself to indexing solutions. There are various approaches to solve multi-armed bandit problems [9]. One category is comprised of semi-uniform strategies. The core of semi-uniform stretegies is pulling the best lever greedily, except when a uniformly random action is taken. For example, as shown in Figure 3, in the epsilon-first strategy, there are distinct exploitation and exploration phases, with exploration for fraction ϵ of trials and exploitation for the rest of the trials. In the exploration phase, a lever is arbitrarily selected with a uniform probability distribution. In the exploitation phase, the best lever is selected.

Epsilon-greedy strategy is a method in which the best lever is selected for a proportion $1 - \epsilon$ of the trials and another lever is randomly selected with uniform probability with proportion ϵ. When the value of ϵ in an epsilon-greedy strategy is decreasing as the number of experiments increases, the strategy is

$$\text{Maximize} \sum_{c=1}^{C} \sum_{p_c^k \in p_c} \sum_{t \geq 1} r_{p_c^k t}^c x_{p_c^k t}^c, \tag{1}$$

subject to

$$\sum_{c=1}^{C} \sum_{p_c^k \in p_c} \sum_{t \geq 1} x_{p_c^k t}^c \leq 1, \tag{2}$$

$$\sum_{p_c^k \in p_c} \sum_{t \geq 1} (x_{p_c^k t}^c + y_{p_c^k t}^c) \leq 1, \quad \forall c = 1, 2, \ldots, C \tag{3}$$

$$x_{p_c^k t+1}^c + y_{p_c^k t+1}^c = y_{p_c^k t}^c, \quad \forall c, p_c^k \in p_c, t \geq 1 \tag{4}$$

$$x_{p_c^k t+1}^c, y_{p_c^k t+1}^c \in [0, 1] \tag{5}$$

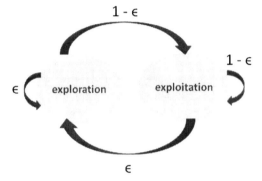

Fig. 3. State machine of semi-uniform multi-armed bandit strategies

called epsilon-decreasing. When the adversary jams different channels in each timeslot, the behavior of rewards are highly changing from the viewpoint of the secondary user. Therefore, a solution can be the adaptive epsilon-greedy strategy, which adjusts ϵ based on reinforcement learning by keeping track of the reward differences during experiments, i.e., high changes in the reward enforce a higher ϵ or more exploration than exploitation [8].

If we denote the set of all available sub-channles by C, each sub-channel by c and the set of jammed sub-channels by $J \subset C$, Equation 6 defines the reward of the secondary user by accessing sub-channel $c \in C$:

$$r = \begin{cases} (S/N)_c & \text{if } c \in C - J \\ 0 & \text{otherwise} \end{cases} \tag{6}$$

where $(S/N)_c$ is the SNR over sub-channel c.

3   Numerical Results

We carried out simulations in MATLAB for various numbers of idle channels provided by the primary user and varying numbers of jammed channels. Our goal here is to show that the proposed method leads to better overall SNR results and helps the secondary user avoid selecting the jammed sub-channels.

Figure 4 shows how using the ϵ-greedy method helps the secondary user to avoid selecting jammed channels and hence obtain more average SNR over selected channels in comparison with choosing sub-channels randomly. It is assumed 30% of the sub-channels are being jammed, zero SNR over jammed channels and the SNR of other sub-channels vary between 5 and 20 dB. The CR user might not have an exact estimate of sub-channels SNR before selecting them, due to reasons such as characteristics of the fading environment, e.g., when the coherence time of the channel is smaller than timeslot duration. The results are averaged over the horizon of 50 trials. Here, ϵ is set to 0.1, which means 10% of trials are dedicated to exploration and the rest to exploitation. However, unlike the ϵ-first method, the exploration and exploitation rounds are interleaved in ϵ-greedy gambling.

An arbitrary proportion of all plays can be dedicated to exploration and the rest to exploitation. To come up with an optimum exploration length, we refer to Figure 5 that shows how the choice of this proportion or ϵ can affect the secondary CR's achieved gain. A horizon containing 50 trials has been considered, with 20% of sub-channels being jammed in random. The experiment, iterated 500 times, demonstrates that increasing the exploration rounds to values beyond 5% does not improve the average SNR.

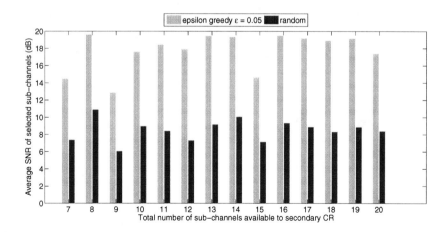

Fig. 4. Comparison of average SNR obtained using ϵ-greedy method and the random scheme

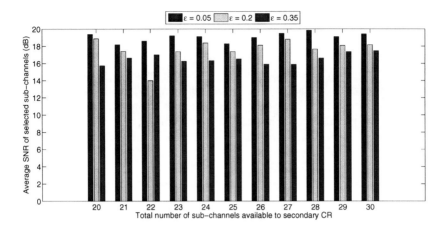

Fig. 5. Effect of exploration phase length on average SNR obtained using ε-greedy method

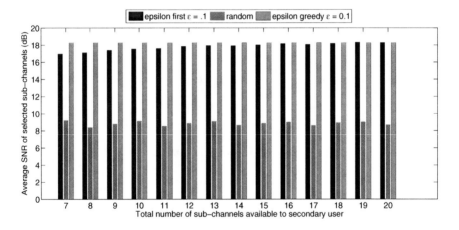

Fig. 6. Comparison of average SNR obtained using ε-first, random and ε-greedy methods

To compare the three methods of ε-first, random, and ε-greedy over a horizon of 50 plays, Figure 6 shows the two multi-armed bandit methods yield more success to the secondary CR in avoding the jammer and accordingly better SNR results than the random method. With almost 1000 instances of completed plays and almost 33% of jammed sub-channels, we observe that the ε-greedy and ε-first methods approach almost the same performance level as the number of sub-channels increases.

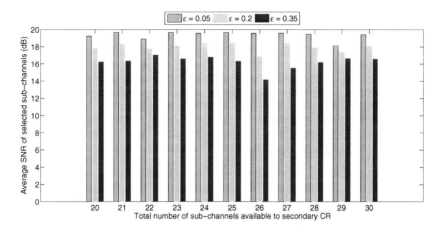

Fig. 7. Effect of exploration phase length on average SNR obtained using ε-first method

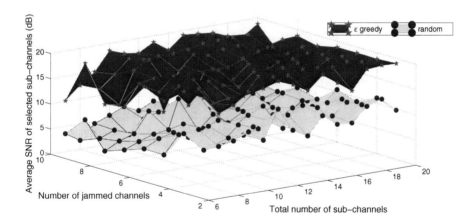

Fig. 8. Average SNR over selected channels in ε-greedy method vs. varying number of sub-channels and jammed sub-channels

In order to get a sense of suitable exploration length with ε-first, we refer to Figure 7, which shows making the exploration phase longer than 5% of rounds degrades the overall results.

Figure 8 compares the performance of ε-greedy ($\epsilon = 0.1$) with random method for different number of sub-channels and different number of jammed channels. As this figure depicts, the proposed method still gives better jammer avoidance results over random channel selection.

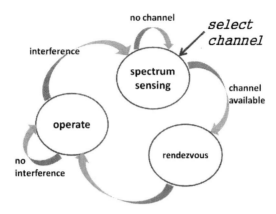

Fig. 9. Dynamic spectrum access by CR

4   Conclusions

In this work, we put forward an anti-jamming strategy for dynamic spectrum access (DSA) by secondary users, in the existence of jamming attacks on some sub-channels that are not revealed to the secondary user until after selection and access. We showed this problem can be cast as a multi-armed bandit processes and carried out simulations for two example multi-armed bandit strategies, ϵ-greedy and ϵ-first. We compared the performance of the proposed jammed channel avoidance schemes with random selection as a benchmark and showed our method always offers better SNR and accordingly more throughput to the secondary user. Applying other learning based multi-armed bandit solutions can be a future topic of research.

4.1   Application to CR

The dynamic channel selection process in CRNs shown in Figure 9 starts with spectrum sensing. If a channel is found, the CR enters the rendezvous state for setting up access to the channel. Operation state is the next step in DSA. The CR remains in that state until it is not interfering with the primary user. The channel selection takes place at spectrum sensing state and the jammer tries to manipulate this phase by injecting interference signals. Our method, basically, offers a defense strategy to the CR at this stage by avoiding the selection of sub-channels under attack.

References

1. Digham, F.F., Alouini, M.-S., Simon, M.: On the energy detection of unknown signals over fading channels. In: Proceedings of the IEEE International Conference on Communications, vol. 55(1), pp. 21–24 (2003)

2. Guha, S., Munagala, K., Shi, P.: Approximation algorithms for restless bandit problems, http://arxiv.org/abs/0711.3861 (accessed June 2011)
3. Gummadi, R., Wetherall, D., Greenstein, B., Seshan, S.: Understanding and mitigating the impact of RF interference on 802.11 networks. ACM SIGCOMM Computer Communication Review 37(4), 385–396 (2007)
4. Kadhe, S., Chandra, M.G., Janakiram, B.: Random Linear Coding Based Anti-Jamming Coding Techniques for Cognitive Radio Systems. In: Proc. IEEE International Conference on Communications, pp. 1–6 (2010)
5. Khattab, S., Mosse, D., Melhem, R.: Modeling of the channelhopping anti-jamming defense in multi-radio wireless networks. In: Proc. 5th Annual International Conference on Mobile and Ubiquitous Systems: Computing, Networking, and Services (MobiQuitous), pp. 1–10 (2008)
6. Navda, V., Bohra, A., Ganguly, S., Rubenstein, D.: Using channelhopping to increase 802.11 resilience to jamming attacks. In: IEEE 26th Conference on Computer Communications (INFOCOM), pp. 2526–2530 (2007)
7. Noubir, G.: On connectivity in ad hoc network under jamming using directional antennas and mobility. In: International Conference on Wired Wireless Internet Communications, pp. 186–200 (2004)
8. Tokic, M.: Adaptive ϵ-Greedy Exploration in Reinforcement Learning Based on Value Differences. In: Dillmann, R., Beyerer, J., Hanebeck, U.D., Schultz, T. (eds.) KI 2010. LNCS, vol. 6359, pp. 203–210. Springer, Heidelberg (2010)
9. Vermorel, J., Mohri, M.: Multi-Armed Bandit Algorithms and Empirical Evaluation. In: Gama, J., Camacho, R., Brazdil, P.B., Jorge, A.M., Torgo, L. (eds.) ECML 2005. LNCS (LNAI), vol. 3720, pp. 437–448. Springer, Heidelberg (2005)
10. Wang, B., Wu, Y., Liu, K.J.R., Clancy, T.C.: An Anti-Jamming Stochastic Game for Cognitive Radio Networks. IEEE Journal on Selected Areas in Communications 29(4), 877–889 (2011)
11. Whittle, P.: Restless bandits: Activity allocation in a changing world. Appl. Prob. 25(A), 287–298 (1988)
12. Wood, A.D., Stankovic, J.A., Zhou, G.: DEEJAM: defeating energyefficient jamming in IEEE 802.15.4-based wireless networks. In: Proc. 4th IEEE Conference on Sensor, Mesh and Ad Hoc Communications and Networks (SECON), pp. 60–69 (2007)
13. Wood, A.D., Stankovic, J.A.: Denial of service in sensor networks. Computer 35(10), 54–62 (2002)
14. Xu, W., Wood, T., Trappe, W., Zhang, Y.: Channel surfing and spatialretreats: defenses against wireless denial of service. In: Proc. 3rd ACM Workshop on Wireless Security (WiSe), pp. 80–89 (2004)

Node Capture Games: A Game Theoretic Approach to Modeling and Mitigating Node Capture Attacks

Tamara Bonaci and Linda Bushnell

Department of Electrical Engineering,
University of Washington, Seattle, WA 98195
{tbonaci,lb2}@uw.edu

Abstract. Unattended wireless sensor networks are susceptible to node capture attacks, where the adversary physically compromises a node, creates functional copies (clones) of it and deploys such clones back into the network, in order to impact the network's functionality. In the absence of a centralized authority, distributed clone detection methods have been developed to mitigate this attack. In this paper, we show that the node capture attack and the network response can be modeled as a simultaneous, noncooperative, two-player game. In developing the game-theoretic framework, we consider a deterministic, linear dynamical model of the attack, as well as a general, stochastic model. For the deterministic model, we develop three games, all of which have quadratic utility for the valid network, whereas the adversary's utility depends on the assumptions about ist abilities. For the stochastic model, we develop a game with convex utility functions. For each game, we prove the existence of a pure strategy Nash Equilibrium and present an efficient way of solving the game. These equilibria can then be used in choosing the appropriate parameters for detecting and responding to the attack. Simulations are provided to illustrate our approach.

Keywords: Node Capture Attack, Distributed Clone Detection Methods, Noncooperative Games, Convex Program.

1 Introduction

Networks of electronic devices, interacting to execute a common task in a distributed fashion, are increasingly becoming systems of choice in military, industrial and medical applications. Examples include target surveillance and tracking, environmental sampling, remote plant control and health monitoring. An important class of such systems is a wireless sensor network (WSN), a collection of wireless sensors deployed over a monitoring area. A typical sensor consists of low-cost hardware components, with constraints on power, communication and computation capabilities. Such sensors are typically expected to operate unattended over an extended period of time. Thus, they have to collaborate in collecting and exchanging data, as well as in preserving their resources.

J.S. Baras, J. Katz, and E. Altman (Eds.): GameSec 2011, LNCS 7037, pp. 44–55, 2011.

Due to their unattended operation, WSNs are vulnerable to *node capture attacks* [6]. In this attack, an adversary physically compromises a sensor, extract data known to it, and using the obtained knowledge, deploys functional copies of captured nodes (*clones*) back into the WSN. Using captured and cloned nodes, referred to as *compromised insiders*, the adversary can mount a slew of efficient attacks.

In the absence of a centralized authority, node capture attacks are typically mitigated through the use of distributed, ad hoc solutions. In [3], we developed a dynamical model of the interaction between the adversary and a network using distributed detection methods. Using a control theoretic methodology, we showed how such a model can be used to control the detection, and consequently, the revocation of compromised insiders. This deterministic model does not, however, characterize a full range of adversarial actions, which may be nonlinear or randomized. Thus, a more general model of the node capture attack is needed.

In this paper, we develop a game-theoretic framework to model the interaction between the adversary and the network under node capture attack. We further show that node capture attack is a Markovian process and introduce a stochastic model to represent it. For this stochastic model, and for the linear dynamical model, we develop two-player, simultaneous, noncooperative games, referred to as *node capture games*. For each game, we prove the existence of a pure-strategy Nash Equilibrium (NE) and present an efficient algorithm of solving for it.

This paper is organized as follows. In Section 2, we state our assumptions about the network, the adversary, and clone detection methods. In Section 3, we present the stochastic model and the linear dynamical model of node capture attack. In Section 4, the game-theoretic framework is introduced and node capture games are developed and analyzed. Simulation analysis is given in Section 5. Section 6 concludes the paper.

2   Background and Preliminaries

2.1   Network Model

We consider a WSN $\mathcal{N}$, consisting of N wireless sensors. All sensors have limited battery life, computation and communication capabilities. Every sensor is on average able to directly communicate with d other nodes, referred to as *neighbors*.

We assume the use of identity-based public key cryptographic methods [10]. Prior to deployment, every sensor is assigned a key pair (PK, SK). The key PK is the node's public key; it is known to all other nodes and it is used as the node's unique identifier (ID). The key SK is the node's secret key. Sensors are further assumed to be able to determine their positions using secure localization mechanisms [7].

2.2   Adversarial Model

We consider a *time-persistent adversary* $\mathcal{A}$, who, over an extended period of time, physically compromises a set of sensors and extracts their data, such as cryptographic secrets, information about the states of network protocols, and sensed data. Using the extracted data, the adversary creates functional copies of captured nodes, denoted as *clones* and deploys them back into the WSN. It is assumed that every captured node is cloned at least once.

The adversary's goal is to compromise the sensing operation. The sensing operation is assumed to be compromised if the adversary can: (1) eavesdrop on the exchanged messages in the targeted WSN, (2) change the content of or tamper with the exchanged messages, (3) delete, drop or destroy messages traversing a compromised node, (4) delay messages long enough to render them invalid, or (5) insert false data into the targeted WSN, in order to cause the network compute incorrect sensing value. We assume that, in order to achieve one of the goals (1) through (5), the adversary has to gain control over a set of at least K nodes, where $K \leq N - 1$.

2.3   Related Work on Distributed Clone Detection Model

Distributed clone detection methods, such as [5,8,9], have been proposed to mitigate node capture attack, in the absence of a centralized monitoring authority. Such methods exploit the idea of collision in detecting compromised insiders. Every node periodically broadcasts a signed message (*location claim*) containing the node's unique identifier (ID) and its current location. Upon receiving a location claim, each neighbor forwards the location claim to a set of nodes, V, with probability p and discards it otherwise. Clone detection methods differ in the way the set V, consisting of g distinct nodes, is chosen: in [9] and [8], the nodes in V are chosen randomly, whereas in [5] the nodes are chosen as a function of the claiming node's ID.

If a node in V receives multiple location claims originating from the same ID, claiming to be at different locations, then that node, denoted as a *witness*, concludes that at least one of the senders is a clone and broadcasts a revocation message for the senders' ID. The effectiveness of a clone detection method can be characterized through *the probability of clone detection* $\mathbb{P}_d$. This probability is a function of the average number of neighbors, d, the number of witnesses, g, and the probability of forwarding the location claim, p [9].

3   Analytical Models of Node Capture Attacks

3.1   Linear Dynamical Model

In [3], a linear dynamical model of a node capture attack was developed. Under this model the adversary is characterized by the parameters $x(t)$, representing *the number of captured nodes* at time t, and *the capture rate*, λ, representing the fraction of valid nodes that the adversary captures per unit of time. Such a modeling of adversarial actions is based on the assumption that the adversary is located in one part of the WSN during one unit of time, and it captures valid, previously uncompromised sensors placed within that same part of the WSN.

The targeted network, using witness-based detection methods, is characterized by *the revocation rate*, representing the fraction of compromised insiders revoked from WSN per unit of time. Based on the observation that in distributed clone detection methods, such as [5,8,9], revocation takes place immediately after a compromised insider is detected, and a valid node cannot falsely be detected as captured (*false alarm*), it follows that the revocation rate is bounded on the interval $[0, 1]$ and it is equal to *the probability of detection*, $\mathbb{P}_d$. The interaction between the adversary and the network is therefore represented by the following dynamical model:

$$\dot{x}(t) = \lambda[N - x(t)] - \mathbb{P}_d x(t) \tag{1}$$

Since compromised insiders are detected after a collision in their location claims occurs, it follows that there is a *delay* between the capture and the revocation processes. This delay causes the number of captured nodes, $x(t)$, to be larger than zero even if the probability of detection $\mathbb{P}_d = 1$. Indeed, from equation (1) it follows that the number of captured nodes stabilizes at $x_{ss} = \frac{\lambda N}{\lambda + \mathbb{P}_d}$.

3.2   Stochastic Model

The model (1) assumes that the process of capturing a node is a linear function of time, whereas both the adversary and the network exhibit nonlinear and possibly stochastic behaviors. Hence a generalized, stochastic model of the attack and response is needed. Any such model should, however, be useful in analyzing the interaction and developing realizable and implementable strategies of the adversary and the network.

In developing such a model, we observe that the targeted network, by expending additional resources on creating *deception*, decreases the efficiency of an intelligent adversary. We do not further explore the ways of creating deception in this paper, but we identify nodes' mobility and redundancy in the nodes' IDs and storage as mechanisms to do so. In the ideal case, by creating deception, the targeted network steers the actions of an intelligent adversary towards the actions of a random adversary, forcing it to capture every node completely independent of all the previous captures. Thus, in a deceptive network, the capture process can be modeled as a Poisson process $X(t)$ with *capture rate* λ. The process $X(t)$ represents *the number of captured nodes* present in the WSN at time t:

$$\mathbb{P}[X(t) = k] = e^{-\lambda t} \frac{(\lambda t)^k}{k!}, \quad k = [0, N - 1] \tag{2}$$

We further observe that the processes of revoking compromised insiders are independent and identically distributed (i.i.d.). In addition, the time required to revoke a compromised insider is the number of independent iterations of the clone detection algorithm until the insider is detected, and is therefore a geometric random variable. As an idealization, for every sensor node we assume that the revocation times $r_i, i = 1, \ldots N - 1$ have exponential distribution with *the revocation rate* equal to the probability of detection, $\mu = \mathbb{P}_d$:

$$\mathbb{P}[r_i \leq T] = 1 - e^{-\mu T}, \ T \geq 0 \tag{3}$$

While there is a non-zero probability that the revocation process fails because the collision-detecting witness has been compromised, and does not report collision, this event is statistically indistinguishable from the case when the witness experiences device failure. As a result, we may assume that the random variable r_i is independent from the times when the nodes were captured.

Taken together, these assumptions imply that the node capture attack on a WSN can be modeled as an M/M/N/N queue [2], where the capture process maps into the arrival process, the number of captured nodes into the number of customers in the system, and the revocation process into the service process. The queue is truncated to the number of nodes in the system, N. The revocation process is further modeled as consisting of N parallel servers, and each server represents the WSN's detection a compromised insiders.

4   A Game Theoretic Approach to Modeling Node Capture Attacks

In this section, a game theoretic framework to model the strategic interaction between the adversary and the network in a node capture attack is presented. We formulate non-cooperative simultaneous games, $\mathscr{G}_i = (\mathcal{P}, \mathcal{S}, \mathcal{U})$, $i = 1, \ldots, 5$, consisting of two players, $\mathcal{P} = \{\mathcal{A}, \mathcal{N}\}$, where $\mathcal{A}$ denotes the adversary and $\mathcal{N}$ the network. For both the linear dynamical model and for the queueing model, the adversary's strategy is to choose the capture rate $\lambda \in [0, \lambda_{\max}]$, and the network's to choose the detection probability, $\mathbb{P}_d \in [0, 1]$.

In mounting the attack, the adversary incurs the following costs: the cost of capture, representing the energy needed to disassemble and to access the information for each captured node, and the cost of revocation, representing the resources wasted if a compromised insider is detected. Similarly, the targeted WSN incurs the cost of captured nodes, quantifying the impact of compromised insiders on the WSN's performance, and the cost of revocation, representing the communication and storage overhead needed to detect and revoke compromised insiders, as well as the effort involved in replacing them with new, secure nodes. The adversary's and the network's optimal strategies depend on how these costs are quantified, as discussed below.

4.1   Games Based on the Linear Dynamical Model

Game $\mathscr{G}_1$: Zero-Sum Game. We first consider a game where both players are greedy and unconstrained in their resources. The utilities of the players are:

$$U_{\mathcal{A}} = -U_{\mathcal{N}} = x(t) \tag{4}$$

In this *zero-sum simultaneous* game, the goal of the adversary $\mathcal{A}$ is to find the capture rate λ that maximizes the number of captured nodes, whereas the goal of the network is to find the revocation rate that minimizes the number of compromised insiders. Both players have a dominant strategy: the adversary always chooses capture rate $\lambda^* = \lambda_{\max}$ and the network always chooses the maximum probability of detection, $\mathbb{P}_d^* = 1$.

Game $\mathscr{G}_2$: Quadratic Cost of the Network. We next formulate a game with cost-oriented players, whose utility functions are defined as:

$$U_{\mathcal{A}}(\lambda, \nu_1, \nu_2) = -\left[\lambda + \nu_1\left(K - \frac{\lambda N}{\lambda + \mathbb{P}_d}\right) + \nu_2(\lambda - \lambda_{\max})\right] \tag{5a}$$

$$U_{\mathcal{N}}(\mathbb{P}_d) = -\frac{1}{T}\int_0^T [q_1 x(t)^2 + r_1(\mathbb{P}_d x(t))^2]dt \tag{5b}$$

where ν_1, ν_2 denote real numbers and T is a positive constant. Under this formulation, the goal of the adversary is to find the minimal capture rate λ, such that the number of captured nodes in the steady state, x_{ss}, is larger than or equal to the number of nodes needed to compromise the WSN, K. The adversary solves the following optimization problem:

$$\min \lambda$$
$$\text{subject to } K \le x_{ss} \tag{6}$$
$$\lambda \le \lambda_{\max}$$

In this game, the network's cost consists of two components: the cost of revocation and the cost (impact) of undetected compromised insiders. For a given λ, the network's best response, $\mathbb{P}_d^*(\lambda)$, can be found by solving the following optimization problem, representing the tradeoff between the two costs:

$$\min_{\mathbb{P}_d} \frac{1}{T}\int_0^T [q_1 x(t)^2 + r_1(\mathbb{P}_d x(t))^2]dt$$
$$\text{subject to } \dot{x}(t) = \lambda[N - x(t)] - \mathbb{P}_d x(t) \tag{7}$$
$$0 \le \mathbb{P}_d \le 1$$

Theorem 1. *In the node capture game $\mathscr{G}_2$, for $\sqrt{\frac{q_1}{r_1}} \le 1$ and T sufficiently large, there exists a unique NE in pure strategies, given as:*

$$(\lambda^*, \mathbb{P}_d^*) = \left(\frac{K}{N}\sqrt{\frac{q_1}{r_1}}, \frac{N - K}{N}\sqrt{\frac{q_1}{r_1}}\right) \tag{8}$$

Proof. The adversary's optimization problem (6) represents an *inequality form* of a linear program (LP) [4] and the optimal capture rate, λ^*, is attained when the constraint $K \le x_{ss}$ holds with strict equality. The optimal capture rate, defined as the best response to the network's optimal strategy, is therefore given as:

$$\lambda^* = BR(\mathbb{P}_d^*) = \frac{K\mathbb{P}_d^*}{N - K} \tag{9}$$

The network's optimization problem (7), after relaxing the constraint $0 \le \mathbb{P}_d \le 1$, represents a Linear Quadratic Regulator (LQR) problem [1]. When T is sufficiently large, the optimal value of $\mathbb{P}_d$, defined as the best response to the adversary's optimal strategy, can be derived as:

$$\mathbb{P}_d^* = BR(\lambda^*) = \frac{P}{r_1} = \lambda^*\left(-1 + \sqrt{1 + \frac{q_1}{r_1}\frac{1}{(\lambda^*)^2}}\right) \tag{10}$$

where P represents a positive solution of the Algebraic Ricatti Equation (ARE): $AP + P^T A + Q - PBR^{-1}B^T P = 0$, with $A = -\lambda$, $B = -1$, $Q = q_1$ and $R = r_1$. The players' optimal strategies (8) are obtained by solving the system of equations (9) and (10) for λ^* and $\mathbb{P}_d^*$. The condition $\sqrt{\frac{q_1}{r_1}} \leq 1$ implies that $0 \leq \mathbb{P}_d$, and hence this equilibrium is feasible.

Game $\mathscr{G}_3$: Quadratic Cost of the Network and of the Adversary. We next consider a game where the adversary incurs two costs in mounting the attack: the cost of capturing a node, $Q_2 = \int_0^\infty r_2(\lambda z(t))^2 dt$ and the cost $R_2 = \int_0^\infty q_2 z(t)^2 dt$, representing the cost of not capturing enough nodes to compromise the WSN. The costs Q_2 and R_2 represent the tradeoff between the invested resources and ability to inflict damage on the targeted system. This tradeoff is modeled through the adversary's cost function:

$$U_A(\lambda) = -\frac{1}{T} \int_0^T [q_2 z(t)^2 + r_2(\lambda z(t))^2] dt \qquad (11)$$

where $z(t)$ represents *the number of valid nodes* remaining in the targeted WSN. In developing this utility function, we make an assumption that all revoked nodes are replenished after revocation. This assumption is valid in target tracking and surveillance applications, where the removal of a node may cause a gap in the monitoring area.

In this node capture game, the network's utility is defined as (5b) and the network's goal as an optimization problem (7). The adversary's goal is to find the capture rate λ such that the cost (11) is minimized:

$$\min_\lambda \frac{1}{T} \int_0^T [q_2 z(t)^2 + r_2(\lambda z(t))^2] dt$$
$$\text{subject to } \dot{z}(t) = \mathbb{P}_d[N - z(t)] - \lambda z(t) \qquad (12)$$
$$0 \leq \lambda \leq 1$$

In the optimization problem (12), the adversary minimizes the cost with respect to the dynamics of the number of remaining valid nodes, $z(t)$. The dynamics of $z(t)$ is derived from equation (1) by noting that $z(t) := K - x(t)$.

Theorem 2. *In the node capture game $\mathscr{G}_3$, under the conditions:*

$$2b^2 + 7b + 4ab \leq 4a^2 + 14a + 9$$
$$4b^2 - 14b - 4ab \leq 2a^2 + 7a - 9 \qquad (13)$$

where $a := \frac{q_1}{r_1}, b := \frac{q_2}{r_2}, c := (2b - 4a), d := (2a - 4b)$ there exists a unique NE in pure strategies, defined as:

$$\{\lambda^*, \mathbb{P}_d^*\} = \left\{ \sqrt{\frac{c + \sqrt{c^2 + 12b^2}}{6}}, \sqrt{\frac{d + \sqrt{(d^2 + 12a^2}}{6}} \right\} \qquad (14)$$

when T is sufficiently large.

Proof. Optimization problems (7) and (12), after relaxing the constraints $0 \leq \lambda \leq 1$ and $0 \leq \mathbb{P}_d \leq 1$, represent LQR problems. From Theorem 1 it thus follows that the players' optimal strategies, defined as the best responses to the opponent's optimal strategy, are defined as:

$$\lambda^* = BR(\mathbb{P}_d^*) = -\mathbb{P}_d^* + \sqrt{(\mathbb{P}_d^*)^2 + b}$$
$$\mathbb{P}_d^* = BR(\lambda^*) = -\lambda^* + \sqrt{(\lambda^*)^2 + a} \tag{15}$$

By solving the system of equations (15) for $\mathbb{P}_d^*$, we obtain:

$$\mathbb{P}_d^* = \sqrt{\lambda^2 + a - b} \tag{16}$$

Now, by substituting equation (16) into the adversary's best response $\lambda^* = BR(\mathbb{P}_d^*)$ we obtain:

$$-3\lambda^4 - \lambda^2(2b - 4a) + b^2 = 0 \tag{17}$$

The optimal capture rate, λ^*, is obtained by solving equation (17). By symmetry of the adversary's and network's optimization problems, the optimal probability of detection, $\mathbb{P}_d^*$, is obtained from the same equation, by substituting $\mathbb{P}_d^*$ and λ^*, and a and b. It can be further shown that the constraints (13) imply that $0 \leq \lambda^* \leq 1$ and $0 \leq \mathbb{P}_d^* \leq 1$, thus confirming that the equilibrium (14) is feasible.

Game $\mathscr{G}_4$: Diminishing Returns in the Adversary's Utility. We next consider an adversary who experiences the effect of *diminishing returns* in the number of captured nodes (i.e., every additional capture contributes to the adversary's utility less than the previous one). This effect occurs, for example, when the adversary influences valid nodes through the compromised insiders within their neighborhood. In such a setup, the number of additional nodes that the adversary is able to influence by capturing a new node decreases with every capture.

In this game formulation, the utility of the network is defined by equation (7) and its goal by the optimization problem (7). The adversary's utility function is given as:

$$U_\mathcal{A}(\lambda) = \log(x_{ss}) - cx_{ss} \tag{18}$$

where the constant c represents the unit cost of capturing a node. The adversary's goal is to find the capture rate λ such that the utility function (18) is maximized:

$$\max_{\lambda} \ \log\left(\frac{\lambda N}{\lambda + \mathbb{P}_d^*}\right) - \frac{c\lambda N}{\lambda + \mathbb{P}_d^*}$$
$$\text{subject to } K \leq x_{ss} \tag{19}$$
$$\lambda \leq \lambda_{\max}$$

In order for the adversary's utility function to be a nonnegative function of λ, the cost of capturing a node, c, satisfies the following conditions: $\frac{\log(K)}{N} \leq c \leq \frac{1}{K}$. The lower bound comes from the observation that the adversary has to capture at least K nodes to compromise the sensing operation, and the upper bound represent the feasibility condition for the optimization problem (19). From these conditions it follows that $K \log(K) \leq N$.

Theorem 3. *In the node capture game $\mathscr{G}_4$, when $\frac{\log(K)}{N} \leq c \leq \frac{1}{K}$ and T sufficiently large, there exists a unique NE in pure strategies, defined as:*

$$\{\lambda^*, \mathbb{P}_d^*\} = \left\{ \sqrt{\frac{q_1}{r_1} \frac{1}{(cN)^2 - 1}}, \sqrt{\frac{q_1}{r_1} \frac{(cN) - 1}{(cN) + 1}} \right\} \tag{20}$$

Proof. The optimal strategy of the network, defined as the best response to the optimal strategy of the adversary, is given by equation (10). Consider now the adversary's optimization problem (19). It represents a constrained convex optimization problem. The optimal capture rate, λ^*, is attained when:

$$\lambda^* = \frac{\mathbb{P}_d^*}{cN - 1} \tag{21}$$

The optimal values of the capture rate, λ^*, and the probability of detection, $\mathbb{P}_d^*$, are obtained solving the system of equations (10) and (21).

4.2 Games Based on the Queueing Model

Game $\mathscr{G}_5$: Maximizing the Average Number of Captured Nodes. A queueing model of a node capture attack allows the steady state analysis of the adversary's and the network's interaction, through the average number of captured nodes in the network, the probability of having zero captured nodes in the WSN, the probability that all (or K) nodes are captured, and through the average time a captured node spends in the WSN before detection. We formulate a two-player noncooperative game, focusing on the average number of captured nodes in the WSN, as defined in [2]:

$$\bar{X} = \sum_{n=0}^{N} n p_n = \sum_{n=0}^{N} n \left(\frac{\lambda}{\mu}\right)^n \frac{p_0}{n!} = \frac{\sum_{n=0}^{N} \left(\frac{\lambda}{\mu}\right)^n \frac{n}{n!}}{\sum_{i=0}^{N} \left(\frac{\lambda}{\mu}\right)^i \frac{1}{i!}} \tag{22}$$

In this game, the adversary, assumed not to know the required number of nodes to capture, K, intends to increase the average number of captured nodes, while the network tries to decrease that number. Both players are cost-oriented, with utility functions defined as:

$$U_{\mathcal{A}}(\lambda) = \bar{X} - c_{\mathcal{A}}\lambda \qquad\qquad U_{\mathcal{N}}(\mu) = -[\bar{X} + c_{\mathcal{N}}\mu]$$

where $c_{\mathcal{A}}$ represents the average capture cost per node, and $c_{\mathcal{N}}$ the average cost of revoking a single node. The goal of the adversary is to find the capture rate, λ, that maximizes the utility function $U_{\mathcal{A}}$. Similarly, the goal of the network is to find the revocation rate, μ, that maximizes the cost function $U_{\mathcal{N}}$. For a large number of sensors, N, the average number of captured nodes, given in equation (22), can be approximated as $\bar{X} \leq \sum_{n=0}^{\infty} n p_n = \frac{\lambda}{\mu}$. Thus, the adversary and the network solve the following optimization problems:

Adversary	**Network**
$\max_\lambda \frac{\lambda}{\mu} - c_{\mathcal{A}}\lambda$	$\min_\mu \frac{\lambda}{\mu} + c_{\mathcal{N}}\mu$
subject to $0 \leq \lambda$	subject to $0 \leq \mu$
$\qquad\qquad \lambda \leq \lambda_{\max}$	$\qquad\qquad \mu \leq \mu_{\max}$
(a)	**(b)**

$$(23)$$

Theorem 4. *In the node capture game $\mathscr{G}_5$ there exist a unique NE in pure strategies, defined as:*

$$\{\lambda^*, \mu^*\} = \begin{cases} \left\{ \lambda_{\max}, \sqrt{\frac{\lambda_{\max}}{c_{\mathcal{N}}}} \right\}, & \text{if } \lambda_{\max} < \frac{c_{\mathcal{N}}}{(c_{\mathcal{A}})^2} \\ \left\{ \frac{c_{\mathcal{N}}}{(c_{\mathcal{A}})^2}, \frac{1}{c_{\mathcal{A}}} \right\}, & \text{if } \lambda_{\max} \geq \frac{c_{\mathcal{N}}}{(c_{\mathcal{A}})^2} \end{cases} \tag{24}$$

Proof. The network's optimization problem (23 (b)) is a constrained convex optimization problem, and the optimal revocation rate $\mu^* = BR(\lambda^*)$ is:

$$\mu^* = BR(\lambda^*) = \sqrt{\frac{\lambda^*}{c_{\mathcal{N}}}} \tag{25}$$

Similarly, the adversary's optimization is a constrained linear program, with the following points of interest:

$$\mu^* < \frac{1}{c_{\mathcal{A}}} \to \lambda^* = \lambda_{\max}, \ \mu^* > \frac{1}{c_{\mathcal{A}}} \to \lambda^* = 0, \ \mu^* = \frac{1}{c_{\mathcal{A}}} \to \lambda^* \in [0, \lambda_{\max}] \tag{26}$$

Now, if $\lambda^* = 0$, it follows from equation (25) that $\mu^* = 0$. This, however, is not a feasible equilibrium point, since $\lambda^* = 0$ only if $\mu^* > \frac{1}{c_{\mathcal{A}}}$. This is a contradiction, implying that $\lambda^* = 0$ is not a NE.

If $\mu^* = \frac{1}{c_{\mathcal{A}}}$, then the adversary is indifferent among different possible strategies. In particular, for $\lambda^* = \frac{c_{\mathcal{N}}}{(c_{\mathcal{A}})^2}$, it follows from equation (25) that $\mu^* = \frac{1}{c_{\mathcal{A}}}$.

Finally, if $\lambda^* = \lambda_{\max}$, it follows from equation (25) that $\mu^* = \sqrt{\frac{\lambda_{\max}}{c_{\mathcal{N}}}}$. This is an equilibrium point only if $\mu^* \leq \frac{1}{c_{\mathcal{A}}}$, thus implying that $\lambda_{\max} < \frac{c_{\mathcal{N}}}{(c_{\mathcal{A}})^2}$.

5   Simulation Results

In this section we provide simulation analysis for different game formulations of node capture attack. We consider two network setups, with simulation parameters provided in Table 1. The optimal values of the capture rate, λ^*, and the probability of detection, $\mathbb{P}_d^*$, obtained as solutions to the games $\mathscr{G}_i, i = 1, \ldots, 5$, are also given in Table 1, where N represents the number of sensors in the WSN,

Table 1. Simulation parameters and optimal solutions of the games $\mathcal{G}_i$, $i = 2,\ldots,5$

Parameters	N	K	$\frac{q_1}{r_1}$	$c_\mathcal{A}$	$c_\mathcal{N}$	$(\lambda_2^*, \mathbb{P}_{d,2}^*)$	$(\lambda_3^*, \mathbb{P}_{d,3}^*)$	$(\lambda_4^*, \mathbb{P}_{d,4}^*)$	$(\lambda_5^*, \mathbb{P}_{d,5}^*)$
Setup I	10^3	100	0.615	1.5	4	(0.12, 0.67)	(0.47, 0.42)	(0.16, 0.64)	(1, 0.5)
Setup II	10^4	1200	0.857	2	6	(0.11, 0.82)	(0.56, 0.49)	(0.13, 0.80)	(1, 0.41)

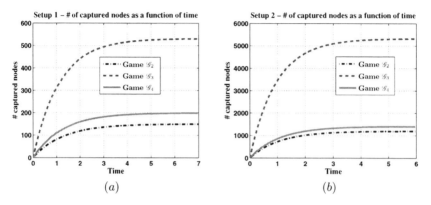

Fig. 1. Comparison of the WSN's performance under node capture attack: (a) Network setup I - $N = 10^3, K = 100, \frac{q_1}{r_1} = 0.615$, (b) Network setup II - $N = 10^4, K = 1200, \frac{q_1}{r_1} = 0.857$. For both network setups, the number of captured nodes, $x(t)$, for games $\mathcal{G}_2, \mathcal{G}_3$ and $\mathcal{G}_4$ is shown.

K the number of sensors the adversary needs to capture to compromise the WSN, $\frac{q_1}{r_1}$ the ratio between the cost of a compromised insider, q_1, and the cost of revocation, r_1, $c_\mathcal{A}$ the average cost of capturing a node, and $c_\mathcal{N}$ the average cost of revoking a node. Figures 1 (a) and (b) depict the number of captured nodes as a function of time, for game formulations $\mathcal{G}_2, \mathcal{G}_3$ and $\mathcal{G}_4$ and network setups I and II. In both figures, we observe that the adversary inflicts the largest impact on the WSN when game $\mathcal{G}_3$ is played. This result reflects the differences in analyzed games: in games $\mathcal{G}_2$ and $\mathcal{G}_4$, the adversary tries to find the minimum capture rate that guarantees at least K captured nodes in the steady state. In the game $\mathcal{G}_3$, however, instead of an explicit constraint on the number of captured nodes, the adversary tries to minimize the tradeoff between capturing nodes and not controlling the required number of nodes in the steady state. This optimization problem, for a given ratio of tradeoff costs, results in a less cost-oriented, but more powerful adversary. Comparing figures (a) and (b), we observe the impact of the ratio $\frac{q_1}{r_1}$ on the WSN performance: in both setups, the number of captured nodes in the steady state stabilizes close to K, for games $\mathcal{G}_2$ and $\mathcal{G}_4$. For setup II, however, this number stabilizes closely below K, while for setup I, it always stabilizes above K. In both setups, the adversary's strategy is the same, but the network strategy is dependant on the ratio, and the higher ratio results in more assertive WSN.

6   Conclusion

In this paper, we studied the node capture attack on wireless sensor networks. We developed a game-theoretic framework to model the interaction between an intelligent, time-persistent adversary and the network. We analyzed this interaction using two models of the attack: a linear dynamical model and a stochastic, queueing model. For both models, noncooperative, simultaneous two-player games were developed. The linear model was explored through a family of games, where the network exhibits a quadratic cost and the adversary's cost varied, based on the assumptions about ist abilities. For the stochastic model, a game with convex utilities were presented. For each game, we proved the existence of a pure strategy Nash Equilibrium. We further presented an efficient way of solving the games.

The developed games are one-stage simultaneous games and it was assumed that both players possess full information about each other. In our future work, we plan to investigate the use of repeated games, as well as the games of incomplete information in characterizing the interaction between the adversary and the network in node capture attacks.

References

1. Andreson, B.D.O., Moore, J.B.: Optimal control: Linear Quadratic Methods. Dover Publications (2007)
2. Bertsekas, D.P., Gallager, R.: Data networks, 2nd edn. Prentice-Hall (1992)
3. Bonaci, T., Bushnell, L., Poovendran, R.: Node capture attacks in wireless sensor networks: A system theoretic approach. In: Proc. of the 49th IEEE Control and Desicion Conference, pp. 6765–6772 (2010)
4. Boyd, S., Vandenberghe, L.: Convex Optimization. Cambridge University Press (2004)
5. Conti, M., Di Pietro, R., Mancini, L.V., Mei, A.: A randomized, efficient, and distributed protocol for the detection of node replication attacks in wireless sensor networks. In: Proc. of the 8th ACM International Symposium on Mobile Ad Hoc Networking and Computing, pp. 80–89 (2007)
6. Eschenauer, L., Gligor, V.D.: A key-management scheme for distributed sensor networks. In: Proc. of the 9th ACM Conference on Computer and Communications Security, pp. 41–47 (2002)
7. Lazos, L., Poovendran, R.: SeRLoc: Robust localization for wireless sensor networks. ACM Trans. on Sensor Networks 1(1), 73–100 (2005)
8. Li, Z., Gong, G.: Randomly directed exploration: An efficient node clone detection protocol in wireless sensor networks. In: Proc. of the 6th International IEEE Conference on Mobile Adhoc and Sensor Systems, pp. 1030–1035 (2009)
9. Parno, B., Perrig, A., Gligor, V.D.: Distributed detection of node replication attacks in sensor networks. In: Proc. of the IEEE Symposium on Security and Privacy, pp. 49–63 (2005)
10. Stinson, D.R.: Cryptography: Theory and Practice. Chapman & Hall/CRC (2002)

Multi-variate Quickest Detection of Significant Change Process

Krzysztof Szajowski

Institute of Mathematics and Computer Science, Wrocław University of Technology,
Wybrzeże Wyspiańskiego 27, 50-370 Wrocław, Poland
Krzysztof.szajowski@pwr.wroc.pl
http://www.im.pwr.wroc.pl/~szajow

Abstract. The paper deals with a mathematical model of a surveillance system based on a net of sensors. The signals acquired by each node of the net are Markovian process, have two different transition probabilities, which depends on the presence or absence of a intruder nearby. The detection of the transition probability change at one node should be confirmed by a detection of similar change at some other sensors. Based on a simple game the model of a fusion center is then constructed. The aggregate function defined on the net is the background of the definition of a non-cooperative stopping game which is a model of the multivariate disorder detection.

Keywords: voting stopping rule, majority voting rule, monotone voting strategy, change-point problems, quickest detection, sequential detection, simple game.

1 Introduction

The aim of this consideration is to construct the mathematical model of a multivariate surveillance system. It is assumed that there is net $\mathfrak{N}$ of p nodes which register (observe) signals modeled by discrete time multivariate stochastic process. At each node the state is the signal at moment $n \in \mathbb{N}$ which is at least one coordinate of the vector $\vec{x}_n \in \mathbb{E} \subset \Re^m$. The distribution of the signal at each node has two forms and depends on *a pure* or *a dirty* environment of the node. The state of the system change dynamically. We consider the discrete time observed signal as $m \geq p$ dimensional process defined on the fixed probability space $(\Omega, \mathcal{F}, \mathbf{P})$. The observed at each node process is Markovian with two different transition probabilities (see [18] for details). In the signal the visual consequence of the transition distribution changes at moment θ_i, $i \in \mathfrak{N}$ is a change of its character. To avoid false alarm the confirmation from other nodes is needed. The family of subsets (coalitions) of nodes are defined in such a way that the decision of all member of some coalition is equivalent with the claim of the net that the disorder appeared. It is not sure that the disorder has had place. The aim is to define the rules of nodes and a construction of the net decision based on individual nodes claims. Various approaches can be found in the recent research

J.S. Baras, J. Katz, and E. Altman (Eds.): GameSec 2011, LNCS 7037, pp. 56–66, 2011.
© Springer-Verlag Berlin Heidelberg 2011

for description or modeling of such systems (see e.g. [24], [17]). The problem is quite similar to a pattern recognition with multiple algorithm when the fusions of individual algorithms results are unified to a final decision. The proposed solution will be based on a simple game and the stopping game defined by a simple game on the observed signals. It gives a centralized, Bayesian version of the multivariate detection with a common fusion center that it has perfect information about observations and *a priori* knowledge of the statistics about the possible distribution changes at each node. Each sensor (player) will declare to stop when it detects disorder at his region. Based on the simple game the sensors' decisions are aggregated to formulate the decision of the fusion center. The sensors' strategies are constructed as an equilibrium strategy in a non-cooperative game with a logical function defined by a simple game (which aggregates their decision).

The general description of such multivariate stopping games has been formulated by Kurano, Yasuda and Nakagami in the case when the aggregation function is defined by the voting majority rule [9] or the monotone voting strategy [25] and the observed sequences of the random variables are independent, identically distributed. It was Ferguson [5] who substituted the voting aggregation rules by a simple game. The Markov sequences have been investigated by the author and Yasuda [22].

The model of detection the disorder at each sensor are presented in the next section. It allows to define the individual payoffs of the players (sensors). It is assumed that the sensors are distributed in homogeneous way in the guarded area and the intruders behavior are well modeled by symmetric random walk. By these assumptions in Section 2 the *a priori* distribution of the disorder moment at each node can be chosen in such a way that it gives the best model of the structure of sensors and the behavior of intruder . Section 3 introduces the aggregation method based on a simple game of the sensors. Section 4 contains derivation of the non-cooperative game and existence theorem for equilibrium strategy. The final decision based on the state of the sensors is given by the fusion center and it is described in Section 6. The natural direction of further research is formulated also in the same section. A conclusion and resume of an algorithm for rational construction of the surveillance system is included in Section 7.

2   Detection of Disorder at Sensors

Following the consideration of Section 1, let us suppose that the process $\{\overrightarrow{X}_n, n \in \mathbb{N}\}$, $\mathbb{N} = \{0, 1, 2, \ldots\}$, is observed sequentially in such a way that each sensor, *e.g.* r (gets its coordinates in the vector $\overrightarrow{X}_n$ at moment n). By assumption, it is a stochastic sequence that has the Markovian structure given random moment θ_r, in such a way that the process after θ_r starts from state $\overrightarrow{X}_{n\,\theta_r-1}$. The objective is to detect these moments based on the observation of $\overrightarrow{X}_n$. at each sensor separately. There are some results on the discrete time case of such disorder detection which generalize the basic problem stated by Shiryaev in [19] (see e.g. Brodsky and Darkhovsky [2], Bojdecki [1],Poor and Hadjiliadis [16], Yoshida [26], Szajowski [21]) in various directions.

Application of the model for the detection of traffic anomalies in networks has been discussed by Tartakovsky et al. [23]. The version of the problem when the moment of disorder is detected with given precision will be used here (see [18]).

2.1   Formulation of the Problem

The observable random variables $\{\vec{X}_n\}_{n\in\mathbb{N}}$ are consistent with the filtration $\mathcal{F}_n$ (or $\mathcal{F}_n = \sigma(\vec{X}_0, \vec{X}_1, \ldots, \vec{X}_n)$). The random vectors $\vec{X}_n$ take values in $(\mathbb{E}, \mathcal{B})$, where $\mathbb{E} \subset \Re^m$. On the same probability space there are defined unobservable (hence not measurable with respect to $\mathcal{F}_n$) random variables $\{\theta_r\}_{r=1}^m$ which have the geometric distributions:

$$\mathbf{P}(\theta_r = j) = p_r^{j-1} q_r, \ q_r = 1 - p_r \in (0, 1), \ j = 1, 2, \ldots. \tag{1}$$

The sensor r follows the process which is based on switching between two, time homogeneous and independent, Markov processes $\{X_{rn}^i\}_{n\in\mathbb{N}}$, $i = 0, 1, r \in \mathfrak{N}$ with the state space $(\mathbb{E}, \mathcal{B})$, both independent of $\{\theta_r\}_{r=1}^m$. Moreover, it is assumed that the processes $\{X_{rn}^i\}_{n\in\mathbb{N}}$ have transition densities with respect to the σ-finite measure μ, i.e., for any $B \in \mathcal{B}$ we have

$$\mathbf{P}_x^i(X_{r1}^i \in B) = \mathbf{P}(X_{r1}^i \in B | X_{r0}^i = x) = \int_B f_x^{ri}(y)\mu(dy). \tag{2}$$

The random processes $\{X_{rn}\}$, $\{X_{rn}^0\}$, $\{X_{rn}^1\}$ and the random variables θ_r are connected via the rule: conditionally on $\theta_r = k$

$$X_{rn} = \begin{cases} X_{rn}^0, & \text{if } k > n, \\ X_{r\,n+1-k}^1, & \text{if } k \leq n, \end{cases}$$

where $\{X_{rn}^1\}$ is started from $X_{r\,k-1}^0$ (but is otherwise independent of $X_{r.}^0$).

For any fixed $d_r \in \{0, 1, 2, \ldots\}$ we are looking for the stopping time $\tau_r^* \in \mathcal{T}$ such that

$$\mathbf{P}_x(|\theta_r - \tau_r^*| \leq d_r) = \sup_{\tau \in \mathfrak{S}^X} \mathbf{P}_x(|\theta_r - \tau| \leq d_r) \tag{3}$$

where $\mathfrak{S}^X$ denotes the set of all stopping times with respect to the filtration $\{\mathcal{F}_n\}_{n\in\mathbb{N}}$. The parameters d_r determines the precision level of detection and it can be different for too early and too late detection. These payoff functions measure the chance of detection of intruder.

2.2   Construction of the Optimal Detection Strategy

In [18] the construction of τ^* by transformation of the problem to the optimal stopping problem for the Markov process $\vec{\xi}$ has been made, such that $\vec{\xi}_{rn} = (\underline{\vec{X}}_{r\,n-1-d_r,n}, \Pi_n)$, where $\underline{\vec{X}}_{r\,n-1-d_r,n} = (\vec{X}_{r\,n-1-d_r}, \ldots, \vec{X}_{r\,n})$ and Π_{rn} is the posterior process:

$$\Pi_{r0} = 0,$$
$$\Pi_{rn} = \mathbf{P}_x(\theta_r \leq n \mid \mathcal{F}_n), \ n = 1, 2, \ldots$$

which is designed as information about the distribution of the disorder instant θ_r. In this equivalent the problem of the payoff function for sensor r is $h_r(\overrightarrow{x}_{r\,d_r+2}, \alpha)$.

3 The Aggregated Decision via the Cooperative Game

There are various methods combining the decisions of several classifiers or sensors. Each ensemble member contributes to some degree to the decision at any point of the sequentially delivered states. The fusion algorithm takes into account all the decision outputs from each ensemble member and comes up with an ensemble decision. When classifier outputs are binary, the fusion algorithms include the majority voting [10], [11], naïve Bayes combination [3], behavior knowledge space [7], probability approximation [8] and singular value decomposition [12].

The majority vote is the simplest. The extension of this method is a simple game.

3.1 A Simple Game

Let us assume that there are many nodes absorbing information and make decision if the disorder has appeared or not. The final decision is made in the fusion center which aggregates information from all sensors. The nature of the system and their role is to detect intrusion in the system as soon as possible but without false alarm.

The voting decision is made according to the rules of *a simple game*. Let us recall that a coalition is a subset of the players. Let $\mathcal{C} = \{C : C \subset \mathfrak{N}\}$ denote the class of all coalitions.

Definition 1. *(see [15], [5])* A simple game *is coalition game having the characteristic function,* $\phi(\cdot) : \mathcal{C} \rightarrow \{0,1\}$.

Let us denote $\mathcal{W} = \{C \subset \mathfrak{N} : \phi(C) = 1\}$ and $\mathcal{L} = \{C \subset \mathfrak{N} : \phi(C) = 0\}$. The coalitions in $\mathcal{W}$ are called the winning coalitions, and those from $\mathcal{L}$ are called the losing coalitions.

Assumptions 2. *By assumption the characteristic function satisfies the properties:*

1. $\mathfrak{N} \in \mathcal{W}$;
2. $\emptyset \in \mathcal{L}$;
3. *(the monotonicity):* $T \subset S \in \mathcal{L}$ *implies* $T \in \mathcal{L}$.

3.2 The Aggregated Decision Rule

When the simple game is defined and the players can vote presence or absence, $x_i = 1$ or $x_i = 0$, $i \in \mathfrak{N}$, of the intruder then the aggregated decision is given by the logical function

$$\pi(x_1, x_2, \ldots, x_p) = \sum_{C \in \mathcal{W}} \prod_{i \in C} x_i \prod_{i \notin C} (1 - x_i). \tag{4}$$

For the logical function π we have (cf [25])

$$\pi(x^1,\ldots,x^p) = x^i \cdot \pi(x^1,\ldots,\overset{i}{1},\ldots,x^p) + \overline{x}^i \cdot \pi(x^1,\ldots,\overset{i}{0},\ldots,x^p).$$

4   A Non-cooperative Stopping Game

Following the results of the author and Yasuda [22] the multilateral stopping of a Markov chain problem can be described in the terms of the notation used in the non-cooperative game theory (see [14], [4], [13], [15]). Let $(\vec{X}_n, \mathfrak{F}_n, \mathbf{P}_x)$, $n = 0, 1, 2, \ldots, N$, be a homogeneous Markov chain with state space $(\mathbb{E}, \mathcal{B})$. The horizon can be finite or infinite. The players are able to observe the Markov chain sequentially. Each player has their utility function $f_i : \mathbb{E} \to \Re$, $i = 1, 2, \ldots, p$, such that $\mathbf{E}_x|f_i(\vec{X}_1)| < \infty$. If process is not stopped at moment n, then each player, based on $\mathfrak{F}_n$, can declare independently their willingness to stop the observation of the process.

Definition 3. (see [25]) *An individual stopping strategy of the player i (ISS) is the sequence of random variables $\{\sigma_n^i\}_{n=1}^N$, where $\sigma_n^i : \Omega \to \{0,1\}$, such that σ_n^i is $\mathfrak{F}_n$-measurable.*

The interpretation of the strategy is following. If $\sigma_n^i = 1$ then player i declares that they would like to stop the process and accept the realization of X_n. Denote $\sigma^i = (\sigma_1^i, \sigma_2^i, \ldots, \sigma_N^i)$ and let $\mathfrak{S}^i$ be the set of ISSs of player i, $i = 1, 2, \ldots, p$. Define

$$\mathfrak{S} = \mathfrak{S}^1 \times \mathfrak{S}^2 \times \ldots \times \mathfrak{S}^p.$$

The element $\sigma = (\sigma^1, \sigma^2, \ldots, \sigma^p)^T \in \mathfrak{S}$ will be called the stopping strategy (SS). The stopping strategy $\sigma \in \mathfrak{S}$ is random matrix. The rows of the matrix are the ISSs. The columns are the decisions of the players at successive moments. The factual stopping of the observation process, and the players realization of the payoffs is defined by the stopping strategy exploiting p-variate logical function. Let $\pi : \{0,1\}^p \to \{0,1\}$. In this stopping game model the stopping strategy is the list of declarations of the individual players. The aggregate function π converts the declarations to an effective stopping time.

Definition 4. *A stopping time $\mathsf{t}_\pi(\sigma)$ generated by the SS $\sigma \in \mathfrak{S}$ and the aggregate function π is defined by*

$$\mathsf{t}_\pi(\sigma) = \inf\{1 \le n \le N : \pi(\sigma_n^1, \sigma_n^2, \ldots, \sigma_n^p) = 1\}$$

$(\inf(\emptyset) = \infty)$. *Since π is fixed during the analysis we skip index π and write* $\mathsf{t}(\sigma) = \mathsf{t}_\pi(\sigma)$.

We have $\{\omega \in \Omega : \mathsf{t}_\pi(\sigma) = n\} = \bigcap_{k=1}^{n-1}\{\omega \in \Omega : \pi(\sigma_k^1, \sigma_k^2, \ldots, \sigma_k^p) = 0\} \cap \{\omega \in \Omega : \pi(\sigma_n^1, \sigma_n^2, \ldots, \sigma_n^p) = 1\} \in \mathfrak{F}_n$, then the random variable $\mathsf{t}_\pi(\sigma)$ is stopping time with respect to $\{\mathfrak{F}_n\}_{n=1}^N$. For any stopping time $t_\pi(\sigma)$ and $i \in \{1, 2, \ldots, p\}$, let

$$f_i(X_{t_\pi(\sigma)}) = \begin{cases} f_i(X_n) & \text{if } t_\pi(\sigma) = n, \\ \limsup_{n \to \infty} f_i(X_n) & \text{if } t_\pi(\sigma) = \infty \end{cases}$$

(cf [20], [22]). If players use SS $\sigma \in \mathfrak{S}$ and the individual preferences are converted to the effective stopping time by the aggregate rule π, then player i gets $f_i(X_{t_\pi(\sigma)})$.

Let $^*\sigma = (^*\sigma^1, ^*\sigma^2, \ldots, ^*\sigma^p)^T$ be fixed SS. Denote

$$^*\sigma(i) = (^*\sigma^1, \ldots, ^*\sigma^{i-1}, \sigma^i, ^*\sigma^{i+1}, \ldots, ^*\sigma^p)^T.$$

Definition 5. (cf. [22]) *Let the aggregate rule π be fixed. The strategy $^*\sigma = (^*\sigma^1, ^*\sigma^2, \ldots, ^*\sigma^p)^T \in \mathfrak{S}$ is an equilibrium strategy with respect to π if for each $i \in \{1, 2, \ldots, p\}$ and any $\sigma^i \in \mathfrak{S}^i$ we have*

$$\mathbf{E}_x f_i(\vec{X}_{t_\pi(^*\sigma)}) \geq \mathbf{E}_x f_i(\vec{X}_{t_\pi(^*\sigma(i))}). \tag{5}$$

The set of SS $\mathfrak{S}$, the vector of the utility functions $f = (f_1, f_2, \ldots, f_p)$ and the monotone rule π define the non-cooperative game $\mathcal{G} = (\mathfrak{S}, f, \pi)$. The construction of the equilibrium strategy $^*\sigma \in \mathfrak{S}$ in $\mathcal{G}$ is provided in [22]. For completeness this construction will be recalled here. Let us define an individual stopping set on the state space. This set describes the ISS of the player. With each ISS of player i the sequence of stopping events $D_n^i = \{\omega : \sigma_n^i = 1\}$ combines. For each aggregate rule π there exists the corresponding set value function $\Pi : \mathfrak{F} \to \mathfrak{F}$ such that $\pi(\sigma_n^1, \sigma_n^2, \ldots, \sigma_n^p) = \pi\{\mathbb{I}_{D_n^1}, \mathbb{I}_{D_n^2}, \ldots, \mathbb{I}_{D_n^p}\} = \mathbb{I}_{\Pi(D_n^1, D_n^2, \ldots, D_n^p)}$. For solution of the considered game the important class of ISS and the stopping events can be defined by subsets $C^i \in \mathcal{B}$ of the state space $\mathbb{E}$. A given set $C^i \in \mathcal{B}$ will be called the stopping set for player i at moment n if $D_n^i = \{\omega : X_n \in C^i\}$ is the stopping event.

For the logical function π we have

$$\pi(x^1, \ldots, x^p) = x^i \cdot \pi(x^1, \ldots, \overset{i}{1}, \ldots, x^p) + \bar{x}^i \cdot \pi(x^1, \ldots, \overset{i}{0}, \ldots, x^p).$$

It implies that for $D^i \in \mathfrak{F}$

$$\Pi(D^1, \ldots, D^p) = \{D^i \cap \Pi(D^1, \ldots, \overset{i}{\check{\Omega}}, \ldots, D^p)\} \tag{6}$$
$$\cup \{\overline{D^i} \cap \Pi(D^1, \ldots, \overset{i}{\check{\emptyset}}, \ldots, D^p)\}.$$

Let f_i, g_i be the real valued, integrable (i.e. $\mathbf{E}_x |f_i(X_1)| < \infty$) function defined on $\mathbb{E}$. For fixed D_n^j, $j = 1, 2, \ldots, p$, $j \neq i$, and $C^i \in \mathcal{B}$ define

$$\psi(C^i) = \mathbf{E}_x \left[f_i(X_1) \mathbb{I}_{iD_1(D_1^i)} + g_i(X_1) \mathbb{I}_{\overline{iD_1(D_1^i)}} \right]$$

where $^iD_1(A) = \Pi(D_1^1, \ldots, D_1^{i-1}, A, D_1^{i+1}, \ldots, D_1^p)$ and $D_1^i = \{\omega : X_n \in C^i\}$. Let $a^+ = \max\{0, a\}$ and $a^- = \min\{0, -a\}$.

Lemma 1. *Let f_i, g_i, be integrable and let $C^j \in \mathcal{B}$, $j = 1, 2, \ldots, p$, $j \neq i$, be fixed. Then the set $^*C^i = \{x \in \mathbb{E} : f_i(x) - g_i(x) \geq 0\} \in \mathcal{B}$ is such that*

$$\psi(^*C^i) = \sup_{C^i \in \mathcal{B}} \psi(C^i)$$

and

$$\psi(^*C^i) = \mathbf{E}_x(f_i(X_1) - g_i(X_1))^+ \mathbb{I}_{{}^iD_1(\Omega)} \tag{7}$$
$$- \mathbf{E}_x(f_i(X_1) - g_i(X_1))^- \mathbb{I}_{{}^iD_1(\Omega)} + \mathbf{E}_x g_i(X_1).$$

Based on Lemma 1 we derive the recursive formulae defining the equilibrium point and the equilibrium payoff for the finite horizon game.

4.1   The Finite Horizon Game

Let horizon N be finite. If the equilibrium strategy $^*\sigma$ exists, then we denote $v_{i,N}(x) = \mathbf{E}_x f_i(X_{t(^*\sigma)})$ the equilibrium payoff of i-th player when $X_0 = x$. For the backward induction we introduce a useful notation. Let $\mathfrak{S}_n^i = \{\{\sigma_k^i\}, k = n, \ldots, N\}$ be the set of ISS for moments $n \leq k \leq N$ and $\mathfrak{S}_n = \mathfrak{S}_n^1 \times \mathfrak{S}_n^2 \times \ldots \times \mathfrak{S}_n^p$. The SS for moments not earlier than n is $^n\sigma = (^n\sigma^1, {}^n\sigma^2, \ldots, {}^n\sigma^p) \in \mathfrak{S}_n$, where $^n\sigma^i = (\sigma_n^i, \sigma_{n+1}^i, \ldots, \sigma_N^i)$. Denote

$$t_n = t_n(\sigma) = t(^n\sigma) = \inf\{n \leq k \leq N : \pi(\sigma_k^1, \sigma_k^2, \ldots, \sigma_k^p) = 1\}$$

to be the stopping time not earlier than n.

Definition 6. *The stopping strategy $^{n*}\sigma = (^{n*}\sigma^1, {}^{n*}\sigma^2, \ldots, {}^{n*}\sigma^p)$ is an equilibrium in $\mathfrak{S}_n$ if*

$$\mathbf{E}_x f_i(X_{t_n(^*\sigma)}) \geq \mathbf{E}_x f_i(X_{t_n(^*\sigma(i))}) \, \mathbf{P}_x - a.e.$$

for every $i \in \{1, 2, \ldots, p\}$, where

$$^{n*}\sigma(i) = (^{n*}\sigma^1, \ldots, {}^{n*}\sigma^{i-1}, {}^n\sigma^i, {}^{n*}\sigma^{i+1}, \ldots, {}^{n*}\sigma^p).$$

Denote

$$v_{i,N-n+1}(X_{n-1}) = \mathbf{E}_x[f_i(X_{t_n(^*\sigma)})|\mathfrak{F}_{n-1}] = \mathbf{E}_{X_{n-1}} f_i(X_{t_n(^*\sigma)}).$$

At moment $n = N$ the players have to declare to stop and $v_{i,0}(x) = f_i(x)$. Let us assume that the process is not stopped up to moment n, the players are using the equilibrium strategies $^*\sigma_k^i$, $i = 1, 2, \ldots, p$, at moments $k = n + 1, \ldots, N$. Choose player i and assume that other players are using the equilibrium strategies $^*\sigma_n^j$, $j \neq i$, and player i is using strategy σ_n^i defined by stopping set C^i. Then the expected payoff $\varphi_{N-n}(X_{n-1}, C^i)$ of player i in the game starting at moment n, when the state of the Markov chain at moment $n-1$ is X_{n-1}, is equal to

$$\varphi_{N-n}(X_{n-1}, C^i) = \mathbf{E}_{X_{n-1}}\left[f_i(X_n)\mathbb{I}_{{}^{i*}D_n(D_n^i)} + v_{i,N-n}(X_n)\mathbb{I}_{\overline{{}^{i*}D_n(D_n^i)}}\right],$$

where $^{i*}D_n(A) = \Pi(^*D_n^1, \ldots, {}^*D_n^{i-1}, A, {}^*D_n^{i+1}, \ldots, {}^*D_n^p)$.

By Lemma 1 the conditional expected gain $\varphi_{N-n}(X_{N-n}, C^i)$ attains the maximum on the stopping set $^*C_n^i = \{x \in \mathbb{E} : f_i(x) - v_{i,N-n}(x) \geq 0\}$ and

$$
\begin{aligned}
v_{i,N-n+1}(X_{n-1}) = &\, \mathbf{E}_x[(f_i(X_n) - v_{i,N-n}(X_n))^+ \mathbb{I}_{i^*D_n(\Omega)}|\mathfrak{F}_{n-1}] \\
&- \mathbf{E}_x[(f_i(X_n) - v_{i,N-n}(X_n))^- \mathbb{I}_{i^*D_n(\emptyset)}|\mathfrak{F}_{n-1}] \\
&+ \mathbf{E}_x[v_{i,N-n}(X_n)|\mathfrak{F}_{n-1}]
\end{aligned} \tag{1}
$$

$\mathbf{P}_x$−a.e.. It allows to formulate the following construction of the equilibrium strategy and the equilibrium value for the game $\mathcal{G}$.

Theorem 1. *In the game $\mathcal{G}$ with finite horizon N we have the following solution.*

(i) *The equilibrium value $v_i(x)$, $i = 1, 2, \ldots, p$, of the game $\mathcal{G}$ can be calculated recursively as follows:*
   1. $v_{i,0}(x) = f_i(x)$;
   2. *For $n = 1, 2, \ldots, N$ we have $\mathbf{P}_x$−a.e.*

$$
\begin{aligned}
v_{i,n}(x) = &\, \mathbf{E}_x[(f_i(X_{N-n+1}) - v_{i,n-1}(X_{N-n+1}))^+ \mathbb{I}_{i^*D_{N-n+1}(\Omega)}|\mathfrak{F}_{N-n}] \\
&- \mathbf{E}_x[(f_i(X_{N-n+1}) - v_{i,n-1}(X_{N-n+1}))^- \mathbb{I}_{i^*D_{N-n+1}(\emptyset)}|\mathfrak{F}_{N-n}] \\
&+ \mathbf{E}_x[v_{i,n-1}(X_{N-n+1})|\mathfrak{F}_{N-n}],
\end{aligned}
$$

   for $i = 1, 2, \ldots, p$.
(ii) *The equilibrium strategy $^*\sigma \in \mathfrak{S}$ is defined by the SS of the players $^*\sigma_n^i$, where $^*\sigma_n^i = 1$ if $X_n \in {}^*C_n^i$, and $^*C_n^i = \{x \in \mathbb{E} : f_i(x) - v_{i,N-n}(x) \geq 0\}$, $n = 0, 1, \ldots, N$.*

We have $v_i(x) = v_{i,N}(x)$, and $\mathbf{E}_x f_i(X_{t(^*\sigma)}) = v_{i,N}(x)$, $i = 1, 2, \ldots, p$.

5   Infinite Horizon Game

In this class of games the equilibrium strategy is presented in Definition 5 but in class of SS

$$
\mathfrak{S}_f^* = \{\sigma \in \mathfrak{S}^* : \mathbf{E}_x f_i^-(X_{t(\sigma)}) < \infty \quad \text{for every } x \in \mathbb{E}, \, i = 1, 2, \ldots, p\}.
$$

Let $^*\sigma \in \mathfrak{S}_f^*$ be an equilibrium strategy. Denote

$$
v_i(x) = \mathbf{E}_x f_i(X_{t(^*\sigma)}).
$$

Let us assume that $^{(n+1)*}\sigma \in \mathfrak{S}_{f,n+1}^*$ is constructed and it is an equilibrium strategy. If players $j = 1, 2, \ldots, p$, $j \neq i$, apply at moment n the equilibrium strategies $^*\sigma_n^j$, player i the strategy σ_n^i defined by stopping set C^i and $^{(n+1)*}\sigma$ at moments $n+1, n+2, \ldots$, then the expected payoff of the player i, when history of the process up to moment $n-1$ is known, is given by

$$
\varphi_n(X_{n-1}, C^i) = \mathbf{E}_{X_{n-1}}\left[f_i(X_n)\mathbb{I}_{i^*D_n(D_n^i)} + v_i(X_n)\mathbb{I}_{\overline{i^*D_n(D_n^i)}}\right],
$$

where $^{i^*}D_n(A) = \Pi(^*D_n^1, \ldots, ^*D_n^{i-1}, A, ^*D_n^{i+1}, \ldots, ^*D_n^p)$, $^*D_n^j = \{\omega \in \Omega : ^*\sigma_n^j = 1\}$, $j = 1, 2, \ldots, p$, $j \neq i$, and $D_n^i = \{\omega \in \Omega : \sigma_n^i = 1\} = 1\} = \{\omega \in \Omega : X_n \in C^i\}$.

By Lemma 1 the conditional expected gain $\varphi_n(X_{n-1}, C^i)$ attains the maximum on the stopping set ${}^*C_n^i = \{x \in \mathbb{E} : f_i(x) \geq v_i(x)\}$ and

$$\varphi_n(X_{n-1}, {}^*C^i) = \mathbf{E}_x[(f_i(X_n) - v_i(X_n))^+\mathbb{I}_{i*D_n(\Omega)}|\mathfrak{F}_{n-1}]$$
$$-\mathbf{E}_x[(f_i(X_n) - v_i(X_n))^-\mathbb{I}_{i*D_n(\emptyset)}|\mathfrak{F}_{n-1}]$$
$$+\mathbf{E}_x[v_i(X_n)|\mathfrak{F}_{n-1}].$$

Let us assume that there exists solution $(w_1(x), w_2(x), \ldots, w_p(x))$ of the equations

$$w_i(x) = \mathbf{E}_x(f_i(X_1) - w_i(X_1))^+\mathbb{I}_{i*D_1(\Omega)} \tag{1}$$
$$-\mathbf{E}_x(f_i(X_1) - w_i(X_1))^-\mathbb{I}_{i*D_1(\emptyset)} + \mathbf{E}_x w_i(X_1),$$

$i = 1, 2, \ldots, p$. Consider the stopping game with the following payoff function for $i = 1, 2, \ldots, p$.

$$\phi_{i,N}(x) = \begin{cases} f_i(x) & \text{if } n < N, \\ v_i(x) & \text{if } n \geq N. \end{cases}$$

Lemma 2. *Let* ${}^*\sigma \in \mathfrak{S}_f^*$ *be an equilibrium strategy in the infinite horizon game* $\mathcal{G}$. *For every* N *we have*

$$\mathbf{E}_x \phi_{i,N}(X_{t^*}) = v_i(x).$$

Let us assume that for $i = 1, 2, \ldots, p$ and every $x \in \mathbb{E}$ we have

$$\mathbf{E}_x[\sup_{n \in \mathbb{N}} f_i^+(X_n)] < \infty. \tag{2}$$

Theorem 2. *Let* $(X_n, \mathfrak{F}_n, \mathbf{P}_x)_{n=0}^\infty$ *be a homogeneous Markov chain and the payoff functions of the players fulfill* (2). *If* $t^* = t({}^*\sigma)$, ${}^*\sigma \in \mathfrak{S}_f^*$ *then* $\mathbf{E}_x f_i(X_{t^*}) = v_i(x)$.

Theorem 3. *Let the stopping strategy* ${}^*\sigma \in \mathfrak{S}_f^*$ *be defined by the stopping sets* ${}^*C_n^i = \{x \in \mathbb{E} : f_i(x) \geq v_i(x)\}$, $i = 1, 2, \ldots, p$, *then* ${}^*\sigma$ *is the equilibrium strategy in the infinite stopping game* $\mathcal{G}$.

6   Determining the Strategies of Sensors

Based on the model constructed in Sections 2–4 for the net of sensors with the fusion center determined by a simple game, one can determine the rational decisions of each nodes. The rationality of such a construction refers to the individual aspiration for the highest sensitivity to detect the disorder without false alarm. The Nash equilibrium fulfills requirement that nobody deviates from the equilibrium strategy because its probability of detection will be smaller. The role of the simple game is to define wining coalitions in such a way that the detection of intrusion to the guarded area is maximal and the probability of false alarm is minimal. The method of constructing the optimum winning coalitions family is not the subject of the research in this article. However, there are some natural methods of solving this problem.

The research here is focused on constructing the solution of the non-cooperative stopping game as to determine the detection strategy of the sensors. To this end, the game analyzed in Section 4 with the payoff function of the players defined by the individual disorder problem formulated in Section 2 should be derived.

The proposed model disregards correlation of the signals. It is also assumed that the fusion center has perfect information about signals and the information is available at each node. The further research should help to qualify these real needs of such models and to extend the model to more general cases. In some type of distribution of sensors, e.g. when the distribution of the pollution in the given direction is observed, the multiple disorder model should work better than the game approach. In this case the *a priori* distribution of disorder moment has the form of sequentially dependent random moments and the fusion decision can be formulated as the threshold one: stop when k^* disorder is detected. The method of a cooperative game was used in [6] to find the best coalition of sensors in the problem of the target localization. The approach which is proposed here shows possibility of modelling the detection problem by multiple agents at a general level.

7   Final Remarks

In a general case the consideration of this paper leads to the algorithm of constructing the disorder detection system.

7.1   Algorithm

1. Define a simple game on the sensors.
2. Describe signal processes and *a priori* distribution of the disorder moments at all sensors. Establish the *a posteriori* processes: $\vec{\Pi}_n = (\Pi_{1n}, \ldots, \Pi_{mn})$, where $\Pi_{kn} = \mathbf{P}(\theta \leq n | \mathcal{F}_n)$.
3. Solve the multivariate stopping game on the simple game to get the individual strategies of the sensors.

References

1. Bojdecki, T.: Probability maximizing approach to optimal stopping and its application to a disorder problem. Stochastics 3, 61–71 (1979)
2. Brodsky, B., Darkhovsky, B.: Nonparametric Methods in Change-Point Problems, Mathematics and Its Applications, vol. 243. Kluwer Academic Publisher, Dordrecht (1993)
3. Domingos, P., Pazzani, M.: On the optimality of the simple bayesian classifier under zero-one loss. Machine Leaning 29, 103–130 (1997)
4. Dresher, M.: The mathematics of games of strategy. Theory and applications. Dover Publications, Inc., New York (1981)
5. Ferguson, T.S.: Selection by committee. In: Nowak, A., Szajowski, K. (eds.) Advances in Dynamic Games, Ann. Internat. Soc. Dynam. Games, vol. 7, pp. 203–209. Birkhäuser, Boston (2005)

6. Gharehshiran, O.N., Krishnamurthy, V.: Coalition formation for bearings-only localization in sensor networks—a cooperative game approach. IEEE Trans. Signal Process. 58(8), 4322–4338 (2010)
7. Huang, Y.S., Suen, C.Y.: A method of combining multiple experts for recognition of unconstrained handwritten numerals. IEEE Transactions on Pattern Analysis and Machine Learning 17, 90–93 (1995)
8. Kang, H.J., Kim, K., Kim, J.H.: Optimal approximation of discrete probability distribution with kth-order dependency and its application to combining multiple classifiers. Pattern Recognition Letters 18, 515–523 (1997)
9. Kurano, M., Yasuda, M., Nakagami, J.: Multi-variate stopping problem with a majority rule. J. Oper. Res. Soc. Jap. 23, 205–223 (1980)
10. Lam, L., Krzyzak, A.: A theoretical analysis of the application of majority voting to pattern recognition, Jerusalem, Israel, pp. 418–420 (1994)
11. Lam, L., Suen, C.Y.: Application of majority voting to pattern recognition: An analysis of its behavior and performance. IEEE Transactions on Systems, Man, and Cybernetics-Part A: Systems and Humans 27(5), 533–568 (1997)
12. Merz, C.: Using correspondence analysis to combine classifiers. Machine Learning 36, 33–58 (1999)
13. Moulin, H.: Game Theory for the Social Sciences. New York University Press, New York (1986)
14. Nash, J.: Non-cooperative game. Annals of Mathematics 54(2), 286–295 (1951)
15. Owen, G.: Game theory, 3rd edn. Academic Press Inc., San Diego (1995)
16. Poor, V.H., Hadjiliadis, O.: Quickest detection. Cambridge University Press, Cambridge (2009)
17. Raghavan, V., Veeravalli, V.V.: Quickest change detection of a Markov process across a sensor array. IEEE Trans. Inform. Theory 56(4), 1961–1981 (2010)
18. Sarnowski, W., Szajowski, K.: Optimal detection of transition probability change in random sequence. Stochastics An International Journal of Probability and Stochastic Processes, 13 (First published on: March 10, 2011 (iFirst))
19. Shiryaev, A.: The detection of spontaneous effects. Sov. Math., Dokl. 2, 740–743 (1961)
20. Shiryaev, A.: Optimal Stopping Rules. Springer, Heidelberg (1978)
21. Szajowski, K.: Optimal on-line detection of outside observations. J. of Statistical Planning and Inference 30, 413–422 (1992)
22. Szajowski, K., Yasuda, M.: Voting procedure on stopping games of Markov chain. In: Christer, A.H., Osaki, S., Thomas, L.C. (eds.) UK-Japanese Research Workshop on Stochastic Modelling in Innovative Manufecuring, July 21-22, 1995. Lecture Notes in Economics and Mathematical Systems, vol. 445, pp. 68–80. Springer, Heidelberg (1996)
23. Tartakovsky, A.G., Rozovskii, B.L., Blažek, R.B., Kim, H.: Detection of intrusions in information systems by sequential change-point methods. Stat. Methodol. 3(3), 252–293 (2006)
24. Tartakovsky, A.G., Veeravalli, V.V.: Asymptotically optimal quickest change detection in distributed sensor systems. Sequential Anal. 27(4), 441–475 (2008)
25. Yasuda, M., Nakagami, J., Kurano, M.: Multi-variate stopping problem with a monoton rule. J. Oper. Res. Soc. Jap. 25, 334–350 (1982)
26. Yoshida, M.: Probability maximizing approach for a quickest detection problem with complocated Markov chain. J. Inform. Optimization Sci. 4, 127–145 (1983)

Interplay between Security Providers, Consumers, and Attackers: A Weighted Congestion Game Approach

Patrick Maillé[1], Peter Reichl[2], and Bruno Tuffin[3]

[1] Institut Telecom, Telecom Bretagne
2 rue de la Châtaigneraie CS 17607
35576 Cesson Sévigné Cedex, France
`patrick.maille@telecom-bretagne.eu`
[2] FTW
Donau-City-Str. 1
A-1220 Wien, Austria
`reichl@ftw.at`
Université européenne de Bretagne
[3] INRIA Rennes - Bretagne Atlantique
Campus universitaire de Beaulieu
35042 Rennes Cedex, France
`bruno.tuffin@inria.fr`

Abstract. Network users can choose among different security solutions to protect their data. Those solutions are offered by competing providers, with possibly different performance and price levels. In this paper, we model the interactions among users as a noncooperative game, with a negative externality coming from the fact that attackers target popular systems to maximize their expected gain. Using a nonatomic weighted congestion game model for user interactions, we prove the existence and uniqueness of a user equilibrium, and exhibit the tractability of its computation, as a solution of a convex problem. We also compute the corresponding Price of Anarchy, that is the loss of efficiency due to user selfishness, and investigate some consequences for the (higher-level) pricing game played by security providers.

Keywords: Security, Game theory, Competition.

1 Introduction

Within the current evolution towards the Future Internet, the provision of appropriate network security is considered to be one of the most difficult as well as most challenging tasks. Among the broad range of related research approaches, the attempt to better understand the mindset of attackers serves for sure as one of the key sources for developing advanced protection mechanisms. Cybercrime concerns colossal amounts of money, and is highly organized so that attacker

J.S. Baras, J. Katz, and E. Altman (Eds.): GameSec 2011, LNCS 7037, pp. 67–86, 2011.

efforts are rationalized to maximize the associated gains. This is why we model here an interesting negative externality effect of security architectures and systems, through the attractiveness for potential attackers: majority products are likely to be larger targets for hackers, and therefore become less attractive for consumers. Then, the choice of a particular system and security protection -that we will call a security provider from now on- by the whole online population can now be considered as a congestion game, where congestion is not considered in the common sense of an excessive demand for a finite resource amount, but more generally as a degradation of the performance on a given choice when it gets too popular. Here the performance degradation is indirect, since it stems from the behavior of attackers.

In the specific context of security, the link between the audience of a system and its attractiveness to attackers can be further described when attacks are intended to steal or damage data: an attacker would be attracted by the potential gain (or damage) of the attack, which depends on the value of the users' data, but that value affects (and is therefore, to some extent, revealed by) the security option users choose. For example, the "safest" solutions may attract users with high-value data to protect, making those solutions an interesting target for an attacker even if their market share is small.

In this paper, we propose a model that encompasses that effect, by considering users with heterogeneous data values making a choice among several security possibilities. The criteria considered in that choice are the security protection level -measured by the likeliness of having one's data stolen or damaged, that is subject to negative externalities- and the price set by the security provider.

The literature on network security involving game-theoretic models and tools is recent and still not very abundant. Some very interesting works have been published regarding the interactions between attacking and defending entities, where the available strategies can consist in spreading effort over the links of a network [6,15] or over specific targets [8], or in selecting some particular attack or defense measures [5,11]. In those references, the security game is a zero-sum game between two players only, and therefore no externalities among several potential defenders are considered.

Another stream of work considers security protection investments, through models that encompass positive externalities among users: indeed, when considering epidemic attacks (like, e.g., worms), the likeliness of being infected decreases with the proportion of neighbors that are protected. Since protection has a cost and users selfishly decide to protect or not without considering the externality they generate, the equilibrium outcome is such that investment is suboptimal [12] and needs to be incentivized through specific measures [17]. For more references on game theory applied to network security contexts, see [1,18].

In contrast, the work presented here considers negative externalities in the choices of security software/procedures. As highlighted above, the negative externality comes from the attractiveness of security solutions for attackers. Such situations can arise when attacks are not epidemic but rather direct, as are attacks targeting randomly chosen IP addresses. The interaction among users can then be modeled as a population game, that is a game where the user payoffs for a given strategy (here, a security solution) change as more users choose that same strategy [10]. Such games are particular cases of so-called *congestion games* where user strategies are subsets of a given set of resources, and the total cost experienced by users is the sum of the costs on each resource [2,22]. Here, users select only one resource, and congestion corresponds to the fact that the more customers, the more likely an attack.

In this paper, we consider a very large population, where the extra congestion created by any individual user is negligible. The set of players can therefore be considered as a continuum; note that such games are called *nonatomic* [29]. The study of nonatomic congestion games has seen recent advances for the case when all users are identical or belong to a finite set of populations [7,14,24,25,26], but we want here to encompass the larger attractiveness to attackers of "rich" users, compared to the ones with no valuable data online. More precisely, we intend to model the heterogeneity in users congestion effects, by introducing a distribution among users valuation for the data to protect. The congestion game is therefore *weighted* in the sense that not all users contribute to congestion in an identical manner. Fewer results exist for those games [4,21], even when user strategies only consist in choosing one resource among a common strategy set.

Moreover, in our model users undergo the congestion cost of the security solution they select - which depends on the congestion as well as on their particular data valuation -, but also the monetary cost associated to that solution - which is the same for all users -. As a result, following [20,21] the game would be called a *weighted congestion game with separable preferences*, and can be transformed into an equivalent *weighted congestion game with player-specific constants* [19] (i.e., the payoffs of users selecting the same strategy only differ through a user-specific additive constant). In general, the existence of an equilibrium is not ensured for such games when the number of users is finite [19,20,21]. In the nonatomic case, the existence of a mixed equilibrium is ensured by [29] and the loss of efficiency due to user selfishness is bounded [4], but the existence of a pure equilibrium in the general case is not guaranteed.

In this paper, we establish the *existence* and essential *uniqueness* of a pure equilibrium for our model, as well as its *tractability* by proving that an equilibrium solves a strictly convex optimization problem. To the best of our knowledge, such proofs for nonatomic games had only been given for unweighted games [27,28], with a finite number of different user populations; here we consider a weighted game with possibly an infinity of different weight values, with the specificity that the differences in user congestion weights are directly linked to their user-specific valuations.

The remainder of the paper is organized as follows. The model is formally introduced in Section 2. We focus on the user equilibrium existence, uniqueness and tractability in Section 3, and give an upper bound on the loss of efficiency due to user selfishness. The results are then applied in Section 4 to give some insights about the prices that profit-oriented security providers should set. We conclude and suggest directions for future work in Section 5.

2   Model

We consider a set $\mathcal{I}$ of security providers (each one on a given architecture), and define $I := |\mathcal{I}|$.

2.1   User Data Valuation

Users differ with the valuation for their data. When an attack is successful over a target user u, that user is assumed to experience a financial loss $v_u \geq 0$, which we call her data valuation. The distribution of valuations over the population is given by a cumulative distribution function F on $\mathbb{R}^+$, where $F(v)$ represents the proportion of users with valuation lower than or equal to v. Since users who do not value their data (i.e., for whom $v_u = 0$) will not play any role in our model, we can ignore them; the distribution function F is therefore such that $F(0) = 0$. The overall total "mass" of users is finite, and through a unit change we can assume it to be 1 without loss of generality.

Equivalently, the repartition F of user preferences among the population can be represented by its corresponding *quantile function* $q : [0, 1) \rightarrow \mathbb{R}^+$. For $x \in [0, 1)$, the quantity $q(x)$ represents the valuation[1] of the (infinitesimal) user at (continuous) position x on a valuation-related increasing ranking. Formally, we have

$$\forall x \in [0, 1), \qquad q(x) = \inf\{v \in \mathbb{R}^+ : F(v) \geq x\}, \qquad (1)$$

$$\forall v \in \mathbb{R}^+, \qquad F(v) = \inf\{x \in [0, 1) : q(x) > v\}, \qquad (2)$$

with the convention $\inf \emptyset := 1$ in the latter equation. Note that F is right-continuous, while the quantile function q is left-continuous. Both functions are nonnegative and nondecreasing.

We may not suppose that the support of F, that we denote by S_v, is bounded, but we assume that the overall value of the data in the population is finite, i.e.,

$$V_{\text{tot}} := \int_{S_v} v \, dF(v) < +\infty.$$

Finally, we define $\mathcal{N}(V)$ as the user mass[2] such that the total data valuation for the $\mathcal{N}(V)$ users with smallest valuation exactly equals V:

$$\forall V \in [0, V_{\text{tot}}), \quad \mathcal{N}(V) := \min\left\{ x : \int_{y=0}^{x} q(y)dy = V \right\}.$$

$\mathcal{N}(V)$ is obtained by inverting the bijective function

$$\mathcal{V} : [0, 1] \mapsto [0, V_{\text{tot}}]$$

$$x \rightarrow \mathcal{V}(x) = \int_{y=0}^{x} q(y)dy. \qquad (3)$$

---

[1] Except, possibly, on a zero-measure set of users.
[2] i.e., proportion since we normalized the total user mass to 1.

Notice that $\mathcal{V}$ is continuous and differentiable on $[0, 1]$, with left-derivative $q(x)$ and right-derivative $q(x^+)$, where $q(x^+) = \lim_{y \to x, y > x} q(y)$. Since q is nondecreasing and strictly positive for $x > 0$, then $\mathcal{V}$ is convex and strictly increasing on $[0, 1]$. As a result, its inverse function $\mathcal{N}$ is concave on $(0, V_{\text{tot}})$, and has left-derivative

$$\mathcal{N}_l'(V) = \frac{1}{q(\mathcal{N}(V))} \tag{4}$$

and right-derivative

$$\mathcal{N}_r'(V) = \frac{1}{q(\mathcal{N}(V)^+)}. \tag{5}$$

The distribution F, the quantity V_{tot} as well as the functions q and $\mathcal{N}$ are illustrated in Figure 1.

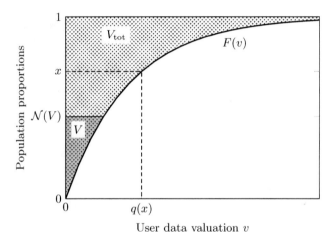

Fig. 1. Values and functions of interest regarding the user valuation distribution F

2.2 Security Systems Performance

In this paper, we focus on direct attacks targeting some specific machines, which may for instance come from an attack-generating robot that randomly chooses IP addresses and launches attacks to those hosts.

The attacks generated by such a scheme have to target a specific vulnerability of a given security system. As a result, the attacker has to select which security system $i \in \mathcal{I}$ to focus on. If an attack is launched to a security system i, we consider that all machines protected by a system $j \neq i$ do not run any risk, while the success probability of the attack is supposed to be fixed, denoted by π_i, on machines with protection system i. In other terms, the parameter π_i measures the effectiveness of the security defense.

2.3 The Attacker Point of View

Successful attacks bring some revenue to the attacker. Be it in terms of damage done to user data, or in terms of stolen data from users, it is reasonable to consider that for a given attack, the gain for the attacker is proportional to the value that the data had to the victim. Indeed, in the case of data steal, more sensitive data (e.g., bank details) are more likely to bring high revenues when used. Likewise, when the objective of the attacker is simply to maximize user damage, then the link between attacker utility and user data valuation is direct.

For a given distribution of the population among providers, let F_i be the (unconditional) distribution of valuations of users associated with provider i, so that $F = \sum_{i \in \mathcal{I}} F_i$. We then define for each provider $i \in \mathcal{I}$ the total value of the protected data, as

$$V_i := \int v \, dF_i(v). \tag{6}$$

For an attacker, the expected benefit from launching an attack targeted at system i (without knowing which users are with provider i) is thus proportional to $\pi_i V_i$. We therefore assume that the likeliness of attacks occurring on system i is a nondecreasing function of $\pi_i V_i$. We discretize time, and denote by $R_i(\pi_i V_i)$ the probability that a particular user is the target of a system-i attack over a time period. Remark that we consider system-specific functions $(R_i)_{i \in \mathcal{I}}$, so that the model can encompass some heterogeneity in the difficulty of creating system-targeted attacks.

To simplify a bit the writing, let us define $T_i(V_i)$ as the risk, for a user, of having one's data compromised when choosing security provider i. Note that it can be written as a function of the total protected data value V_i:

$$T_i(V_i) := \pi_i R_i(\pi_i V_i) = \pi_i R_i\left(\pi_i \int v dF_i(v)\right). \tag{7}$$

We will often make use of the assumption below.

Assumption A *For all $i \in \mathcal{I}$, T_i is a continuous and strictly increasing function of V_i, and $T_i(0) = 0$.*

For T_i functions of the form given in (7), Assumption A is equivalent to

- $\pi_i > 0$ for all $i \in \mathcal{I}$ (no provider offers a perfect protection against attacks),
- R_i is a continuous and strictly increasing function with $R_i(0) = 0$, for all $i \in \mathcal{I}$ (attackers do not target providers not protecting valuable data).

2.4 User Preferences

For a user u with data valuation v_u, the *total expected cost* at provider i depends on the risk of being (successfully) attacked, and on the price p_i charged by the security provider. That total cost is therefore given by

$$v_u T_i(V_i) + p_i.$$

To ensure that all users select one option, we can assume that there exists a provider i with $p_i = 0$, which would correspond to security solutions offered by free software communities (e.g., avast!®[3]). Indeed, if $p_i = 0$, the total cost is the valuation times a product of probabilities, and therefore less than the valuation itself, so that this choice of a free service is always a valuable option[4].

Remark that we consider risk-neutral users here, as may be expected from large entities, while private individuals should rather be considered risk-averse. Nevertheless, one can imagine some extra mechanisms (e.g., insurance [17]) to reach a risk-neutral equivalent formulation.

3 User Equilibrium

In this section, we investigate how demand is split among providers, when their prices p_i and security levels π_i are fixed. Recall we assumed that users are infinitely small: their individual choices do not affect the overall user distribution among providers (and therefore the total values $(V_i)_{i \in \mathcal{I}}$).

The outcome from such user interactions should be determined by user selfishness: demand should be distributed in such a way that each user u chooses one of the cheapest providers (in terms of perceived price) with respect to her valuation v_u and the current risk values $(T_i(V_i))_{i \in \mathcal{I}}$. Such a distribution of users among providers, if it exists, will be called a *user equilibrium*. In other words, if provider $i \in \mathcal{I}$ is chosen by some users u, then it is cheaper for those users (in terms of total expected cost) than any other provider $j \in \mathcal{I}$, otherwise they would be better off switching to j. Formally,

$$i \in \arg \min_{j \in \mathcal{I}} \ v_u T_j(V_j) + p_j.$$

We use here the nonatomicity assumption: each user u considers the values $(V_j)_{j \in \mathcal{I}}$ as fixed when making her individual choice.

3.1 Structure of a User Equilibrium

We now investigate the existence and uniqueness of a user equilibrium, for fixed values of prices and attack success probabilities. To do so, we first define the notion of *user repartition*.

Definition 1. *Denote by $\mathcal{P}_{\mathcal{I}}$ the set of probability distributions over providers in $\mathcal{I}$, i.e., $\mathcal{P}_{\mathcal{I}} := \{(y_1, ..., y_I) \geq 0, \sum_{i \in \mathcal{I}} y_i = 1\}$. For a given price profile $p = (p_1, ..., p_I)$, a user repartition is a mapping $A : S_v \mapsto \mathcal{P}_{\mathcal{I}}$, that is interpreted as follows:*

For all $v \in S_v$, among users with valuation v, a proportion $A_i(v)$ chooses provider i, where $A(v) = (A_1(v), ..., A_I(v))$.

---

[3] http://www.avast.com
[4] We implicitly assume here that each user u is willing to pay at least v_u to benefit from the online service.

Therefore, to a given user repartition A corresponds a unique distribution $\mathbf{V} = (V_i)_{i \in \mathcal{I}}$ of the total data valuation V_{tot} among providers, given by

$$V_i(A) = \int_{v \in S_v} v\, A_i(v)\, dF(v) \qquad \forall i \in \mathcal{I}. \tag{8}$$

Remark also that $F_i(v) = \int_{w \le v} A_i(w) dF(w)$.

Reciprocally, we say that a distribution $\mathbf{V} = (V_i)_{i \in \mathcal{I}}$ of the data valuation is *feasible* if $V_i \ge 0$ for all i, and $\sum_{i \in \mathcal{I}} V_i = V_{\text{tot}}$. For a feasible distribution $\mathbf{V}$, when providers are sorted such that $p_1 \le \dots \le p_I$, we define for each $i \in \mathcal{I} \cup \{0\}$ the quantity

$$V_{[i]} := \sum_{j=1}^{i} V_j,$$

with $V_{[0]} = 0$. $V_{[i]}$ therefore represents the total value of the data protected by the i cheapest providers.

We now formally define the outcome that we should expect from the interaction of users, i.e., an *equilibrium* situation.

Definition 2. *A user equilibrium is a user repartition A^{eq} such that no user has an interest to switch providers. In other words, for any value $v \in S_v$, a user with valuation v cannot do better than following the provider choice given by $A^{eq}(v)$. Formally, A^{eq} is a user equilibrium if and only if* $\forall v \in S_v$,

$$A_i^{eq}(v) > 0 \quad \Rightarrow \quad i \in \arg\min_{j \in \mathcal{I}} v T_j(V_j(A^{eq})) + p_j, \tag{9}$$

where $V_j(A^{eq})$ is given by (8).

We now establish some monotonicity properties that should be verified by a user equilibrium: if a user y values her data strictly less than another user x, then she selects cheaper (in terms of price) providers than x.

Lemma 1. *Consider a user equilibrium A^{eq}. Then user choices -in terms of price of the chosen provider(s)- are monotone in their valuation: for any two users x and y with respective valuations v_x and v_y, and any providers i and j,*

$$(v_x - v_y) \cdot A_i^{eq}(v_x) \cdot A_j^{eq}(v_y) > 0 \quad \Rightarrow \quad p_i \ge p_j. \tag{10}$$

Proof. Let us write $V_i := V_i(A^{eq})$ and $V_j := V_j(A^{eq})$. From (9) applied to users x and y, the left-hand inequality of (10) implies

$$v_x T_i(V_i) + p_i \le v_x T_j(V_j) + p_j$$
$$\text{and} \qquad v_y T_i(V_i) + p_i \ge v_y T_j(V_j) + p_j. \tag{11}$$

Subtracting those inequalities gives $T_i(V_i) \le T_j(V_j)$ since $(v_x - v_y) > 0$. Then (11) yields the right-hand side of (10).

We then use that result to prove that for a given value repartition $(V_i)_{i \in \mathcal{I}}$ over the providers, there can be only one equilibrium repartition if all providers set different prices.

Lemma 2. *Assume that all providers set different prices. If a user equilibrium exists, it is completely characterized (unless for a zero-measure set of users) by the total values* $(V_i)_{i \in \mathcal{I}}$ *of protected data for each provider* $i \in \mathcal{I}$, *provided that* $\sum_{i \in \mathcal{I}} V_i = V_{\text{tot}}$.

Proof. Without loss of generality, assume that provider prices are sorted, such that $p_1 < p_2 < ... < p_I$.

From Definition 1 and (8), to a given equilibrium corresponds a unique set of values $(V_i)_{i \in \mathcal{I}}$.

Reciprocally, consider a feasible data value repartition $\mathbf{V} = (V_i)_{i \in \mathcal{I}}$, and assume it corresponds to a user equilibrium A^{eq}. Since we do not differentiate users with similar valuations, we can sort them -still without loss of generality- in an increasing order of the price of their chosen provider: if $x < y$ and $q(x) = q(y)$ then we can impose that $p_{i_x} \le p_{i_y}$, where i_x (resp. i_y) would be the (unique) provider chosen by user at position x (resp. y) in the user valuation ranking. Therefore from Lemma 1, at the user equilibrium A^{eq}, provider prices can be considered as sorted in a increasing order of user valuations among *all* users. Thus, user choices are uniquely (unless on a zero-measure user set) determined by their position $x \in [0, 1]$ in the user valuation ranking, and given by

$$\mathcal{V}(x) \in (V_{[i-1]}, V_{[i]}) \quad \Rightarrow \text{user } x \text{ selects provider } i, \tag{12}$$

where $\mathcal{V}$ is defined in (3).

3.2 The Case of Several Providers with the Same Price

In this subsection, we establish a way to consider several providers with the same price as one single option from the user point of view. Let us consider a common price p, and define $\mathcal{I}_p := \{i \in \mathcal{I} : p_i = p\}$.

First, if one such provider i gets positive demand (i.e., $V_i > 0$), then at a user equilibrium all providers with the same price also get positive demand: indeed, Assumption A implies that $T_i(V_i) > 0$, and thus the total cost of a user u with positive valuation choosing provider $i \in \mathcal{I}_p$ is $v_u T_i(V_i) + p > p$. Therefore each provider $j \in \mathcal{I}_p$ necessarily has a strictly positive T_j, otherwise it would have cost $v_u T_j(0) + p = p$ for user u, who would be better off switching from i to j. Consequently, at a user equilibrium we necessarily have $T_i(V_i) = T_j(V_j)$.

When the set of users choosing one of the providers with price p is fixed, so is the total valuation $V_{\mathcal{I}_p}$ of those users' data. Consequently, the distribution of users among all providers in $\mathcal{I}_p$ should be such that

$$\begin{cases} i, j \in \mathcal{I}_p \Rightarrow T_i(V_i) = T_j(V_j) \\ \sum_{i \in \mathcal{I}_p} V_i = V_{\mathcal{I}_p}. \end{cases} \tag{13}$$

Following [2], we reformulate (13) as a minimization problem:

$$(V_i)_{i \in \mathcal{I}_p} \in \arg\min_{(x_i)_{i \in \mathcal{I}_p} \ge 0} \sum_{i \in \mathcal{I}_p} \int_{y=0}^{x_i} T_i(y) dy \tag{14}$$

$$\text{s.t.} \sum_{i \in \mathcal{I}_p} x_i = V_{\mathcal{I}_p}.$$

Under Assumption A, there exists a unique vector of values $(V_i)_{i \in \mathcal{I}_p}$ satisfying the above system. In the following, we will denote by $T_{\mathcal{I}_p}(V)$ the corresponding common value of $T_i(V_i)$. Interestingly, remark that the function $T_{\mathcal{I}_p}$ that we have defined also satisfies Assumption A. As a result, in the rest of the analysis of user equilibria, we will associate providers with the same price p and consider them as a single choice $\mathcal{I}_p$ that we assimilate as a single provider k, with corresponding risk function $T_k(V) := T_{\mathcal{I}_p}(V)$ satisfying Assumption A.

3.3   Game Equilibrium as a Solution of an Optimization Problem

Based on the reasoning in Subsection 3.2, we assume that all providers submit a different price, and we sort them such that $p_1 < \ldots < p_I$. Now let us consider the following measure:

$$\mathcal{L}(\mathbf{V}, \mathbf{p}) := \sum_{i \in \mathcal{I}} \left(\int_{y=0}^{V_i} T_i(y)dy + p_i \underbrace{\left(\mathcal{N}(V_{[i]}) - \mathcal{N}(V_{[i-1]}) \right)}_{\text{Market share of prov. } i} \right) \tag{15}$$

$$= \sum_{i=1}^{I} \int_{y=0}^{V_i} T_i(y)dy + p_I - \sum_{i=1}^{I-1} (p_{i+1} - p_i)\mathcal{N}(V_{[i]}), \tag{16}$$

with $p_0 := 0$. Remark that the first part of the quantity $\mathcal{L}(\mathbf{V}, \mathbf{p})$ in (15) is the potential function usually associated to unweighted congestion games (see, e.g., [2]), while the second part stands for the total price paid by all users.

The expression (16) highlights the fact that $\mathcal{L}$ is a strictly convex function of $\mathbf{V}$, since $\mathcal{N}$ is concave and under Assumption A, T_i is strictly increasing. It thus admits a unique minimum $\mathbf{V}^*$ on the (convex) domain of feasible value shares; and $\mathbf{V}^*$ is completely characterized by the first-order conditions. We now prove that this valuation repartition $\mathbf{V}^*$ actually corresponds to a user equilibrium.

Proposition 1. *Let Assumption A hold. For any price profile* $\mathbf{p}$*, there exists a user equilibrium, that is completely characterized by the valuation repartition* $\mathbf{V}^*$*, unique solution of the convex optimization problem*

$$\min_{\mathbf{V} \text{ feasible}} \mathcal{L}(\mathbf{V}, \mathbf{p}). \tag{17}$$

Proof. We first consider the feasible directions consisting in switching some infinitesimal amount of value from $i > 1$ to $j < i$, when $V_i^* > 0$. The optimality condition in (16) then yields

$$0 \leq T_j(V_j^*) - T_i(V_i^*) - \sum_{k=j}^{i-1} (p_{k+1} - p_k)\mathcal{N}'_r(V_{[k]}^*)$$

$$\leq T_j(V_j^*) - T_i(V_i^*) - (p_i - p_j)\mathcal{N}'_r(V_{[i-1]}^*), \tag{18}$$

where the second line comes from the concavity of $\mathcal{N}$.

Notice that since $p_j < p_i$ and $\mathcal{N}$ is nondecreasing,(18) and Assumption A imply that $V_j^* > 0$. Consequently, if we define $i^* := \max\{i \in \mathcal{I} : V_i^* > 0\}$, then

$$V_i^* > 0 \Leftrightarrow i \leq i^*. \tag{19}$$

As a result, since $V_i > 0$ and $i > 1$ in (18), then $0 < V_{[i-1]}^* < V_{\text{tot}}$. Thus, from (5), $\mathcal{N}_r'(V_{[i-1]}^*) = \frac{1}{q(\mathcal{N}(V_{[i-1]}^*))}$ is strictly positive. (18) is then equivalent to

$$\underline{v}_i^* T_i(V_i^*) + p_i \leq \underline{v}_i^* T_j(V_j^*) + p_j, \tag{20}$$

with $\underline{v}_i^* := q(\mathcal{N}(V_{[i-1]}^*)^+) = \inf\{v : \int_{u=0}^{v} u dF(u) > V_{[i-1]}^*\}$. Remark that necessarily from (20), $T_i(V_i^*) < T_j(V_j^*)$ since $p_i > p_j$.

For $i < I$ such that $V_i^* > 0$ (i.e., $i \leq i^*$), we now investigate the possibility of switching some value from i to $j > i$. Still applying the optimality condition for $\mathbf{V}^*$, we get

$$0 \leq T_j(V_j^*) - T_i(V_i^*) + \sum_{k=i}^{j-1}(p_{k+1} - p_k)\mathcal{N}_l'(V_{[k]}^*)$$
$$\leq T_j(V_j^*) - T_i(V_i^*) + (p_j - p_i)\mathcal{N}_l'(V_{[i]}^*), \tag{21}$$

where we used again the concavity of $\mathcal{N}$.

Applying (4), Relation (21) is equivalent to

$$\bar{v}_i^* T_i(V_i^*) + p_i \leq \bar{v}_i^* T_j(V_j^*) + p_j, \tag{22}$$

with $\bar{v}_i^* = q(\mathcal{N}(V_{[i]}^*)) = \inf\{v : \int_{u=0}^{v} u dF(u) \geq V_{[i]}^*\}$.

Relations (20) and (22) can be interpreted as users with valuation $v \in [\underline{v}_i^*, \bar{v}_i^*]$ preferring provider i over any other one, for the repartition value $\mathbf{V}^*$. Formally,

$$v \in [\underline{v}_i^*, \bar{v}_i^*] \quad \Rightarrow \quad i \in \arg\min_{j \in \mathcal{I}} vT_j(V_j^*) + p_j. \tag{23}$$

Now, consider the provider choices induced by the value repartition $\mathbf{V}^*$ as given in (12). We prove here that this repartition is a user equilibrium: no user has an interest to change providers. Take a provider $i \in \mathcal{I}$. Then for $x \in [0,1]$,

$$\mathcal{V}(x) \in (V_{[i-1]}^*, V_{[i]}^*) \Leftrightarrow V_{[i-1]}^* < \int_{y=0}^{x} q(y)dy < V_{[i]}^*$$
$$\Leftrightarrow \mathcal{N}(V_{[i-1]}^*) < x < \mathcal{N}(V_{[i]}^*)$$
$$\Rightarrow \underline{v}_i^* \leq q(x) \leq \bar{v}_i^*.$$

The last line and (23) imply that the considered user, that is at position x in the population when it is ranked according to valuations, cannot do better than choosing the provider suggested by (12). In other words, each user is satisfied with her current provider choice, i.e., we have a user equilibrium.

We now establish the uniqueness of the equilibrium value repartition $\mathbf{V}^*$ (and thus, of the user equilibrium due to Lemma 2 when all prices are different).

Proposition 2. *Under Assumption A, the value repartition at a user equilibrium necessarily equals* $\mathbf{V}^* = \arg\min_{\mathbf{V} \text{ feasible}} \mathcal{L}(\mathbf{V}, \mathbf{p})$. *Consequently, there exists a unique value equilibrium value repartition, and the user equilibrium is unique (unless for a zero-measure set of users) when all providers set different prices.*

Proof. We consider a user equilibrium, and prove that the corresponding value repartition $\tilde{\mathbf{V}}$ satisfies the first-order conditions of the convex optimization problem (17), that has been shown to have a unique solution $\mathbf{V}^*$.

We actually only need to show the counterpart of Relation (18) (resp., (21)) for $j = i-1$ (resp., $j = i+1$), since the other cases immediately follow. From (12), at a user equilibrium we should have for all $x \in (0,1)$ and all $i, j \in \mathcal{I}$,

$$x \in \left(\mathcal{N}(\tilde{V}_{[i-1]}), \mathcal{N}(\tilde{V}_{[i]})\right) \quad \Rightarrow \quad q(x)(T_i(\tilde{V}_i) - T_j(\tilde{V}_j)) + p_i - p_j \leq 0. \quad (24)$$

Consider $i \in \mathcal{I}$ such that $\tilde{V}_i > 0$.

- If $j = i - 1$, then $T_i(\tilde{V}_i) < T_j(\tilde{V}_j)$. When x tends to $\mathcal{N}(\tilde{V}_{[i-1]})$, (24) yields

$$\underbrace{q(\mathcal{N}(\tilde{V}_{[i-1]})^+)}_{=\mathcal{N}'_r(\tilde{V}_{[i-1]})}(T_i(\tilde{V}_i) - T_j(\tilde{V}_j)) + p_i - p_j \leq 0,$$

  which is exactly the counterpart of (18).
- Likewise for $j = i + 1$, from (24) for x tending to $\mathcal{N}(\tilde{V}_{[i]})$ we get the counterpart of (21) (using the fact that q is left-continuous)

$$\underbrace{q(\mathcal{N}(\tilde{V}_{[i]}))}_{=\mathcal{N}'_l(\tilde{V}_{[i]})}(T_i(\tilde{V}_i) - T_j(\tilde{V}_j)) + p_i - p_j \leq 0.$$

The repartition $\tilde{\mathbf{V}}$ satisfies the first-order conditions of the convex optimization problem (17) and is feasible, therefore $\tilde{\mathbf{V}} = \mathbf{V}^*$, the unique solution of (17).

The second claim of the proposition is a direct application of Lemma 2.

Note that the uniqueness of the equilibrium value repartition $\mathbf{V}^*$ implies that even when several user equilibria exist, for all users the cost of each provider at equilibrium is unique; the user equilibrium is then said *essentially unique* [2].

Note also that it was not compulsory to aggregate providers with the same price p: at the minimum of $\mathcal{L}(\cdot, \mathbf{p})$ we notice from (14) that the term $\int_0^{V_{\mathcal{I}_p}} T_{\mathcal{I}_p}$ involving the aggregated function coincides with $\sum_{i \in \mathcal{I}_p} \int_{y=0}^{x_i} T_i(y)dy$. Therefore, the equilibrium value distribution $\mathbf{V}^*$ can directly be found by solving the potential minimization problem (17). Nevertheless, the interpretation of the potential is changed, since the terms $\mathcal{N}(V_{[i]}) - \mathcal{N}(V_{[i-1]})$ of (15) do not necessarily correspond anymore to provider i's market share.

The next result shows some continuity properties of the user equilibrium.

Proposition 3. *The (unique) equilibrium value repartition $\mathbf{V}^*$ is continuous in the price profile. Moreover, at any price profile such that all prices are different, the provider market shares are continuous in the price profile.*

Proof. Remark that $\mathcal{L}(\mathbf{V}, \mathbf{p})$ is jointly continuous in $\mathbf{V}$ and $\mathbf{p}$, and that the set of feasible value repartitions is compact. Therefore, from the Theorem of the Maximum (see [3]) applied to the minimization problem (17), the set of equilibrium distributions is upper hemicontinuous in $\mathbf{p}$. It is actually continuous due to the uniqueness of the equilibrium distribution $\mathbf{V}^*$.

For a given price profile $\bar{\mathbf{p}}$ where all prices differ, the strict order of prices is maintained within a vicinity of $\bar{\mathbf{p}}$, where the market share of provider i is $\mathcal{N}(V_{[i]}^*) - \mathcal{N}(V_{[i-1]}^*)$, which is jointly continuous in $\mathbf{V}$ and $\mathbf{p}$ since $\mathcal{N}$ is continuous.

Note that while the equilibrium value repartition $\mathbf{V}^*$ is continuous for all price profiles, that is not the case of provider market shares. Indeed, market shares $(\theta_i)_{i \in \mathcal{I}}$ strongly depend on the *order* of prices through the expression $\mathcal{N}(V_{[i]}^*) - \mathcal{N}(V_{[i-1]}^*)$, that holds when prices are sorted in an increasing order. Since $\mathcal{N}$ is a concave function, then the market share of a provider may drastically decrease when a slight price modification changes his position from k to $k+1$ in the price ranking. This effect is more prominent when $\mathcal{N}$ is more concave, i.e., when user valuations are heterogeneous.

3.4   Price of Anarchy of the User Game

In non-cooperative games, the Price of Anarchy measures the loss of efficiency due to user selfishness [16]. This metric is usually defined as the worst-case ratio of the total cost at an equilibrium to the minimal feasible total cost, and has been extensively studied in the last years [7,24,25,26]. The results closest to the one presented in this subsection come from [4]: the authors consider weighted congestion games, where the cost experienced by each user would correspond to the situation where all prices are set to 0 in our model. Then the authors prove that the upper bound for the Price of Anarchy is not greater for the weighted game than for its unweighted counterpart. We actually establish the same kind of result for any value of the provider price profile $\mathbf{p}$, except that in our case the total user cost (sum of the costs perceived by all users) for any feasible user valuation repartition $\mathbf{V}$ is

$$C_{\mathrm{u}} := \sum_{i \in \mathcal{I}} \left(V_i T_i(V_i) + p_i(\mathcal{N}(V_{[i]}) - \mathcal{N}(V_{[i-1]})) \right). \tag{25}$$

Proposition 4. *Assume that the risk functions $(T_i)_{i \in \mathcal{I}}$ belong to a family $\mathcal{C}$, and define as in [7] the quantity $\beta(\mathcal{C}) := \sup\limits_{T \in \mathcal{C}, (x,y) \in [0, V_{\mathrm{tot}}]^2} \dfrac{x(T(y) - T(x))}{y T(y)}$. Then for any nonnegative price profile $\mathbf{p}$,*

$$\frac{C_{\mathrm{u}}^*}{C_{\mathrm{u}}^{\mathrm{opt}}} \le \frac{1}{1 - \beta(\mathcal{C})}, \tag{26}$$

where C_{u}^* (resp. $C_{\mathrm{u}}^{\mathrm{opt}}$) is the total user cost at the user equilibrium (resp. the minimum total user cost) for the price profile $\mathbf{p}$.

Proof. We apply a variational inequality that is satisfied by the user equilibrium value repartition $\mathbf{V}^*$, and that directly stems from the fact that users only select their preferred provider: for any feasible value repartition $\mathbf{V}$, we have

$$\sum_{i \in \mathcal{I}} \left(V_i^* T_i(V_i^*) + p_i (\mathcal{N}(V_{[i]}^*) - \mathcal{N}(V_{[i-1]}^*)) \right) \leq \sum_{i \in \mathcal{I}} \left(V_i T_i(V_i^*) + p_i (\mathcal{N}(V_{[i]}) - \mathcal{N}(V_{[i-1]})) \right).$$

This yields

$$C_{\mathrm{u}}^* \leq C_{\mathrm{u}} + \sum_{i \in \mathcal{I}} V_i (T_i(V_i^*) - T_i(V_i)) \leq C_{\mathrm{u}} + \beta(\mathcal{C}) \sum_{i \in \mathcal{I}} V_i^* T_i(V_i^*) \leq C_{\mathrm{u}} + \beta(\mathcal{C}) C_{\mathrm{u}}^*,$$

which establishes the proposition.

It is shown in [7] that if $\mathcal{C}$ is the set of affine risk functions the bound $1/(1 - \beta(\mathcal{C}))$ equals $4/3$, resulting in a moderate loss of efficiency due to selfishness. Values 1.626 and 1.896 have also been found respectively for the sets of quadratic and cubic cost risk functions, and $\beta(\mathcal{C}) = d/(d+1)^{1+1/d}$ for the set of polynomials of degree at most d with non-negative coefficients.

As in [4], we find that the introduction of weights among user congestion effects (and here, in addition, among user perceived costs) does not worsen the Price of Anarchy. The bound given in Proposition 4 can indeed be attained, when $\mathcal{C}$ includes the constant functions, with a simple 2-provider instance with prices set to zero, and all users having the same weight.

4   Pricing Decisions of Security Providers

We now focus on the decisions made by security providers when choosing their charging price. We consider that providers are able to anticipate user reactions when fixing their prices. We then have a two-stage game, where at a first step (larger time scale) providers compete on setting their prices so as to maximize revenue, considering that at a second step (smaller time scale) users selfishly select their provider.

The utility of provider i is given by his revenue $r_i := p_i \theta_i$, where θ_i is the market share of provider i. When all providers propose different prices and providers are ranked such that $p_1 < p_2 < ... < p_I$, from Proposition 2 the user equilibrium exists and is unique, and we simply have $\theta_i = \mathcal{N}(V_{[i]}^*) - \mathcal{N}(V_{[i-1]}^*)$, where $\mathbf{V}^*$ is the equilibrium value repartition. On the other hand, if several providers in a set $\mathcal{I}_p$ propose the same price p, then the equilibrium valuation repartition $\mathbf{V}^*$ is unique, but the user equilibrium choices need not be unique: indeed, any price-monotone user repartition consistent with $\mathbf{V}^*$ is a user equilibrium, and several such repartitions may exist. For those special cases, a reasonable assumption could be that users make their provider choice independently of their valuation when they have several equally preferred providers. As a result, the total market share of providers in $\mathcal{I}_p$ would be split among them proportionally to the data value V_i^* that they attract, yielding

$$\theta_i = \frac{V_i^*}{\sum_{j:p_j = p_i} V_j^*} \left(\mathcal{N}\left(\sum_{j:p_j \leq p_i} V_j^* \right) - \mathcal{N}\left(\sum_{j:p_j < p_i} V_j^* \right) \right).$$

We now establish that, when there exists a bounded price alternative, the revenue of any provider tends to zero if he increases his price to infinity. In practice, such a bounded-price option always exists, even if it has bad performance: one just needs to consider any free security possibility. Therefore, prices will not be arbitrarily high when providers want to maximize revenue.

Proposition 5. *Assume that there exists a provider i_0 with price $p_{i_0} \leq \bar{p}_{i_0} < \infty$. Then for any provider $j \neq i_0$, the revenue $r_j = p_j \theta_j$ tends to 0 when $p_j \to \infty$.*

Proof. Let us consider a user with valuation v, for whom provider j is among the favorite providers. In particular, that user prefers j over i_0, thus at a user equilibrium we have

$$v(T_{i_0}(V_{i_0}) - T_j(V_j)) \geq p_j - p_{i_0} \geq p_j - \bar{p}_{i_0}. \tag{27}$$

Therefore if $p_j > \bar{p}_{i_0}$ then $T_j(V_j) < T_{i_0}(V_{i_0})$ and

$$v \geq \frac{p_j - \bar{p}_{i_0}}{T_{i_0}(V_{i_0}) - T_j(V_j)} \geq \frac{p_j - \bar{p}_{i_0}}{T_{i_0}(V_{\text{tot}})} := v_{\min}.$$

The revenue $r_j = p_j \theta_j$ of provider j can then be upper bounded:

$$r_j \leq p_j \int_{v=v_{\min}}^{+\infty} dF(v) = T_{i_0}(V_{\text{tot}}) \frac{p_j - \bar{p}_{i_0}}{T_{i_0}(V_{\text{tot}})} \underbrace{\int_{v=\frac{p_j-\bar{p}_{i_0}}{T_{i_0}(V_{\text{tot}})}}^{+\infty} dF(v)}_{\xrightarrow[p_j \to \infty]{} 0} + \bar{p}_{i_0} \underbrace{\int_{v=\frac{p_j-\bar{p}_{i_0}}{T_{i_0}(V_{\text{tot}})}}^{+\infty} dF(v)}_{\xrightarrow[p_j \to \infty]{} 0},$$

where the two terms tend to zero since $\int_0^\infty v dF(v) = V_{\text{tot}} < \infty$.

4.1 Licensed versus Free Security Provider

We consider here a simple situation with two providers, but only one trying to maximize his profit through subscription benefits. The other provider (or, more likely, a community of developers) offers the security service for free.

Denote by 0 and 1 the freeware provider and the licensed provider, respectively. From Proposition 1, there exists a unique value repartition $(V_0(p), V_{\text{tot}} - V_0(p))$ at the user equilibrium, for any price p set by provider 1. Likewise, for any $p > 0$ the equilibrium market share of provider 1 is unique and given by $\theta_1 = 1 - \mathcal{N}(V_0(p))$; the profit maximization problem of provider 1 can therefore be written as

$$\max_{p \geq 0} \quad p \cdot (1 - \mathcal{N}(V_0(p))). \tag{28}$$

Note that provider 1 gets demand as soon as his price is strictly below $\sup(S_v) \times T_0(V_{\text{tot}})$, therefore by choosing $p \in (0, \sup(S_v)T_0(V_{\text{tot}}))$ he can ensure a positive revenue. Therefore from Propositions 3 and 5, the provider revenue optimization problem (28) has a solution, that is finite.

Corollary 1. *When a profit-oriented provider faces only a competitor with null price, then under Assumption A there exists a finite price $\bar{p} > 0$ that maximizes his revenue, whose maximum value is strictly positive.*

4.2   Competition among Providers: The Risk of Price War

Competitive contexts where providers play on price to attract customers often lead to *price war* situations, i.e., situations where each provider has an interest in decreasing his price below the price of his competitor. The outcome then corresponds to providers making no profit, and possibly not surviving.

With the model presented in this paper, not all demand goes to the cheapest provider because of the congestion effect due to attackers' behavior. However, some threshold effect still exist, as illustrated by the non-continuity of provider market shares when provider prices cross each other.

Let us for example consider two identical profit-oriented providers and a free alternative. Due to the symmetry of the game, one would expect a situation where both providers set their price to the same level, say $p > 0$. As a result, again from symmetry arguments both providers would be chosen by users to protect, at equilibrium, the same value $V_1^* = V_2^* := V^*$ of data each, while the free provider covers a total data value V_0. Then, if provider 1 sets his price to $p - \varepsilon$ for a small $\varepsilon > 0$, the market share repartition is such that when $\varepsilon \to 0$,

$$\theta_0 = \mathcal{N}(V_0^*),$$
$$\theta_1 = \mathcal{N}(V_0^* + V^*) - \mathcal{N}(V_0^*),$$
$$\theta_2 = \mathcal{N}(V_0^* + 2V^*) - \mathcal{N}(V_0^* + V^*).$$

When users choosing provider 1 or 2 are not all homogeneous in their data valuations (which is for example the case if the valuation distribution F admits a density), then $\theta_1 > \theta_2$. In other words, provider 1 strictly improves his market share (and thus his revenue) by setting his price just below the price of his competitor. But provider 2 can make the exact same reasoning, resulting in a price war situation.

Consequently, there can be no symmetric Nash equilibrium (i.e., a price profile such that no provider can improve his revenue by a unilateral change) where $p_1 = p_2 > 0$, despite the symmetry of the pricing game. Furthermore, the price profile where all prices are set to 0 is not an equilibrium either: both providers would get no revenue, which each one could strictly improve by a small price increase as stated in Corollary 1.

Remark that this reasoning does not rule out the possibility of the pricing game having a (non-symmetric) Nash equilibrium, however we cannot always guarantee that such an equilibrium exists. An explanation to the existence of stable price profiles can nevertheless still be found from game-theoretic arguments, since the pricing game among providers is not played only once but repeatedly over time. When considering *repeated games* (i.e., where players take into account not only their current payoff but also a discounted sum of the future ones), the set of Nash equilibria is indeed much larger than for their one-shot counterpart, as evidenced by the *Folk theorem* [23]. The stability of prices can then stem from the threat of being sanctioned by competitors for an (immediate-profit) price change.

We illustrate those results when user valuations are distributed according to an exponential law with average value $1/\lambda = 10$ monetary units. Such a distribution models an unbounded continuum of valuations among the population, where a large majority of users have limited valuations, but there exist few people with extremely high value data to protect. The risk function considered in our numerical computations is $R_i(x) = 1 - e^{-x}$ for each provider i, which models the fact that systems with no valuable data are not targeted while successful systems are very likely to attract attacks.

In our numerical illustration, we consider here three providers: a provider 0 with performance parameter $\pi_0 = 0.05$, that is always free: $p_0 = 0$; and two profit-oriented providers, namely 1 and 2, with respective performance values $\pi_1 = 0.01$ and $\pi_2 = 0,005$. Providers protected data values and market shares are shown in Figures 2 and 3, and the revenue of provider 2 is displayed in Figure 4. The curves illustrate the continuity results of Proposition 3. Interestingly, we remark in Figure 4 that despite the discontinuity in revenue when prices cross each other, provider 2 actually has a revenue-maximizing price $p_2^{\mathrm{BR}}(p_1)$ strictly below the price of his competitor. That last figure shows the price war situation: if providers engage in successive best-reply price adaptations to the competition, then prices tend to very low values, which jeopardizes the viability of security providers. However, a situation with strictly positive prices from both providers could be stable in a repeated game context. Consider a price profile (p_1, p_2) such that each provider obtains at least what he could obtain with an aggressive competitor (i.e., a competitor that tries to minimize the provider revenue); when providers value the future almost as much as the present (i.e., when the discount factor that relates current prices to future prices is close to 1), that price profile can be maintained as a subgame-perfect equilibrium of the repeated game [9].

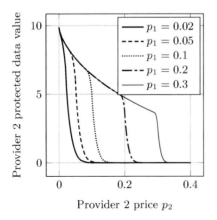
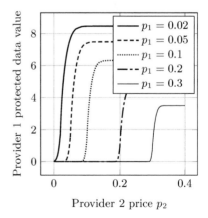

Fig. 2. Protected data values when provider 2 varies his price

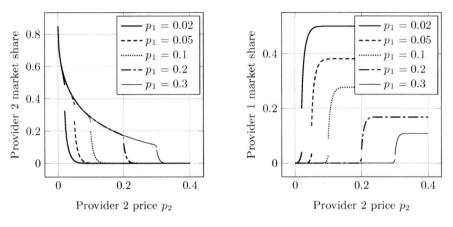

Fig. 3. Market share of provider 1, when p_1 varies

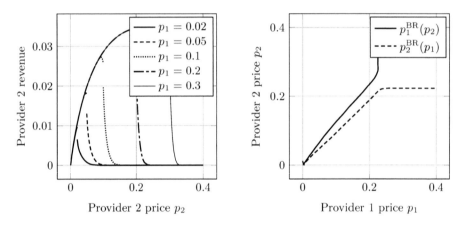

Fig. 4. Revenue of provider 2 ($\pi_2 = 0.005$) facing provider 1 ($\pi_1 = 0.01$) and free provider 0 ($\pi_0 = 0.05$) *(left)*, and best-reply functions of providers 1 and 2 *(right)*

5 Conclusions

The model introduced in this paper takes into account the attractiveness that successful security systems represent to profit-minded attackers. This constitutes a negative externality among users: their (selfish) security choices then form a noncooperative congestion game. We have considered heterogeneity among user valuations for data protection, which affects both the externality level and the user cost functions. The corresponding game is therefore a weighted congestion game with user-specific payoffs. We have studied that game for the case of a continuum of infinitesimal users, and have proved that it admits a potential and therefore an equilibrium, that is unique when providers submit different prices.

The study of the user selection game has helped us understand the interaction among security providers, who have to attract customers but are then subject to quality degradation due to more attacks, hence a trade-off. Our analysis shows that providers will keep their prices low, and that competition may lead to price war situations, unless providers consider long-term repeated interactions.

Future work can focus on the information asymmetry and uncertainty among actors: we have studied the interactions in a complete information context, whereas users may not have a perfect knowledge of the performance level of the different providers, or of their total protected data value. Likewise, attackers can only estimate the potential gain from targeting a given system.

Another interesting direction for future research concerns the investment strategies that security providers should implement: indeed, improving the protection performance has a cost, that has to be compensated by the extra revenue due to user subscription decisions. While there exist references for this kind of problem when users are homogeneous [13], the case when users have different weights deserves further attention.

Acknowledgements. The authors acknowledge the support of European initiative COST IS0605, Econ@tel. Part of this work has been supported by the Austrian government and the city of Vienna in the framework of the COMET competence center program, and by the French research agency through the CAPTURES project.

References

1. Alpcan, T., Başar, T.: Network Security: A Decision and Game Theoretic Approach. Cambridge University Press (2011)
2. Beckmann, M., McGuire, C.B., Winsten, C.B.: Studies in the economics of transportation. Yale University Press, New Heaven (1956)
3. Berge, C.: Espaces topologiques. Fonctions multivoques, Collection Universitaire de Mathématiques, Dunod, Paris, vol. III (1959)
4. Bhawalkar, K., Gairing, M., Roughgarden, T.: Weighted Congestion Games: Price of Anarchy, Universal Worst-Case Examples, and Tightness. In: de Berg, M., Meyer, U. (eds.) ESA 2010. LNCS, vol. 6347, pp. 17–28. Springer, Heidelberg (2010)
5. Bistarelli, S., Dall'Aglio, M., Peretti, P.: Strategic Games on Defense Trees. In: Dimitrakos, T., Martinelli, F., Ryan, P.Y.A., Schneider, S. (eds.) FAST 2006. LNCS, vol. 4691, pp. 1–15. Springer, Heidelberg (2007)
6. Bohacek, N., Hespanha, J.P., Lee, J., Lim, C., Obraczka, K.: Game theoretic stochastic routing for fault tolerance and security in computer networks. IEEE Transactions on Parallel and Distributed Systems 18(9), 1227–1240 (2007)
7. Correa, J.R., Schulz, A.S., Stier-Moses, N.: A geometric approach to the price of anarchy in nonatomic congestion games. Games and Economic Behavior 64(2), 457–469 (2008)
8. Cremonini, M., Nizovtsev, D.: Understanding and influencing attackers' decisions: Implications for security investment strategies. In: Proc. of 5th Workshop on the Economics of Information Security (WEIS), Cambridge, UK (2006)

9. Fudenberg, D., Maskin, E.: The folk theorem in repeated games with discounting or with incomplete information. Econometrica 54(3), 533–554 (1986)
10. Fudenberg, D., Tirole, J.: Game Theory. MIT Press (1991)
11. Ganesh, A., Gunawardena, D., Jey, P., Massoulié, L., Scott, J.: Efficient quarantining of scanning worms: Optimal detection and co-ordination. In: Proc. of IEEE INFOCOM 2006, Barcelona, Spain (2006)
12. Jiang, L., Anantharam, V., Walrand, J.: Efficiency of selfish investments in network security. In: Proc. of 3rd Workshop on the Economics of Networks, Systems, and Computation, Seattle, WA, USA (2008)
13. Johari, R., Weintraub, G.Y., Van Roy, B.: Investment and market structure in industries with congestion. Operations Research 58(5), 1303–1317 (2010)
14. Karakostas, G., Kolliopoulos, S.G.: Edge pricing of multicommodity networks for heterogeneous selfish users. In: Proc. of FOCS, pp. 268–276 (2004)
15. Kodialam, M., Lakshman, T.V.: Detecting network intrusions via sampling: A game theoretic approach. In: Proc. of IEEE INFOCOM, San Francisco, CA, USA (2003)
16. Koutsoupias, E., Papadimitriou, C.: Worst-Case Equilibria. In: Meinel, C., Tison, S. (eds.) STACS 1999. LNCS, vol. 1563, pp. 404–413. Springer, Heidelberg (1999)
17. Lelarge, M., Bolot, J.: Economic incentives to increase security in the internet: The case for insurance. In: Proc. of IEEE INFOCOM, Rio de Janeiro, Brazil (2009)
18. Maillé, P., Reichl, P., Tuffin, B.: Of threats and costs: A game-theoretic approach to security risk management. In: Gülpınar, N., Harrison, P., Rüstem, B. (eds.) Performance Models and Risk Management in Communication Systems. Springer, Heidelberg (2010)
19. Mavronicolas, M., Milchtaich, I., Monien, B., Tiemann, K.: Congestion Games with Player-Specific Constants. In: Kučera, L., Kučera, A. (eds.) MFCS 2007. LNCS, vol. 4708, pp. 633–644. Springer, Heidelberg (2007)
20. Milchtaich, I.: Congestion games with player-specific payoff functions. Games and Economic Behavior 13(1), 111–124 (1996)
21. Milchtaich, I.: Weighted congestion games with separable preferences. Games and Economic Behavior 67(2), 750–757 (2009),
http://www.sciencedirect.com/science/article/B6WFW-4WOSK2D-1/
2/63c8ccc38f57d26ba8db4ed12c1596d3
22. Monderer, D., Shapley, L.S.: Potential games. Games and Economic Behaviour 14, 124–143 (1996)
23. Osborne, M.J., Rubinstein, A.: A Course in Game Theory. MIT Press (1994)
24. Perakis, G.: The "Price of Anarchy" under nonlinear and asymmetric costs. Mathematics of Operations Research 32(3), 614–628 (2007)
25. Roughgarden, T.: Selfish Routing and the Price of Anarchy. MIT Press (2005)
26. Roughgarden, T., Tardos, E.: Bounding the inefficiency of equilibria in nonatomic congestion games. Games and Economic Behavior 47(2), 389–403 (2004)
27. Sandholm, W.H.: Potential games with continuous player sets. Journal of Economic Theory 97(1), 81–108 (2001)
28. Sandholm, W.H.: Large population potential games. Journal of Economic Theory 144(4), 1710–1725 (2009)
29. Schmeidler, D.: Equilibrium points of nonatomic games. Journal of Statistical Physics 7(4), 295–300 (1973)

Network Games with and without Synchroneity

Ahmad Termimi Ab Ghani and Kazuyuki Tanaka

Mathematical Institute
Tohoku University
Sendai 980-8578 Japan
{sa9d12,tanaka}@math.tohoku.ac.jp

Abstract. To formulate a network security problem, Mavronicholas et al. [6] introduced a strategic game on an undirected graph whose nodes are exposed to infection by attackers, and whose edges are protected by a defender. Subsequently, MedSalem et al. [9] generalized the model so that they have many defenders instead of a single defender. Then in [1], we introduced a new network game with the roles of players interchanged, and obtained a graph-theoretic characterization of its pure Nash equilibria. In this paper we study mixed Nash equilibria for stochastic strategies in this new game, and then we generalize our network game to an asynchronous game, where two players repeatedly execute simultaneous games. Although the asynchronous game is formally an infinite game, we show that it has a stable solution by reducing it to a finite game.

Keywords: Network Game, Asynchronous Game, Nash Equilibrium.

1 Introduction

Following a series of studies by Mavronicholas et al. [6], we propose a new network game where attackers aim to damage the network by attacking an edge, and defenders aim to protect the network by securing a vertex. Both the attackers and defenders make individual (but simultaneous) decisions for their placements in the network, seeking to maximize their objectives. The defenders seek to protect the network as much as possible, while each attacker wishes to avoid being caught so as to be able to damage the network. Inspired by the work of MedSalem et al. [9], we suppose that to increase the number of defenders will improve the quality of protection of the network. We model this network game as a non-cooperative strategic game on a graph, and analyze the characteristics of its Nash equilibrium [10], which is a stable solution immune from unilateral deviations, that is, each player has no incentive to deviate from his/her strategy given that other players do not deviate from theirs. Nash [11] proved that a finite game has a Nash equilibrium in mixed strategies.

We then generalize the previous network game to a new game, called an asynchronous game, which consists of sequential executions of simultaneous games with two players. The most important property of asynchronous game is that after one round of the game, the base graph may be reduced by removing an

J.S. Baras, J. Katz, and E. Altman (Eds.): GameSec 2011, LNCS 7037, pp. 87–103, 2011.

edge damaged by the attack. Then, the final goal of the attacker is to vanish the graph as quickly as possible, while that of the opponent is to keep it as long as possible. Actually, an asynchronous game can be viewed as a special Blackwell game [2]. In 1998, Martin proved that a Blackwell game with Borel payoff is determinate, i.e., it has a stable solution [5]. Thus, our game with a suitable payoff is also determinate and each player has a stable strategy. However, we show this by reducing it to a finite game without referring to the Blackwell determinacy.

1.1   Motivation

In this paper, we consider a network security problem concerning the protection of a network from harmful entities. It is said that the more widely networks grows, the more vulnerable they become to security risks. Thus, the challenge is to invent a proper theoretical model for understanding the mathematical aspects of network security. We formulate a network security problem as a strategic graph-theoretic game and study its associated Nash equilibria. More specifically, we view a network as an undirected graph whose edges are exposed to virus infection disseminated by attackers, and nodes can be protected, for instance, by a system security software. In this model, attackers and defenders over the network security have oposing aims that seek to maximize damage and protection, respectively. An attacked edge is destroyed unless one of its end nodes is protected by the security software.

Our work continues the study of the network games introduced by Mavronicholas et al. [6]. In particular, their study focused on the network security model as a graph whose vertices are exposed to infection by attackers and whose edges can be protected by a defender. Their model has been further studied in [3],[7] and [8]. Especially, it models a risk scenario for a synchronous issue of network attacks and a limited power security mechanism.

In this paper, our final aim is to study the dynamism of attack-protection effects in the network. In other words, we consider a new game where a synchronous game is executed repeatedly (unbounded many times). To invent such a game, we realize that a network game with edge-attackers and node-defenders can be more naturally repeated. In a previous game, a hub node is easier to avoid damage since it has many edges for protection, but once it is damaged, the network might have a terminal breakdown. Contrastively, in our new game, even if a trunk edge is damaged, another rooting could take its role, and thus the network could maintain major functions after the effective attack.

Our work is also motivated from the study by MedSalem et al. [9]. They establish that the increased number of the defender results in better protection of the network. Their work can be seen as a generalization of [6] where the games with many non-centralized defenders are investigated in terms of the complexity of pure Nash equilibria.

In the following sections, we first give all necessary backgrounds of our new network game (Sect. 2) and then review our previous results of pure strategies (Sect. 3). Next, we turn to the mixed strategies (Sect. 4) and discuss various conditions for computing mixed Nash equilibria, and particularly we work with games on bipartite graphs. In the final section (Sect. 5) we present a new network game where the players repeatedly make stochastic moves, that is, an asynchronous game.

2 A New Strategic Game

2.1 Basic Notions of Graph Theory

We consider an undirected graph $G = (V, E)$ where V denotes a set of vertices and E denotes a set of edges. If v is adjacent to an edge e, then we write $v \in e$. A *vertex cover* of G is a vertex set $C_V \subseteq V$ such that for each edge $e \in E$, there is a $v \in C_V$ such that $v \in e$. A *minimum* vertex cover is one that has the minimum size. An *edge cover* of G is an edge set $C_E \subseteq E$ such that $\forall v \in V, \exists e \in C_E, v \in e$. A *matching* M of G is a subset of E such that no vertex is incident to more than one edge in M (i.e. no two edges in M have a common vertex). The two ends of an edge in M are said to be *matched* under M. A matching M is said to be *maximum* if for any other matching M', $|M| \geq |M'|$. A vertex set $I_V \subseteq V$ is an *independent set* of G if for all pairs of vertices $u, v \in I_V, (u, v) \notin E$. Given $\bar{E} \subset E$ and $\bar{V} \subset V$, define a set $V(\bar{E}) = \{v \in V : v \in e \text{ for some } e \in \bar{E}\}$, and $E(\bar{V}) = \{e \in E : \exists v \in \bar{V}, v \in e\}$. We write $V(e)$ for $V(\{e\})$ and $E(v)$ for $E(\{v\})$. Let $n_V(G)$ and $n_E(G)$ denote the numbers of vertices and edges in G, respectively. Whenever no confusion arises we write n_V and n_E instead of $n_V(G)$ and $n_E(G)$, respectively.

2.2 Definitions of Network Games and Profits

Definition 1. *Let $G = (V, E)$ be an undirected graph with no isolated vertices. Fix integers α and δ with $\alpha, \delta \geq 1$. A strategic game $\Gamma_{\alpha,\delta}(G) = \langle \mathcal{N}, \mathcal{S} \rangle$ on G is defined as follows:*

- *$\mathcal{N} = \mathcal{N}_A \cup \mathcal{N}_D$ is the set of players, where $\mathcal{N}_A$ and $\mathcal{N}_D$ are disjoint and $\mathcal{N}_A$ is a finite set of attackers a_i, where $1 \leq i \leq \alpha$ $\mathcal{N}_D$ is a finite set of defenders d_j, where $1 \leq j \leq \delta$*
- *$\mathcal{S} = E^\alpha \times V^\delta$ is the strategy set of $\Gamma_{\alpha,\delta}(G)$*

An element $\langle e_1, ..., e_\alpha, v_1, ..., v_\delta \rangle$ of $\mathcal{S}$ is also called a *profile* of the game, and e_i, v_j strategies of a_i, d_j, respectively. Note that all players make their choice simultaneously. Now fix a profile $\boldsymbol{s} = \langle e_1, ..., e_\alpha, v_1, ..., v_\delta \rangle$ of the game $\Gamma_{\alpha,\delta}(G)$. We define a profit (income) of the players as follows.

- The individual *profit* of attacker a_i, $1 \leq i \leq \alpha$, is given by

$$P_{\boldsymbol{s}}(a_i) = \begin{cases} 0 & \text{if } v_j \in e_i \text{ for some } j, 1 \leq j \leq \delta \\ 1 & \text{if } v_j \notin e_i \text{ for all } j, 1 \leq j \leq \delta \end{cases}$$

In other words, attacker a_i receives 0 if it is caught by a defender d_j, and 1 otherwise.

– The individual *profit* of defender d_j, $1 \leq j \leq \delta$, is given by

$$P_s(d_j) = |\{i : 1 \leq i \leq \alpha, v_j \in e_i\}|$$

representing the number of attackers captured by d_j.

3 Pure Nash Equilibria

Definition 2. *A profile s is a Nash equilibrium if for any player $r \in \mathcal{N}$, $P_s(r) \geq P_{\bar{s}}(r)$ for any profile $\bar{s}$ which differs from s only on the strategy of r.*

In other words, in a Nash equilibrium no player can improve his individual profit by changing his strategy unilaterally. Before we proceed to the theorem, we define the following sets:

$$A_s = \{e \in E : \exists i, 1 \leq i \leq \alpha, \text{ where } e = e_i\},$$

$$D_s = \{v \in V : \exists j, 1 \leq j \leq \delta \text{ such that } v = v_j\},$$

where $s = \langle e_1, ..., e_\alpha, v_1, ..., v_\delta \rangle$.

Theorem 1. *The game $\Gamma_{\alpha,\delta}(G)$ has a Nash equilibrium if and only if there exist $D \subset V$ and $A \subset E$ such that*
(1) $|D| \leq \delta$ and $|A| \leq \alpha$
(2) D is a vertex cover of G
(3) $\forall v \in D, |A \cap E(v)| = \max_{\bar{v} \in V} |A \cap E(\bar{v})|$.

Proof. Suppose $\Gamma_{\alpha,\delta}(G)$ has a Nash equilibrium, say s. Let $A = A_s$ and $D = D_s$. Then, (1) is straightforward. To prove (2), suppose to the contrary that there exists $\bar{e} \in E$ such that $v \notin \bar{e}$ for all $v \in D$. Then, any attacker can receive 1 by switching to $\bar{e}$. Since s is a Nash equilibrium, each attacker must already get 1, which means that all defenders receive 0. However, any defender can get at least 1 by switching to a vertex incident to an attacked edge, which contradicts the assumption that s is Nash equilibrium. Similarly if (3) does not hold, there would be a $v_j \in D$ and $\bar{v}$ such that $|A \cap E(v_j)| < |A \cap E(\bar{v})|$. Thus, defender j would find it beneficial to change his choice from v_j to $\bar{v}$, which contradicts the fact that s is a Nash equilibrium. Conversely, suppose there exist A and D satisfying condition (1), (2) and (3). By (1), let s be a profile so that each element of A (resp. D) is chosen by at least one attacker (resp. defender). By (2), no matter how an attacker changes his strategy, he will always get 0. Thus, an attacker has no incentive to change his strategy. By (3), if a defender changes his choice, it won't increase his profit, since the number of protected edges is already maximum. □

Theorem 2. *If α is the size of a maximum matching in G and $\delta = 2\alpha$, then the game $\Gamma_{\alpha,\delta}(G)$ has a Nash equilibrium.*

Proof. Let A be a maximum matching in G and D be the set of vertices incident to an edge in A. Then obviously conditions (1) and (3) of Theorem 1 hold. For (2), assume that there were $\bar{e} \in E$ such that $v \notin \bar{e}$ for all $v \in D$. Then, $A \cup \{\bar{e}\}$ would be also a matching, which contradicts with the maximality of A. □

In general, it is difficult to determine whether a network game has a Nash equilibrium. The problem of finding a Nash equilibrium is related to the problem of finding a maximum matching in a graph, which is also known to be computationally hard to treat. But for a bipartite graph, such a problem is tractable. Moreover, bipartite graphs have many nice properties studied in the past research. For instance, König's theorem states that, in a bipartite graph, the size of a minimum vertex cover is equal to the size of a maximum matching, which in fact leads to a result that a minimum vertex cover and a maximum independent set can be found in polynomial time for a bipartite graph.

Definition 3. *The graph G is bipartite if $V = V_0 \cup V_1$ for some disjoint vertex sets $V_0, V_1 \subseteq V$ so that for each edge $(u, v) \in E$, $u \in V_0$ and $v \in V_1$ (or $u \in V_1$ and $v \in V_0$).*

Theorem 3. *For a bipartite graph G, the game $\Gamma_{\alpha,\delta}(G)$ has a Nash equilibrium if and only if $\alpha, \delta \geq m$, where m is the size of a maximum matching in G.*

Proof. The proof easily follows from König's duality theorem. For a bipartite graph G, if M is a maximum matching and C_V^{min} is a minimum vertex cover, then $\forall e \in M, \exists! v \in C_V^{min}$ such that $v \in e$. On the other hand, $\forall v \in C_V^{min}, \exists e \in M$ such that $v \in e$. So, $A = M, D = C_V^{min}$ satisfy the three conditions of Theorem 1. The other direction is similar. □

Theorem 4. *For a bipartite graph G, the existence of pure Nash equilibrium on $\Gamma_{\alpha,\delta}(G)$ can be determined in $O(\sqrt{n_V}n_E)$.*

Proof. The proof is based on an augmenting path algorithm (see [12]). The overall complexity of finding a maximum matching by such an algorithm is $O(n_V n_E)$. This can be improved to $O(\sqrt{n_V}n_E)$ by augmenting along several augmenting paths simultaneously. □

In sum, Theorem 3 reduces the problem of finding a Nash equilibrium to that of finding a maximal matching in the case of a bipartite graph. Then, Theorem 4 shows that the existence of a pure Nash equilibrium on a bipartite graph can be determined in almost linear time by computing an augmenting path through the graph.

4   Mixed Nash Equilibria

In this section, we introduce a mixed strategy for our network game, and investigate the existence conditions of a mixed Nash equilibrium by applying some results in the previous sections. We start with basic notations and definitions.

4.1 Mixed Strategies

A *mixed strategy* for an attacker (resp. defender) is a probability distribution over edges (resp. vertices) of G. A *mixed profile* $s = \langle \sigma_1, ..., \sigma_\alpha, \tau_1, ..., \tau_\delta \rangle$ is a collection of mixed strategies, one for each player. So, $\sigma_i(e)$ is the probability that attacker a_i chooses edge e and $\tau_j(v)$ is the probability that defender d_j chooses vertex v. Thus, for each $i \leq \alpha$,

$$\sigma_i : E \to [0, 1] \text{ satisfies } \sum_{e \in E} \sigma_i(e) = 1,$$

and for each $j \leq \delta$,

$$\tau_j : V \to [0, 1] \text{ satisfies } \sum_{v \in V} \tau_j(v) = 1.$$

The support of a player $r \in \mathcal{N}$ in a profile s, denoted by $S_s(r)$, is the set of edges or vertices to which r assigns positive probability in s. Finally, let

$$S_s(A) = \bigcup_{a_i \in \mathcal{N}_A} S_s(a_i)$$

and

$$S_s(D) = \bigcup_{d_j \in \mathcal{N}_D} S_s(d_j).$$

We say that s is *uniform* if for any i,

$$S_s(a_i) = S_s(A) \text{ and } \sigma_i(e) = \frac{1}{|S_s(A)|} \text{ for } e \in S_s(A)$$

and for any j,

$$S_s(d_j) = S_s(D) \text{ and } \tau_j(v) = \frac{1}{|S_s(D)|} \text{ for } v \in S_s(D).$$

Now, fix a mixed profile s. For an edge $e \in E$, let $\mathsf{Save}(e)$ denote the event that at least one end $v \in e$ is protected by a defender in s. For a vertex $v \in V$, let $\mathsf{Save}(v)$ denote the event that at least one defender protects the node v in s. The probability of $\mathsf{Save}(v)$, denoted by $\pi_s(\mathsf{Save}(v))$, is defined by

$$\sum_j \tau_j(v) - \sum_{j \neq k} \tau_j(v) \cdot \tau_k(v) + \sum_{j \neq k, k \neq l, j \neq l} \tau_j(v) \cdot \tau_k(v) \cdot \tau_l(v) - \cdots .$$

Thus for $e = (u, v)$, the probability of $\mathsf{Save}(e)$, denoted by $\pi_s(\mathsf{Save}(e))$, is defined as follows:

$$\pi_s(\mathsf{Save}(e)) = \pi_s(\mathsf{Save}(u)) + \pi_s(\mathsf{Save}(v)) - \pi_s(\mathsf{Save}(u))\pi_s(\mathsf{Save}(v)).$$

4.2   Expected Profits

A mixed profile s induces an expected individual profit $P_s(r)$ for each player $r \in \mathcal{N}$, which is the expectation, according to s, of its corresponding individual profit (defined in Section 2.2). We proceed to define the expected profit for each player.

For a defender $d_j \in \mathcal{N}_D$,

$$P_s(d_j) = \sum_{\substack{v \in V \\ i \leq \alpha \\ e \in E(v)}} \tau_j(v)\sigma_i(e) = \sum_{v \in V} \tau_j(v) \sum_{\substack{i \\ e \in E(v)}} \sigma_i(e).$$

For an attacker $a_i \in \mathcal{N}_A$,

$$P_s(a_i) = \sum_{e \in E} \sigma_i(e) \cdot (1 - \pi_s(\mathsf{Save}(e))).$$

Definition 4. *A mixed profile s is a mixed Nash equilibrium if for each player $r \in \mathcal{N}$, it maximizes P_s over all profiles $\bar{s}$ that differ from s only with respect to the mixed strategy of player r.*

Intuitively, no player can gain more by a unilateral change of his strategy. We proceed to study the characterization of a mixed Nash equilibrium.

4.3   Properties of Mixed Nash Equilibria

In this section, we show that two covering properties, that is, the existence of a vertex cover and an edge cover, characterize a necessary condition for mixed Nash equilibria. In fact, we show that the supports of defenders and attackers are a vertex cover and an edge cover of the graph, respectively.

Theorem 5. *In any mixed Nash equilibrium s of $\Gamma_{\alpha,\delta}(G)$, $S_s(D)$ is a vertex cover of G.*

Proof. Assume $S_s(D)$ is not a vertex cover. Let E_{NC} be a nonempty set of edges of G not covered by $S_s(D)$. Then, any attacker $a_i \in \mathcal{N}_A$, by setting $S_s(a_i) \subseteq E_{NC}$, cannot be caught by a defender. That is, any attacker is not caught by a defender according to s, since it is a mixed Nash equilibrium. This implies that for any j, $P_s(d_j) = 0$, which contradicts with the property of a mixed Nash equilibrium since a defender can choose a vertex incident to an attacked edge. $\qquad\square$

Theorem 6. *In any mixed Nash equilibrium s of $\Gamma_{\alpha,\delta}(G)$, $S_s(A)$ is an edge cover of the subgraph of G obtained by restricting to $S_s(D)$.*

Proof. Assume the contrary, and let $v \in S_s(D)$ such that v is not covered by $S_s(A)$. Since $v \in S_s(D)$, there is a defender d_j who chooses v with positive probability. Now, if he chooses a vertex incident to an attacked edge with the same probability instead of v, he can increase his profit, which is impossible in a mixed Nash equilibrium. $\qquad\square$

The following characterization is useful for checking whether or not it is a mixed Nash equilibrium.

Proposition 1. *Given a graph G, a mixed profile s is a Nash equilibrium if and only if:*

1. $\forall j \leq \delta, \forall v \in S_s(d_j)$

$$\sum_{\substack{i \\ e \in E(v)}} \sigma_i(e) = \max_{\bar{v} \in V} \sum_{\substack{i \\ e \in E(\bar{v})}} \sigma_i(e)$$

2. $\forall i \leq \alpha, \forall e \in S_s(a_i)$

$$\pi_s(Save(e)) = \min_{\bar{e} \in E} \pi_s(Save(\bar{e}))$$

Proof. It is easy to see from the definition of a mixed Nash equilibrium and the expected profit. □

Remark 1. *Given a profile s, the condition that $S_s(A)$ and $S_s(D)$ are an edge cover and vertex cover respectively, does not necessarily imply that s is a Nash equilibrium.*

For example, let $G = \{\{v_0, v_1, v_2\}, \{e_0, e_1\}\}$ where $e_0 = (v_0, v_1)$ and $e_1 = (v_1, v_2)$. Let $\alpha = \delta = 1$, and also $s = \langle \sigma_1, \tau_1 \rangle$ such that $\sigma_1(e_0) = 0.9$, $\sigma_1(e_1) = 0.1$ and $\tau_1(v_0) = 0.9$, $\tau_1(v_1) = 0$, $\tau_1(v_2) = 0.1$.

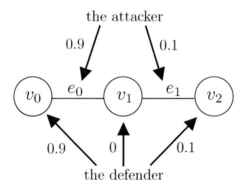

Fig. 1. An illustration of the example

Then, clearly $S_s(A) = \{e_0, e_1\}$ and $S_s(D) = \{v_0, v_2\}$ are an edge cover and vertex cover, respectively. However,

$$\sum_{\substack{i \\ e \in E(v_0)}} \sigma_i(e) = 0.9 \neq \max_{\bar{v} \in V} \sum_{\substack{i \\ e \in E(\bar{v})}} \sigma_i(e) = 1$$

Hence, s is not a Nash equilibrium by Proposition 1.

4.4   Perfect Covering Profiles

In this section, we introduce a new notion, called a *perfect covering* profile, and show it is a sufficient condition for the existence of a Nash equilibrium in a bipartite graph. These characterization enables an almost linear time algorithm to compute a perfect Nash equilibrium on a bipartite graph with a given independent vertex cover and a maximal matching.

Definition 5. *A mixed profile $s = (\sigma_1, ..., \sigma_\alpha, \tau_1, ..., \tau_\delta)$ is said to be a perfect covering profile if it is uniform (see Remark 2) and satisfies the following conditions.*

1. $S_s(D)$ *is an independent vertex cover.*
2. $S_s(A)$ *is a perfect matching.*

Remark 2 (cf. Section 4.1). *Recall that if s is uniform, then $\sigma_i(e) = \sigma_{i'}(e')$ for any $e, e' \in S_s(A)$ and $\tau_j(v) = \tau_{j'}(v')$ for any $v, v' \in S_s(D)$.*

Lemma 1. *In a bipartite graph, a perfect covering profile is a mixed Nash equilibrium.*

Proof. We shall use Proposition 1 to prove this lemma. We first show condition (1). Fix $j \leq \delta$ and $v \in S_s(d_j)$, and take $\bar{v} \in V$ arbitrary. Since $S_s(A)$ is a perfect matching, for any $w \in V$, there is exactly one $e_w \in S_s(A)$ such that $w \in e_w$. Hence, we have

$$\sum_{\substack{i \\ e \in E(v)}} \sigma_i(e) = \sum_i \sigma_i(e_v) = \sum_i \sigma_i(e_{\bar{v}}) = \sum_{\substack{i \\ e \in E(\bar{v})}} \sigma_i(e)$$

by uniformity of s. Thus,

$$\sum_{\substack{i \\ e \in E(v)}} \sigma_i(e) = \max_{\bar{v} \in V} \sum_{\substack{i \\ e \in E(\bar{v})}} \sigma_i(e).$$

To show that condition (2) holds, fix $i \leq \alpha, e \in S_s(a_i)$ and take $\bar{e} \in E$ arbitrary. Since $S_s(D)$ is an independent set and vertex cover, for any $f \in E$, there is exactly one $v_f \in S_s(D)$ such that $v_f \in f$ for each $j \leq \delta$. Hence, we have

$$\pi_s(\mathsf{Save}(e)) = \pi_s(\mathsf{Save}(v_e))$$

$$= \sum_j \tau_j(v_e) - \sum_{j \neq k} \tau_j(v_e) \cdot \tau_k(v_e) + \cdots$$

$$= \sum_j \tau_j(v_{\bar{e}}) - \sum_{j \neq k} \tau_j(v_{\bar{e}}) \cdot \tau_k(v_{\bar{e}}) + \cdots$$

$$= \pi_s(\mathsf{Save}(v_{\bar{e}}))$$

$$= \pi_s(\mathsf{Save}(\bar{e}))$$

by uniformity of s. Thus,

$$\pi_s(\mathsf{Save}(e)) = \min_{\bar{e} \in E} \pi_s(\mathsf{Save}(\bar{e})).$$

By Proposition 1, it follows that s is a Nash equilibrium.   □

4.5   Perfect Nash Equilibria

A perfect covering profile is called a *perfect Nash equilibrium* if it is a Nash equilibrium. A subgraph of G is said to be an *odd cycle* if it is a cycle with an odd number of vertices. Similarly for an *even cycle* with an even number of vertices. Then, we can easily show the following fact.

Remark 3. *If G has a perfect matching and no odd cycle, then G has an independent vertex cover.*

The following theorem provides necessary and sufficient conditions for a game $\Gamma_{\alpha,\delta}(G)$ to have a perfect Nash equilibrium.

Theorem 7. *The game $\Gamma_{\alpha,\delta}(G)$ has a perfect Nash equilibria if and only if G has a perfect matching and no odd cycle.*

Proof. Suppose first that $\Gamma_{\alpha,\delta}(G)$ has a perfect Nash equilibrium, say s. By definition, $S_s(D)$ is an independent vertex cover, which imply $V \setminus S_s(D)$ is independent as well. Suppose that G had an odd cycle $\mathcal{C}_{odd}$. Then, there would be an edge e contained in $\mathcal{C}_{odd}$ with endpoints both in $S_s(D)$ or both in $V \setminus S_s(D)$, which is a contradiction. By definition, G has a perfect matching. Conversely, assume G has a perfect matching and no odd cycle. By Remark 3, G has an independent vertex cover. Since now we have a perfect matching M and an independent vertex cover I_{C_V}, we construct a perfect covering profile $s = (...\sigma_i..., ...\tau_j...)$ as follows:

$$\sigma_i(e) = \begin{cases} \frac{1}{|M|} & \text{if } e \in M \\ 0 & \text{if otherwise} \end{cases}$$

and

$$\tau_j(v) = \begin{cases} \frac{1}{|I_{C_V}|} & \text{if } v \in I_{C_V} \\ 0 & \text{if otherwise} \end{cases}$$

where $M := S_s(A)$ and $I_{C_V} := S_s(D)$. Clearly, a perfect covering profile s is a mixed Nash equilibrium, as needed. Therefore, the game $\Gamma_{\alpha,\delta}(G)$ has a perfect Nash equilibrium.   □

For a subset $U \subseteq V$ of a graph G, denote $N_G(U) = \{v \mid (u,v) \in E \text{ for some } u \in U\}$, the *neighbour* set of U in G. If G is (V_0, V_1)-bipartite, and $U \subseteq V_0$, then $N_G(U) \subseteq V_1$. The graph G is a V_0-*Expander* graph if for each set $U \subseteq V_0$, $|U| \leq |N_G(U)|$. The *degree* of a vertex v is the number $|E(v)|$ of edges at v, this is equal to the number of neighbours of v. If all the vertices of G have the same degree k, then G is k-*regular*, or simply *regular*.

Theorem 8 (Hall's Theorem). *Let G be a bipartite graph with bipartition (V_0, V_1). Then, G contains a matching M that matches all the vertices in V_0 if and only if $|N_G(U)| \geq |U|$ for all $U \subseteq V_0$.*

Corollary 1. *If G is a k-regular bipartite graph with $k > 0$, then the game $\Gamma_{\alpha,\delta}(G)$ has a perfect Nash equilibrium.*

Proof. Let G be a k-regular bipartite graph $(V_0 \cup V_1; E)$. Thus, we have $k|V_0| = n_E = k|V_1|$ by regularity, and hence $|V_0| = |V_1|$ since $k > 0$. Now, let $U \subseteq V_0$. Let $E_0 = \{e \in E : v \in e,$ for some $v \in U\}$ and $E_1 = \{e \in E : v \in e,$ for some $v \in N_G(U)\}$. Clearly, $E_0 \subseteq E_1$. Therefore, $k|N_G(U)| = |E_1| \geq |E_0| = k|U|$, and so $|N_G(U)| \geq |U|$. By Theorem 8, G has a matching M that matches every vertex in V_0. Since $|V_0| = |V_1|$, this matching is necessarily perfect. By Theorem 7, G admits a perfect Nash equilibrium.   □

Note that the example for Remark 1 is bipartite but not regular, and that it has an independent vertex cover but no perfect matching.

4.6   Computing a Nash Equilibrium

Theorem 9. *For a bipartite graph G, it can be computed in $O(\sqrt{n_V} n_E)$ whether or not $\Gamma_{\alpha,\delta}(G)$ has a perfect Nash equilibrium, and if any, one can obtain in $O(\sqrt{n_V} n_E)$.*

Proof. Define $S_{\boldsymbol{s}}(A) := M$ and $S_{\boldsymbol{s}}(D) := I_{C_V}$ such that

$$
\sigma_i(e) = \begin{cases} \frac{1}{|M|} & \text{if } e \in M \\ 0 & \text{if otherwise} \end{cases}
$$

and

$$
\tau_j(v) = \frac{1}{|I_{C_V}|}
$$

Clearly, $\boldsymbol{s}$ is a Nash equilibrium. Thus, the problem of deciding the game $\Gamma_{\alpha,\delta}(G)$ has a perfect Nash equilibrium is equivalent to finding a maximum matching M. The algorithmic problem of finding such a matching thus can be reduces to finding an augmenting path which can be compute in $O(\sqrt{n_V} n_E)$.   □

5   An Asynchronous Game

5.1   Introduction

So far, we have only discussed network games where all the attackers and defenders make individual decisions simultaneously. We now start considering a new game where a pair of an attacker and a defender executes a simultaneous game in turn. Or we may think that there is exactly one attacker and one defender, who repeatedly execute simultaneous games. As in previous games, the attacker aims to damage the network by attacking an edge, and the defender aims to protect the network by choosing a vertex. The important characteristic of our new game is that after one round of the game, the base graph may be reduced by removing an edge damaged by the attack but not (adjacent to a vertex) protected in the round. Then, the final objective of the attacker is to vanish the graph as quickly as possible, while that of the opponent is to keep it as long as possible. Technical definitions will be given in the next section.

Although there may be many possible variations of this games, we are particularly interested in our formulation. Actually, to consider iterations of old simultaneous games, we have thought it more natural that an edge is to be attacked and a vertex protected rather than the opposite. In a previous game, a hub node is easier to avoid damage since it has many edges for protection, but once it is damaged, the network might have a terminal breakdown. Contrastively, in our new game, even if a trunk edge is damaged, another rooting could take its role, and thus the network could maintain major functions after the effective attack. Since we have just started this line of research, we must leave detailed analysis of the variations of this game to the future study.

We would like to point out some important facts. First of all, our new game can be viewed as a special Blackwell game [2]. Blackwell games are infinite games of imperfect information, which is introduced by D. Blackwell in 1969. It has been shown in the standard set theory that in any Blackwell game with a Borel payoff, the players have optimal strategy, namely the game is determinate. Thus, our game (with a suitable payoff) is also determinate, and each player has a stable strategy so that the estimation of the duration of the game is determined.

The Blackwell game is defined as follows:

Definition 6 (Blackwell Games). *Let X and Y be two finite nonempty sets, and put $W = (X \times Y)^{\mathbb{N}}$. Let $f : W \to \mathbb{R}$ be a bounded Borel-measurable function. The Blackwell game $\Gamma(f)$ with payoff function f is the two-person zero-sum infinite game of imperfect information played as follows:*

- *Player I selects an element $x_1 \in X$, and simultaneously, Player II selects an element $y_1 \in Y$. Then both players are told $z_1 = (x_1, y_1)$, and the game has reached position (z_1). Then,*
- *Player I selects $x_2 \in X$, and simultaneously, Player II selects $y_2 \in Y$. Then, both players are told $z_2 = (x_2, y_2)$, and the game is at position (z_1, z_2).*

Then both players simultaneously selects $x_3 \in X$ and $y_3 \in Y$, etc. Thus they produce a play $w = (z_1, z_2, ...)$. Finally, player II pays player I the amount $f(w)$, ending the game.

5.2   Definitions of Asynchronous Games and Profits

We are ready to define an asynchronous game. First of all, we define partial plays and corresponding sequences of subgraphs. Let $G = (V, E)$ be an undirected graph. The set $\mathcal{P} \subset (E \times V)^{<\mathbb{N}}$ of partial plays is defined recursively as follows. Put the empty sequence $\lambda \in \mathcal{P}$ and $E_\lambda = E$ and $V_\lambda = V$. Now assume that $\eta \in \mathcal{P}$, and $E_\eta \subset E$ and $V_\eta \subset V$ have been defined. We put $\rho := \eta^\frown \langle (e, v) \rangle$ into $\mathcal{P}$, if $e \in E_\eta$ and $v \in V_\eta$. Then, we define

$$E_\rho := \begin{cases} E_\eta & \text{if } v \in e \\ E_\eta - \{e\} & \text{if } v \notin e \end{cases}$$

$$V_\rho := V(E_\rho).$$

Finally, let $[\mathcal{P}] = \{w \in (E \times V)^{\mathbb{N}} : \text{each finite initial segment of } w \text{ belongs to } \mathcal{P}\}$. A payoff function needs to be defined on $[\mathcal{P}]$. But it is often regarded as a function on $W = (E \times V)^{\mathbb{N}}$ by setting $f(w) = 0$ for $w \in W - [\mathcal{P}]$.

Definition 7. *Let $f : W \to \mathbb{R}$ be a payoff function. An asynchronous game $\Gamma(G; f)$ with payoff function f is an infinite game of imperfect information played as follows: Attacker a selects an element $e^1 \in E$ and simultaneously defender d chooses an element $v^1 \in V$. Then both the players are told $z_1 = (e^1, v^1)$. Next, the attacker selects $e^2 \in E$, and simultaneously the defender selects $v^2 \in V$, and $z_2 = (e^2, v^2)$, etc. In this manner they produce an infinite sequence $w = (z_1, z_2, ...) \in (E \times V)^{\mathbb{N}}$. Finally, the defender pays the attacker the amount $f(w)$, ending the game.*

A mixed strategy in an asynchronous game is defined as follows.

Definition 8. *A mixed strategy for attacker a in $\Gamma(G; f)$ is a function σ^* assigning to each position ρ a probability distribution on E_ρ, that is, $\sigma^* : \mathcal{P} \to [0,1]^E$ satisfying $\sum_{e \in E_\rho} \sigma^*(\rho)(e) = 1$. Similarly, a mixed strategy for defender d is given by $\tau^* : \mathcal{P} \to [0,1]^V$ such that $\sum_{v \in V_\rho} \tau^*(\rho)(v) = 1$.*

Definition 9. *Let σ^* and τ^* be strategies for the attacker and the defender in $\Gamma(G; f)$, respectively. The probability measure μ_{σ^*, τ^*} on $(E \times V)^\omega$ is given by*

$$\mu_{\sigma^*, \tau^*}([\rho]) = \prod_{n < |\rho|} \left[\sigma^*(\rho_{|n})(e^n) \tau^*(\rho_{|n})(v^n) \right]$$

where $[\rho] = \{w \in W : \rho \text{ is an initial segment of } w\}$ and $\rho_{|n}$ is an initial segment of ρ with the length n.

Given a pair of mixed strategies $s = (\sigma^*, \tau^*)$, the expected profit of the attacker in $\Gamma(G; f)$, according to σ^* and τ^*, is the expectation of $f(w)$ under this probability measure:

$$P_s(\Gamma(G; f)) = \int f(w) d\mu_{\sigma^*, \tau^*}(w)$$

The value of a strategy σ^* for the attacker in $\Gamma(G; f)$ is the expected income the attacker can guarantee if they plays according to σ^*. Similarly, the value of a strategy τ^* for the defender in $\Gamma(G; f)$ is the amount to which the defender can restrict the attacker's profit if he plays according to τ^*, that is,

$$\mathrm{val}_{\sigma^*}(\Gamma(G; f)) = \inf_{\tau^*} P_s(\Gamma(G; f)),$$

$$\mathrm{val}_{\tau^*}(\Gamma(G; f)) = \sup_{\sigma^*} P_s(\Gamma(G; f)).$$

Definition 10. *The lower value of $\Gamma(G; f)$ is the smallest upper bound on the income that the attacker can guarantee. Similarly, the upper value of $\Gamma(G; f)$ is* the largest lower bound on the restrictions the defender can put on the attacker's income, that is,

$$\mathsf{val}_{\downarrow}(\Gamma(G;f)) = \sup_{\sigma^*} \inf_{\tau^*} P_{\boldsymbol{s}}(\Gamma(G;f)),$$

$$\mathsf{val}_{\uparrow}(\Gamma(G;f)) = \inf_{\tau^*} \sup_{\sigma^*} P_{\boldsymbol{s}}(\Gamma(G;f)).$$

If $\mathsf{val}_{\uparrow}(\Gamma(G;f)) = \mathsf{val}_{\downarrow}(\Gamma(G;f))$, then $\Gamma(G;f)$ is called determined.

A strategy σ^* for the attacker in $\Gamma(G;f)$ is optimal if

$$\mathsf{val}_{\sigma^*}(\Gamma(G;f)) = \mathsf{val}_{\downarrow}(\Gamma(G;f)),$$

and similarly, a strategy τ^* for the defender in $\Gamma(G;f)$ is optimal if

$$\mathsf{val}_{\tau^*}(\Gamma(G;f)) = \mathsf{val}_{\uparrow}(\Gamma(G;f)).$$

5.3   One-Round Games and Profits

We are going to analyze our asynchronous game with a concrete payoff. Since the objective of the attacker is to vanish the graph as quickly as possible, his strategy at each round can not be evaluated simply by the expected numbers of edges damaged by it. So, we start with the following one-round game.

Given a finite graph G and a state evaluation $f : \{G' : G'$ is a proper subgraph of $G\} \to \mathbb{R}$, we define a one-round game $\Gamma(G;f)$. A mixed strategy σ for attacker is a probability distribution over E. Similarly for defender, a mixed strategy τ is a probability distribution over V. So, $\sigma(e)$ is the probability that attacker a chooses edge e and $\tau(v)$ is the probability that defender d chooses vertex v. We define a function $h : E \times V \to \{$subgraphs of $G\}$ by

$$h(e,v) := \begin{cases} G & \text{if } v \in e \\ G(E \setminus \{e\}) & \text{if } v \notin e \end{cases}$$

Now, we define the expected profit of the game $\Gamma(G;f)$.

Definition 11. Let $\boldsymbol{s} = (\sigma, \tau)$ be a pair of mixed strategies. The expected profit of the attacker with the delay constant c is given by

$$P_{\boldsymbol{s}}(\Gamma(G;f)) := \sum_{v \notin e} \sigma(e)\tau(v)\{1 + f(h(e,v))\} + \sum_{v \in e} \sigma(e)\tau(v)cP_{\boldsymbol{s}}(\Gamma(G;f)),$$

i.e.,

$$P_{\boldsymbol{s}}(\Gamma(G;f)) = \sum_{v \notin e} \sigma(e)\tau(v)\{1 + f(h(e,v))\}/(1 - c\sum_{v \in e} \sigma(e)\tau(v)).$$

To understand the above definition, suppose that the attacker chooses edge e and the defender chooses vertex v. If v is not adjacent to e, the attack is successful and so one point is given to the attacker. Since the new round starts with the graph $h(e, v)$, it follows that the attacker receives $1 + f(h(e, v))$ in this case. If v is adjacent to e, the next round starts with the same graph, but the expected profit is evaluated as $cP_s(\Gamma(G; f))$ with the delay constant c. For simplicity, we will assume $c = \frac{1}{2}$ from now on, though any $c \in (0, 1)$ works as the same.

Now we investigate the existence of a mixed Nash equilibria in the one-round game. The theorem is stated as follows.

Theorem 10. *Given a graph G and function $f : \{G' \subsetneq G : G'$ a subgraph$\} \to \mathbb{R}$. Then, the one-round game $\Gamma(G; f)$ is determined with a stable solution.*

The proof follows from a combination of the mini-max theorem and Brouwer's fixed point theorem (or the intermediate value theorem for this particular case). First of all, we recall Brouwer's fixed point theorem.

Theorem 11 (Brouwer's Fixed Point Theorem). *Let S be a subset of some space $\mathbb{R}^n$ that is convex and compact, and let ϕ be a continuous function from S to S. Then, ϕ has at least one fixed point, that is, a point s in S such that $\phi(s) = s$.*

Brouwer's theorem has been generalized in numerous ways, e.g., Schauder's and Tychonoff's fixed point theorems. Kakutani also proposed a multifunction ana-logue to Brouwer's theorem, and then show that this generalized theorem implies the famous von Neumann's mini-max theorem.

Theorem 12 (Mini-Max Theorem). *Let K, L be two bounded closed convex sets in $\mathbb{R}^m, \mathbb{R}^n$. Let $f(x, y)$ be a continuous real function on $K \times L$ such that for any $x_0 \in K$ and $\alpha \in \mathbb{R}$, $\{y \in L : f(x_0, y) \le \alpha\}$ is convex, and for any $y_0 \in L$ and $\beta \in \mathbb{R}$, $\{x \in K : f(x, y_0) \ge \beta\}$ is convex. Then we have*

$$\max_{x \in K} \min_{y \in L} f(x, y) = \min_{y \in L} \max_{x \in K} f(x, y).$$

Proof of the Theorem 10. Suppose $f : \{G' \subsetneq G\} \to \mathbb{R}$, and $f(G) = x \in \mathbb{R}$ (for the next round with the same G) are given. The expected profit according to a profile (σ, τ) is following:

$$P(\sigma, \tau, x) := \sum_{v \notin e} \sigma(e)\tau(v)\{1 + f(h(e, v))\} + \sum_{v \in e} \sigma(e)\tau(v)\frac{1}{2}x.$$

By the mini-max theorem, we have

$$\max_{\sigma} \min_{\tau} P(\sigma, \tau, x) = \min_{\tau} \max_{\sigma} P(\sigma, \tau, x)$$

for all $x \in \mathbb{R}$. Note that the set of strategy σ's (similar for τ's) is bounded closed convex, and so it is easy to see that $M(x) = \max_{\sigma} \min_{\tau} P(\sigma, \tau, x)$ is a continuous function. Now put $m = 1 + \max f$. Clearly, $M : [0, m] \to [0, m]$. So it has a fixed point $\hat{x}$. Then $M(\hat{x})$ serves as a stable solution.   □

Note that in the above proof, we cannot use the second formula of the expected profit in Definition 11, since the conditions of the mini-max theorem do not hold for that formula.

5.4   Games with a Natural Payoff Function

Now, we are going to define a function $f : W \to \mathbb{R}$, which plays the role of a natural payoff for the attacker of our asynchronous game. Recall that $[\mathcal{P}] \subseteq W$ is defined to be the set of infinite sequences whose initial segments are all included in $\mathcal{P}$. For $w = ((e^1, v^1), (e^2, v^2), (e^3, v^3), ...) \in [\mathcal{P}]$, we set

$$b_i(w) := \begin{cases} 0 & \text{if } v^i \in e^i, \\ 1 & \text{if } v^i \notin e^i. \end{cases}$$

Then we define

$$f(w) = \sum_{i>0} \frac{b_i(w)}{2^i}$$

Thus $f(w)$ does not only evaluate the number of damaged edges through w, but also evaluate the promptness of attacks. Finally we have,

Theorem 13. *The asynchronous game $\Gamma(G; f)$ is determined.*

Proof. We show the existence of a stable solution in the game without referring to the Blackwell determinacy. We call a subgraph G' of G a terminal graph if the graph can not be reduced to a smaller graph whatever the players play on it. Let $\mathcal{G}_0$ be the set of terminal graphs of G. Now consider a subgraph G' of G, which is reduced to a terminal graph or unchanged after a one-round game on G' is executed. By assigning 0 to each terminal graph, we can compute the profit or the state value of this graph by Theorem 10. We define $\mathcal{G}_1$ to be the set of such subgraphs of G. Next consider a subgraph G'' of G which is either reduced to a graph in $\mathcal{G}_0 \cup \mathcal{G}_1$, or unchanged in a one-round game. By assigning 0 to each terminal graph, and to a graph G' in $\mathcal{G}_1$, its state value, we can compute the state value of the new graph by Theorem 10. By continuing this process (at most $n_E(G)$ times), we finally obtain the profit of the attacker for G.   □

We should notice that in the above proof, the infinite game $\Gamma(G; f)$ is reduced to a finite-round game in a constructive way.

6   Concluding Remarks

In this paper, we presented a new network game with many attackers and defenders. We have considered various conditions for computing mixed Nash equilibria for this game. Then we generalized it to an asynchronous game, which may be viewed as an infinite Blackwell game. Finally, we have shown that an asynchronous game can be constructively reduced to a finite-round game. We believe that this reduction method is quite useful to analyze the computational contents of this game, which we will develop in the future literature.

Acknowledgments. The authors would like to thank Mr. Kojiro Higuchi, for carefully reading a draft of this paper and offering many useful comments that resulted in an improved paper.

References

1. Ahmad Termimi, A.G., Tanaka, K.: Network Games with Many Attackers and Defenders. In: Proceedings of Research Institute for Mathematical Sciences (RIMS) Kôkyûroku, vol. 1729, pp. 146–151. Kyoto University (2011)
2. Blackwell, D.: Infinite G_δ games with imperfect information. In: Zastosowania Matematyki Applicationes Mathematicae, Hugo Steinhaus Jubilee, vol. X, pp. 99–101 (1969)
3. Gelastou, M., Mavronicholas, M., Papadopoulou, V.G., Philippou, A., Spirakis, P.: The Power of the Defender. In: CD-ROM Proceedings of the 2nd International Workshop on Incentive-Based Computing (July 2006)
4. Hall, P.: On representatives of subsets. J. London Math. Soc. 10, 26–30 (1935)
5. Martin, D.A.: The determinacy of Blackwell games. Journal Symbolic Logic 63(4), 1565–1581 (1998)
6. Mavronicholas, M., Papadopoulou, V.G., Philippou, A., Spirakis, P.G.: A Network Game with Attacker and Protector Entities. In: Deng, X., Du, D.-Z. (eds.) ISAAC 2005. LNCS, vol. 3827, pp. 288–297. Springer, Heidelberg (2005)
7. Mavronicolas, M., Papadopoulou, V.G., Philippou, A., Spirakis, P.G.: A Graph-Theoretic Network Security Game. In: Deng, X., Ye, Y. (eds.) WINE 2005. LNCS, vol. 3828, pp. 969–978. Springer, Heidelberg (2005)
8. Mavronicolas, M., Michael, L., Papadopoulou, V.G., Philippou, A., Spirakis, P.G.: The Price of Defense. In: Královič, R., Urzyczyn, P. (eds.) MFCS 2006. LNCS, vol. 4162, pp. 717–728. Springer, Heidelberg (2006)
9. Ould-MedSalem, M., Manoussakis, Y., Tanaka, K.: A Game on Graphs with Many non-Centralized Defenders. In: 22nd European Conference on Operations Research (EURO XXII), Prague, July 8-11 (2007)
10. Nash, J.: Equilibrium Points in n-Person Games. Proceedings of the National Academy of Sciences 36, 48–49 (1950)
11. Nash, J.F.: Noncooperative Games. Annals of Maths 54, 286–295 (1951)
12. West, D.B.: Introduction to Graph Theory, 2nd edn. Prentice Hall (2001)

An Asymptotic Solution of Dresher's Guessing Game

Robbert Fokkink and Misha Stassen

Delft Institute of Applied Mathematics,
TU Delft, P.O.Box 5031, 2600 GA Delft, Netherlands
r.j.fokkink@ewi.tudelft.nl, m.j.stassen@student.tudelft.nl

Abstract. In his 1961 monograph on Game Theory, Melvin Dresher considered a high-low guessing game on N numbers. The game was solved for $N \leq 11$ by Selmer Johnson but solutions for higher values of N have never been reported in the literature. In this paper we derive an asymptotic formula for the value of the game as $N \to \infty$ and we present an algorithm that allows us to numerically solve the game for $N \leq 256$.

Keywords: search game, binary search, asymptotic analysis.

We consider the following zero-sum game:

Alice secretly writes down one of the numbers $1, 2, \ldots, N$ *and Bob must repeatedly guess this number until he gets it, losing* 1 *for each guess. After each guess, Alice must say whether the guess is too high, too low or correct.*

Apparently this game first appeared in a monograph on game theory by Dresher [8, p. 33] and that it why it is called *Dresher's guessing game*. It is a classic game which often serves as an instructive example of a zero-sum game [9,18]. If Alice is allowed to lie a certain number of times, then it is called Ulam's game, which is intimately related to problems in coding theory and has accumulated a fair amount of research [17]. Dresher's game is very similar to games that are used in economics to model bargaining situations, see [4,19] and the references therein. Dresher's game is also related to the study of coordination games and spatial dispersion [3,6]. The following example is illustrative. A group of N people commutes to a part of town that lies across a river. There are two bridges that cross the river and the commuters can take either one, but they have to find a way to coordinate their choice since the total capacity of the two bridges is equal to N. The commuters will get stuck in a jam until they finally divide themselves according to the capacities of the bridges. In other words, the commuters coordinate their choice based on a daily high-low feedback.

Despite the fact that Dresher's game is a classic game describing a common situation, it has only been solved for $N \leq 11$ and that was done a long time ago [15].[1] The reason for this is, and this is typical for games involving deci-

---

[1] Outside the research literature, the game has been solved for slightly larger values. In a recent problem solving contest, IPSC 2011, one of the problems was to solve the guessing game for $N = 16$. A solution can be found in http://ipsc.ksp.sk/contests/ipsc2011/

J.S. Baras, J. Katz, and E. Altman (Eds.): GameSec 2011, LNCS 7037, pp. 104–116, 2011.
© Springer-Verlag Berlin Heidelberg 2011

sions [2], that the number of pure strategies grows exponentially with N so that standard algorithms such as the simplex method quickly become numerically infeasible. In this paper we present a new algorithm, using a delayed column generation, that allows a computational solution for $N \leq 256$. The computational time increases exponentially at each consecutive power of 2 and we were unable to push the computation to 512 on a standard PC.

Dresher's game is an example of a high-low search game. Such games have accumulated a fair amount of literature and many of them have been solved [2, Chapter 5]. Gal solved a very similar guessing game in which Alice must say whether the guess is too high or not too high [10]. The optimal strategies for this version of the guessing game, and its value, can be described explicitly. For reasons given below, an explicit solution of Dresher's guessing game is probably too much to hope for. However, we are able to derive an asymptotic formula for the value of the game:

Theorem 1. *Let* $\mathcal{V}(N)$ *be the value of Dresher's guessing game and let* $\lg(N)$ *be the logarithm of* N *to base 2. Suppose that* $2^k \leq N \leq 2^{k+1}$ *for a non-negative integer* k *and* $x(N) = \frac{N}{2^k} - 1$. *In particular* $0 \leq x \leq 1$ *and if we let* N *go to infinity in such a way that* $x(N)$ *converges to a fixed limit* x *in* $[0, 1]$, *then*

$$\lim_{N \to \infty} \lg(N + 1) - \mathcal{V}(N) = \lg(1 + x) + \frac{1 - x}{1 + x}$$

In other words, $\lg(N + 1) - \mathcal{V}(N)$ has no proper limit and its value oscillates, depending on the relative position of N between consecutive powers of two. Such oscillations are a well known phenomenon in the analysis of binary search trees, see for instance [14].

We use the numerical results from our algorithm to estimate the convergence of $\lg(N + 1) + \mathcal{V}(N)$ to $\lg(1 + x) + \frac{1-x}{1+x}$.

Theorem 2. *Suppose that* $N \geq 256$ *and we write* $x = x(N) = \frac{N}{2^k} - 1$, *then*

$$\lg(1 + x) + \frac{1 - x}{1 + x} - 0.1 < \lg(N + 1) - \mathcal{V}(N) < \lg(1 + x) + \frac{1 - x}{1 + x}.$$

The paper is organized as follows. We first show in section 1 that $\mathcal{V}(N)$ varies periodically around $\lg N$ as N goes to infinity. In section 2 we calculate the limit of $\lg(N+1) - \mathcal{V}(N)$ and obtain Theorem 1. In section 3 we supply our algorithm to solve the game and discuss our numerical results.

1   Asymptotic Periodicity of $\mathcal{V}(N)$

We briefly describe the strategy spaces of Dresher's guessing game. A more detailed discussion can be found in [12,15,16]. Dresher's game is finite and zero-sum, so it has a well-defined value $\mathcal{V}(N)$. After each incorrect guess, the game can continue in two ways depending on Alice' secret number, so the course of the game can be described by a tree. For instance, if $N = 7$ and if Bob plays by bisection, then his initial guess is 4, and his consecutive guess is 2 if the guess is too high or 6 if it is too low, etc. The corresponding tree is depicted below. Note that the top node is divisible by 4, the nodes at the next level are divisible by 2, and the odd numbers are down below. It is not hard to show that a pure strategy

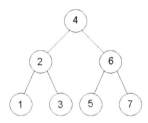

for Bob corresponds to a binary search tree on N nodes [16]. The payoff of the pure strategy depends on Alice's secret number and is equal to the depth of the node containing the secret number. Bob's strategy space corresponds to the family of all binary trees, so its cardinality is equal to the N-th Catalan number $c(N)$, which is a number that grows exponentially with N. Alice's strategy space is relatively small, as it consists of N elements only: the matrix of the game has size $N \times c(N)$.

Lemma 1. $\lg(N + 1) - 1 < \mathcal{V}(N) \le \lg(N + 1)$.

Proof. Note that $\mathcal{V}(N) \le \mathcal{V}(M)$ if $N \le M$, because the game on N numbers can be seen as a version of the game on M numbers in which Alice has a restricted choice. If $N = 1$ then it takes one guess to get the secret number so $\mathcal{V}(1) = 1$. We prove by induction that $\mathcal{V}(N) \le \lg(N+1)$. Suppose that N is odd and that Bob's initial guess is $(N + 1)/2$. Then either the game is immediately over or Bob loses one and play continues on $(N-1)/2$ numbers. In the game on $(N-1)/2$ numbers, Bob has a strategy that guarantees an average loss of $\mathcal{V}((N-1)/2)$, and it follows that $\mathcal{V}(N) \le 1 + \mathcal{V}((N-1)/2)$. According to our induction hypothesis is $1 + \mathcal{V}((N-1)/2) \le \lg(N+1)-1$. So if N is odd, then $\mathcal{V}(N) \le \lg(N+1)$. Now suppose that N is even and that Bob initially guesses $N/2$ or $N/2 + 1$ equiprobably. If the game is not immediately over, then it continues equiprobably on $N/2 - 1$ or on $N/2$ numbers. Once again by induction $\mathcal{V}(N) \le 1 + \frac{1}{2}\left[\mathcal{V}\left(\frac{N}{2} - 1\right) + \mathcal{V}\left(\frac{N}{2}\right)\right]$ and by the concavity of the logarithm we find that this is $\le \lg(N + 1)$. This concludes the proof of the upper bound on $\mathcal{V}(N)$.

To prove the lower bound, consider Alice's mixed strategy in which she chooses a secret number uniformly randomly. Then the payoff of Bob's pure strategy is equal to the average depth of a node in the binary search tree. It has one node of depth 1 - at most two nodes of depth 2 - at most four nodes of depth 3, etc. If $2^k \le N \le 2^{k+1} - 1$ then the average depth of a binary search tree is at least

$$\frac{1}{N}\left(1 + 2.2 + 4.3 + \cdots + 2^{k-1}k + (N - 2^k + 1)(k + 1)\right) \qquad (1)$$

Since this is Alice's minimal payoff if she uses the uniform strategy, it puts a lower bound on the value of the game. We will come back to this lower bound below. For now we observe that it is equal to

$$\frac{1}{N} \sum_{j=1}^{N} \lceil \lg(j+1) \rceil = \frac{1}{N} \int_0^N \lceil \lg(x+1) \rceil dx > \frac{1}{N} \int_0^N \lg(x+1) dx,$$

which is equal to $\frac{(N+1)\lg(N+1)-N}{N}$ and we conclude that the lower bound holds.

Lemma 2. *The function $c(N) = \lg(N+1) - V(N)$ satisfies the inequalities $c(2N+1) \geq c(N)$ and $c(2N) \geq \frac{c(N-1)+c(N)}{2}$.*

Proof. In the proof above we found that $V(2N+1) \leq 1+V(N)$ which is equivalent to $c(2N+1) \geq c(N)$. We also found that $V(2N) \leq 1 + \frac{1}{2}V(N) + \frac{1}{2}V(N-1)$, which implies that $c(2N) > \frac{c(N)+c(N-1)}{2}$.

To describe the way that $V(N)$ varies between consecutive powers of two, we need to compare $c(N)$ to $c(N/2)$. Therefore it is convenient to write $N+1 = 2^k(1+x)$ for $x \in [0,1]$ and to define $c_k(x) = c(N)$. So instead of a single function $c(N)$ we obtain a sequence of functions $c_k(x)$. We extend the domain of c_k to the entire unit interval by linear interpolation. Note that if $N = 2^k - 1$ then $c_k(0) = c_{k-1}(1) = c(N)$. Also note that Lemma 2 can be rewritten as $c_{k+1}(x) \geq c_k(x)$ so the limit function $c_\infty(x) = \lim_{k\to\infty} c_k(x)$ exists and it satisfies $c_\infty(1) = c_\infty(0)$. In other words, the value of Dresher's guessing game is asymptotically periodic.

2   Determining $c_\infty(x)$

In the proof of Lemma 1 we encountered equation (1) which gives a lower bound on $V(N)$. Therefore, it gives an upper bound on the functions $c_k(x)$ and in this section we show that this upper bound is asymptotically sharp.

Lemma 3.
$$c_\infty(x) \leq \lg(1+x) + \frac{1-x}{1+x}$$

Proof. By equation (1) we know that

$$V(N) \geq \frac{(k+1)(N-2^k+1) + \sum_{j=1}^{k} j2^{j-1}}{N}.$$

Now using that $\sum_{j=1}^{k} j2^{j-1} = (k-1)2^k + 1$ we find

$$V(N) \geq \frac{(k+1)(N+1) - 2^{k+1} + 1}{N} = k+1 + \frac{k+2-2^{k+1}}{N}. \tag{2}$$

Note that the formula at the right hand side of this inequality gives the average payoff if Alice chooses her secret number equiprobably. We come back to this below. As before we write $N+1 = 2^k(1+x)$ and find that

$$c_k(x) = \lg(N+1) - \mathcal{V}(N) \le \lg(1+x) + \frac{1-x}{1+x} - \frac{k+2}{2^k(1+x)}.$$

and the last term vanishes as $k \to \infty$.

Bob's strategy space consists of binary search trees on N nodes. Each nodes corresponds to a secret number. One tree that is particularly easy to describe is β, the bisecting tree for $N = 2^k - 1$. The depth of a node n in β is equal to $k - j$ if and only if $j = \mathrm{ord}_2(n)$. If $N = 2^k - 2$ then Bob may still use the bisecting tree β, which now has one node less as it loses the end node $2^k - 1$. Let $\beta - 1$ be the binary search tree in which compared to β all nodes reduce by 1. The trees β and $\beta - 1$ on $2^k - 2$ nodes are illustrated below for $k = 3$.

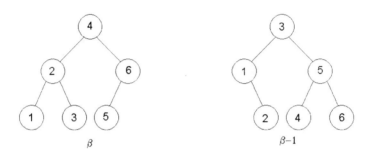

β $\beta-1$

If $N = 6$ and if Bob picks one of the two trees β and $\beta - 1$ equiprobably, then on average he has to pay Alice no more than $2\frac{1}{2}$. The next lemma extends this observation.

Lemma 4. *Let $d < k$ be a natural number. Then $\mathcal{V}(2^k - 2^d) \le k - 1 + \frac{1}{2^d}$.*

Proof. Suppose that Bob picks one of the trees $\beta - i$ equiprobably for $i = 0, \ldots, 2^d - 1$. Let $n \in \{1, \ldots, 2^k - 2^d\}$ be an arbitrary secret number. The depth of node n in $\beta - i$ is equal to $k - \mathrm{ord}_2(n + i)$. Consider the numbers $n + i$ for $i = 0, \ldots, 2^d - 1$. Half of the are odd and half of them are even. Of the even numbers, half are divisible by 4 and half of them are not, etc. More succinctly, $n + i$ runs through all residues modulo 2^d. If $n + i = 0 \bmod 2^d$, then $\mathrm{ord}_2(n + i) \ge d$. All the other orders are determined by the residue. Since Bob takes one of the i equiprobably, the average number of guesses against secret number n is

$$\frac{1}{2^d} \sum_{i=0}^{2^d-1} (k - \mathrm{ord}_2(n+i)) \le k - \frac{d}{2^d} - \frac{1}{2^d} \sum_{j=0}^{d-1} (d-1-j)2^j.$$

We rewrite $\frac{1}{2^d} \sum_{j=0}^{d-1} (d-1-j)2^j$ by substituting $i = d - 1 - j$ as $\sum_{i=0}^{d-1} \frac{i}{2^{i+1}}$, which is equal to $1 - \frac{d+1}{2^d}$. Hence, if Bob uses this strategy than the average number of guesses against any secret number is bounded by $k - 1 + \frac{1}{2^d}$.

Suppose that τ is a binary search tree on $\{1, \ldots, N\}$ and suppose that $M > N$. Let $S \subset \{1, \ldots, M\}$ be a subset of cardinality N. In particular, there is an order preserving bijection $\phi \colon \{1, \ldots, N\} \to S$. Define the binary search tree τ_S on $\{1, \ldots, M\}$ as follows. Replace each node j by $\phi(j)$. Once Bob guesses a number that is an end node of this tree without guessing the secret number, which happens if the secret number is not in S, then he guesses consecutive numbers. If his guess is j and Alice says 'too low', then Bob guesses $j + 1$. If Alice says 'too high', then Bob guesses $j - 1$. This is a well defined pure strategy which corresponds to a binary search tree τ_S. To illustrate this, consider the search tree β for $N = 6$ that we considered above. Let $M = 10$ and let $S = \{1, 3, 5, 6, 9, 10\}$, then β_S is equal to the binary search tree in the figure below.

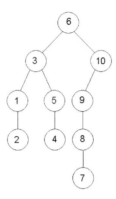

Note that it takes two additional guesses to get the secret number 7. The depth of the tree has increased by two. This is because the complement of S contains consecutive numbers, namely 7 and 8. If there are no consecutive numbers in the complement of S, then the depth of τ_S exceeds the depth of τ by no more than one.

Lemma 5. *Let* $N = 2^k - 2^d$ *and let* $y \leq N$ *be any natural number. Then*

$$\mathcal{V}(N + y) \leq k - \frac{N - y}{N + y} + \frac{N}{N + y} \cdot \frac{1}{2^d}$$

Proof. We say that a subset $C \subset \{1, \ldots, N\}$ contains no consecutive numbers if whenever $j \in C$ then $j - 1 \notin C$ and $j + 1 \notin C$. Here we calculate modulo N, so we consider the numbers 1 and N to be consecutive. For instance, $F = \{1, 3, \ldots, 2y+1\}$ is a subset of $\{1, \ldots, N+y\}$ of cardinality y without consecutive numbers. Also $j + F$, which is a subset of $\{1, \ldots, N + y\}$ if we calculate modulo $N + y$, contains no consecutive numbers,.

Let S be the complement of $j+F$. Suppose Bob takes a number $j \in \{1, \ldots, N+y\}$ equiprobably and chooses the search tree β_S. If the secret number is in S, which has probability $N/(N + y)$, then the expected number of guesses is $\leq k - 1 + \frac{1}{2^d}$ by the previous lemma. If not, then the number of guesses is $\leq k + 1$

since the depth of β is equal to k. So the average number of guesses of this strategy against any secret number is bounded by

$$\frac{N}{N+y}\left(k-1+\frac{1}{2^d}\right)+\frac{y}{N+y}(k+1) = k - \frac{N-y}{N+y} + \frac{N}{N+y}\cdot\frac{1}{2^d}.$$

Theorem 3.

$$c_\infty(x) = \lg(1+x) + \frac{1-x}{1+x}$$

Proof. We already proved that $c_\infty(x) \leq \lg(1+x) + \frac{1-x}{1+x}$, so it suffices to establish the reverse inequality. By the previous lemma we have for $N = 2^k - 2^d + y$ that $\mathcal{V}(N) \leq k - \frac{2^k-2^d-y}{N} + \frac{N-y}{N}\cdot\frac{1}{2^d}$ which is bounded by $k - \frac{2^k-y}{N} + \frac{2^d}{N} + \frac{1}{2^d}$. In particular if we put $N + 1 = 2^k(1+x)$ then

$$c(N) = \lg(N+1) - \mathcal{V}(N) \geq \lg(1+x) + \frac{2^k-y}{N} - \frac{2^d}{N} - \frac{1}{2^d}$$

Now we may let d go to infinity while being much smaller than k, so we can see to it that $\frac{2^d}{N} - \frac{1}{2^d}$ is arbitrarily small. Furthermore, as $N = 2^k - 2^d - y = 2^k(1+x) - 1$ it follows that asymptotically, $\frac{2^k-y}{N} = \frac{2^k-2^kx}{N} = \frac{1-x}{1+x}$.

3 Solving Dresher's Guessing Game by Backward Induction

Dresher's game is a zero-sum finite game, so it can be solved by linear programming. Unfortunately, the matrix of the game is of dimension $N \times c(N)$, where $c(N)$ denotes the N-th Catalan number, see [16]. Johnson pointed out that the size of the matrix can be reduced somewhat by symmetry considerations, but this only allowed a solution of the game up to $N = 11$. The twelfth Catalan number is equal to 208012 and since it roughly quadruples with each increment of N, storing the entire matrix quickly becomes infeasible. Therefore, the simplex algorithm can only be used to solve the game for very small values of N. In each iteration of the simplex algorithm, the matrix of the game (the tableau) is used only once, to determine a certain column vector, see [8, chapter 2.8]. Our main idea is to compute this column by backward induction rather than to determine it from the matrix. In this way, we circumvent the problem of storing the matrix in a way that has some similarity to the implicit tableaux representation that is known from linear programming [7].

The solution of the storage problem does not solve all our computational problems. We also need to deal with the computational time. It is a well known fact that the simplex method often produces the solution in a relatively short computational time, but for Dresher's guessing game it does not. It turns out that the computational time gets problematic if N is close to a power of 2. We are able to push the computations to 256. It is perhaps possible to extend the results to 512, but 1024 seems to be infeasible.

We briefly recall the simplex method in terms of Dresher's guessing game. It starts with a selection of pure strategies γ for Bob (the base). Let V denote the value of the game if Bob uses only pure strategies in the base and let α be an optimal strategy for Alice in this game. Now check the tableau to see if there is a pure strategy γ' outside the base that has payoff less than V against α. If such a γ' exists then it is brought in the base and some other strategy is removed from the base. This procedure is iterated until no such γ' exists, and then the game is solved. In our approach, instead of checking the tableau we *compute* the optimal pure strategy γ' by dynamic programming.

Theorem 4. *Suppose Alice uses a mixed strategy α, in which she chooses the secret number i with probability α_i. It is possible to construct an optimal binary search tree against α in time $\mathcal{O}(N^3)$.*

This theorem also follows from a result of Gottinger, but we include a proof to keep our paper self-contained. A fully worked out example of how to construct the optimal tree can be found in [13]. We introduce some notation. First observe that a binary search tree τ on $\{1, \ldots, N\}$ can be represented by an N-dimensional vector, in which the i-th entry of the vector is equal to the number of guesses if the secret number is equal to i. If Alice uses the mixed strategy α and if Bob uses the pure strategy τ, then the average payoff for Alice is equal to the inner product $\alpha \cdot \tau$. By $\tau^{m,n}$ we denote a binary search tree on a subinterval $\{m, \ldots, n\} \subset \{1, \ldots, N\}$. We represent this by an N-dimensional vector which has entries equal to zero for all coordinates that lie outside $\{m, \ldots, n\}$. We denote the inner product of this vector with α by

$$I(\alpha, \tau^{m,n}) = \alpha \cdot \tau^{m,n}$$

and $\tau_\star^{m,n}$ denotes the search tree on $\{m, \ldots, n\}$ that minimizes $I(\alpha, \tau^{m,n})$. Each binary tree is a concatenation of two subtrees. If i is the top node (initial guess) of the search tree $\tau^{m,n}$, then we denote its concatenation by $\tau^{m,n} = \tau^{m,i-1} * \tau^{i+1,n}$.

Lemma 6. $\tau_\star^{m,n} = \tau_\star^{m,i-1} * \tau_\star^{i+1,n}$ *for the i that minimizes $I(\alpha, \tau_\star^{m,i-1}) + I(\alpha, \tau_\star^{i+1,n})$.*

Proof. Let $\tau^{m,n} = \tau^{m,i-1} * \tau^{i+1,n}$ be a search tree with top node i. Then $\tau^{m,n}$ uses one guess more than the subtrees it is concatenated from. So

$$I(\alpha, \tau^{m,n}) = I(\alpha, \tau^{m,i-1}) + I(\alpha, \tau^{i+1,n}) + \sum_{j=m}^{n} \alpha_j.$$

If we minimize this expression we might as well minimize $I(\alpha, \tau^{m,i-1}) + I(\alpha, \tau^{i+1,n})$ since $\sum_{j=m}^{n} \alpha_j$ is constant. If i is fixed, then the minimum is $I(\alpha, \tau_\star^{m,i-1}) + I(\alpha, \tau_\star^{i+1,n})$, so the minimizing search tree with initial guess i is $\tau_\star^{m,i-1} * \tau_\star^{i+1,n}$. Minimizing this expression over i gives the search tree $\tau_\star^{m,n}$.

Bob's best strategy against a given α is $\tau_*^{1,N}$, and we can construct this search tree from the lemma combined with standard backward induction, as follows. In the first stage we construct the trivial search trees $\tau_*^{i,i+1}$. At stage k the search trees $\tau_*^{j,j'}$ are known for all $j' - j \leq k$ so we can construct the trees in stage $k + 1$ from the recursion $\tau_*^{j,j+k+1} = \tau_*^{j,i-1} * \tau_*^{i+1,k+1}$ for the minimizing initial guess i. During this backward induction we calculate and store each of the $\binom{N}{2}$ search trees $\tau_*^{m,n}$, minimizing over the $m - n$ possible initial guesses. So we store $O(N^3)$ numbers and we perform $O(N^3)$ computational operations. This proves Theorem 4.

Now we have a dynamic pivoting rule all that remains is to decide how to start the iteration. We choose the initial base from binary search trees that are optimal against a uniform strategy, i.e., the coordinate sum is minimal for vectors in the initial base.

```
V(1) := 1

For n = 2 to N do

    Create the initial base B

    Compute the maximin strategy α and the minimax
    value V for B

    Compute an optimal search tree τ against α
    by backward induction

    While α · τ < V do

        { Use the simplex method to compute the new maximin strategy
          α for the matrix that consists of the base plus τ

        Determine the exit strategy and delete it from the matrix

        Compute an optimal search tree τ against α
        by backward induction }

    V(n) := V

End
```

This algorithm has been stylized. If many simplex iterations are necessary, computing an optimal search tree gets expensive and it is faster to first check a limited tableau. In our implementation of the algorithm, we have kept a tableau with a number of $\mathcal{O}(N^2)$ strategies. In this way we were able to compute the value of the game for N up to 256. The running time to solve the game for N close to 256 is less than 10 minutes on a standard PC (which should be compared to a running time of just a few seconds if N is close to 128). We were unable to push the computations further. The size of the denominator of $\mathcal{V}(N)$ grows out of bounds for N close to 512 and so does the number of iterations in our algorithm.

We have given the values of the game for N up to 128 in Table 1, the horizontal lines mark intervals $N \in \{2^k + 1, 2^{k+1}\}$. The values are rational since the game matrix is integral. The main point of the table is to illustrate that most values have a denominator in the order of N, but the denominator increases enormously if N gets close to a power of 2. For instance, if $N = 251$ then the value is equal to

$$\frac{26536065379642766512509321060411885534161463563222201}{3773691101808015326761679535801450411664055784123701}$$

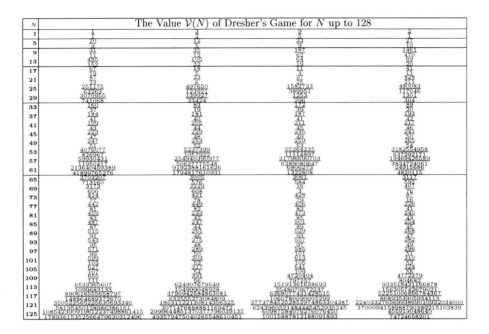

The Value $\mathcal{V}(N)$ of Dresher's Game for N up to 128

Even if it is possible to find an explicit formula for the value of the game, then the size of the numerator and the denominator indicates that such a formula will be very complicated. It is possible, however, to find formulas that hold for a substantial part of the numbers N. Many values are equal to $\frac{(k+1)(N+5)-2^{k+1}-4}{N+4}$ or $\frac{(k+1)(N+3)-2^{k+1}-1}{N+2}$. It may be possible to obtain an explicit solution of the game for numbers N that are not close to a power of two.

Since we have computed the game for $N \leq 256$, we can determine the functions c_k from section 2 for $k \leq 7$. Since c_k converges to c_∞ monotonically, for $N \geq 256$ we have that

$$c_7(x) \leq \log(N+1) - \mathcal{V}(N) \leq c_\infty(x).$$

The graph of $c_\infty(x) - c_7(x)$ is depicted in the figure below. Its maximum is < 0.1 and so we have the following result.

Theorem 5. $\lg(1+x) + \frac{1-x}{1+x} - 0.1 < \lg(N+1) - \mathcal{V}(N) < \lg(1+x) + \frac{1-x}{1+x}$, for $N \geq 256$.

3.1 Brief Remarks on Alice's Optimal Strategy

So far we have not discussed optimal strategies of the player. We conclude our exposition with some very brief comments on Alice's optimal strategies. In equation 2 we found that Alice's average payoff is equal to $\mathcal{E}(N) = k+1+\frac{k+2-2^{k+1}}{N}$ if she chooses a secret number equiprobably. It follows from our asymptotic formula for the value of the game that $\lim_{N\to\infty} \mathcal{V}(N) - \mathcal{E}(N) = 0$. Since $\mathcal{V}(N) - \mathcal{E}(N)$

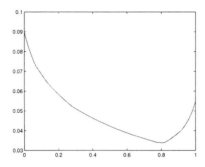

Fig. 1. The graph of $c_\infty(x) - c_7(x)$

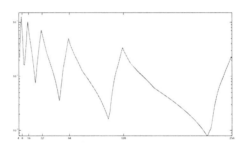

Fig. 2. $\mathcal{V}(N) - \mathcal{E}(N)$

converges to zero, one could say that the equiprobable strategy is 'asymptotically optimal'. In the figure below we have depicted $\mathcal{V}(N) - \mathcal{E}(N)$ on a logarithmic scale for $N \leq 256$, demonstrating that this difference is relatively large if N is close to a power of two, but also showing that the absolute difference is rather small and in the order of 0.01 once $N > 100$.

By saying that the equiprobable strategy is 'asymptotically optimal', we do not mean that Alice's optimal strategy converges to the uniform distribution as N goes to infinity. Indeed, that is not true. Gilbert [12] conjectured that Alice ought to avoid the secret numbers $N/2$ as well as $N/4$ and $3N/4$ and the like. This seems to be true if N is close to a power of two, as illustrated in the figure below, which depicts Alice's optimal strategy for $N = 249$. However, if N is 'in between' powers of two, then Alice plays almost equiprobably: only the secret numbers $1, 2, N - 1, N$ get a slightly enlarged probability mass - all other numbers are equiprobable.

Johnson [15] made the following conjectures for Alice's optimal strategies for $N > 4$:

1. The probability that Alice's secret number is 1 is equal to the probability that the secret number is 2 or 3.

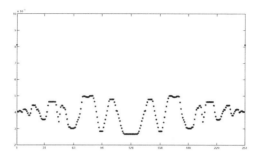

Fig. 3. Probability mass of Alice's optimal strategy for $N = 251$

2. The probability that Alice's secret number is 1 is twice the probability that the secret number is 2.

The second conjecture turns out to be wrong for $N = 13$ already, but our numerical results confirm the other conjecture. However, we are unable to prove it.

References

1. Alpern, S.: Search for point in interval, with high-low feedback. Math. Proc. Cambr. Phil. Soc. 98, 569–578 (1985)
2. Alpern, S., Gal, S.: The theory of search games and rendezvous. Kluwer Academic Publishers (2003)
3. Alpern, S., Reyniers, D.: Spatial dispersion as a dynamic coordination problem. Theory and Decision 53(1), 29–59 (2002)
4. Alpern, S., Snower, D.J.: High-low search in product and labor markets. American Economic Review 78, 356–362 (1988)
5. Baston, V.J., Bostock, F.A.: A high-low search game on the interval. Math. Proc. Cambr. Phil. Soc. 97, 345–348 (1985)
6. Crawford, V.P., Heller, H.: Learning How to Cooperate: Optimal Play in Repeated Coordination Games. Econometrica 58, 571–595 (1990)
7. Dantzig, G., Wolfe, P.: Decomposition principle for linear programs. Operations Res. 8, 101–111 (1960)
8. Dresher, M.: The Mathematics of Games of Strategy. Dover Publications (1981)
9. Ferguson, T.S.: Game theory, part II,
   `http://www.math.ucla.edu/~tom/Game_Theory`
10. Gal, S.: A discrete search game. SIAM J. Appl. Math. 27(4), 641–648 (1974)
11. Garnaev, A.Y.: Search games and other applications of game theory. Springer, Heidelberg (2000)
12. Gilbert, E.: Games of identification and convergence. SIAM Rev. 4(1), 16–24 (1962)
13. Gottinger, H.W.: On a problem of optimal search. Z. Operations Res. 21(5), 223–231 (1977)
14. Grabner, P.: Searching for losers. Random Struct. Algor. 4, 99–110 (1993)
15. Johnson, S.: A search game. In: Dresher, M., Shapley, L.S., Tucker, A.W. (eds.) Advances in Game Theory. Annals of Mathematics Studies, vol. 52, pp. 39–48 (1964)

16. Konheim, A.: The number of bisecting strategies for the game (N). SIAM Rev. 4(4), 379–384 (1962)
17. Pelc, A.: Searching games with errors, fifty years of coping with liers. Theor. Comput. Sci. 270(1-2), 71–109 (2002)
18. Raghavan, T.E.S.: Zero-sum two-person games. In: Aumann, R.J. (ed.) Handbook of Game Theory with Economic Applications, vol. 2, pp. 735–757. North Holland Publishers (1995)
19. Reyniers, D.J.: A dynamic model of collective bargaining. Comp. Economics 11(3), 205–220 (1998)

Security Games with Market Insurance

Benjamin Johnson[1], Rainer Böhme[2], and Jens Grossklags[3]

[1] Department of Mathematics, UC Berkeley
[2] Department of Information Systems, University of Münster
[3] College of Information Sciences and Technology, Penn State University
benjamin@math.berkeley.edu,
rainer.boehme@wi.uni-muenster.de,
jensg@ist.psu.edu

Abstract. Security games are characterized by multiple players who strategically adjust their defenses against an abstract attacker, represented by realizations of nature. The defense strategies include both actions where security generates positive externalities and actions that do not. When the players are assumed to be risk averse, market insurance enters as a third strategic option. We formulate a one-shot security game with market insurance, characterize its pure equilibria, and describe how the equilibria compare to established results. Simplifying assumptions include homogeneous players, fair insurance premiums, and complete information except for realizations of nature. The results add more realism to the interpretation of analytical models of security games and might inform policy makers on adjusting incentives to improve network security and foster the development of a market for cyber-insurance.

Keywords: Game theory, Security, Externalities, Protection, Self-insurance, Market insurance.

1 Introduction

It is widely accepted that network security has properties of a public good. A series of works on security games has led to a set of formal tools to analyze the provision of network security by individual agents who control nodes on the network. One distinctive feature of this work over the traditional literature on the provisioning of public goods is the distinction of two types of security technologies, *protection*, which exhibits externalities, and *self-insurance*, which does not. This combination frames network security as a hybrid between a public and a private good, modulated by the relative costs of the two security technologies.

Another distinctive feature of network security over other public goods problems is the existence of uncertainty. An agent's security investment at present only pays off if an attack occurs in a future state of the world. In the existing body of literature, this uncertainty is treated by considering the expected loss as decision variable, largely for the sake of tractability. By contrast, decision science has a rich variety of more realistic models of human and organizational

J.S. Baras, J. Katz, and E. Altman (Eds.): GameSec 2011, LNCS 7037, pp. 117–130, 2011.

decision making under uncertainty. One key concept is the notion of risk aversion, typically expressed in a concave utility function. Introducing risk aversion into security games and revisiting the equilibrium strategies is interesting of its own. Even more so because risk aversion naturally leads to a third strategic option, namely agents seeking *market insurance* as a means to transfer the financial risk of uncertain future outcomes.

The question of market insurance for network security has attracted the attention of practitioners, policy makers, and researchers who contributed to a meanwhile sizable body of literature on cyber-insurance.

This paper, to the best of our knowledge for the first time, tries to merge the two streams of research and formally analyzes security games with optional market insurance. To do so, we extend the basic setup of security games with complete information—except for the uncertainty of future states—by a utility function with risk aversion. We discuss this case as an intermediate result before we advance to the analysis of market insurance. The analysis here focuses on the existence of an insurance market, an unanswered question that is relevant to inform policy makers on endeavors to bootstrap a market for cyber-insurance for their assumed positive effects on other frictions to network security not captured by our present formal model, such as information asymmetries and negligence.

This paper is organized as follows: Section 2 recalls the broader context of security games and cyber-insurance within the field of economics of information security. Section 3 presents our model, which is then analyzed in Section 4. The final Section 5 wraps up with discussion and conclusion.

2   Background

An increasing amount of evidence about the economic and technical underpinnings of cybercrime highlight the need for thorough security measures. A growing number of specialized measurement studies demonstrate the professionalism of miscreants concerning a variety of nefarious business models. For example, Holz *et al.* [14] document the elicit trade with payment credentials that have been previously stolen through keylogging malware. Even relatively benign activities such as spam distribution now depend on sophisticated infrastructures existing in the form of botnets and their command-and-control centers [15,21].

The devastating success of these threats frequently depends on interdependencies in computer networks that inhibit the deployment of effective countermeasures. For example, botnets as vehicles behind almost all volume crime on the internet can only exist because some nodes connected to the network apply lower-than-optimal security standards. Similarly, for targeted attacks, a single breach of a corporate perimeter may allow an attacker to harvest resources from all machines located within its confines.

To better understand the implications of these interdependencies for individual defenders, Varian [22] conducted an analysis of system reliability within a public goods game-theoretical framework. He discusses the best effort, weakest-link, and total effort games, as originally analyzed by Hirshleifer [12]. In Varian's model, security investments take the form of protection effort (e. g., patching system vulnerabilities) where aggregate investments have a decreasing marginal contribution to network security. Further, the security of an individual defender depends on her own effort and on the contributions by all her peers.

Grossklags *et al.* extend this framework by treating security as a hybrid between public and private goods. This is highly interesting because the characteristic as a public or private good is not only determined by the available technology (i. e., its cost) and the architecture of the network (i. e., the functional form of interdependencies). Moreover, individual agents decide strategically on how to split their security investments between protection and self-insurance [7,8]. Self-insurance only affects the investing defender directly, and is consequently a private good (e. g., having good backups). This is different to Ehrlich and Becker's [6] terminology, who use self-insurance to denote loss protection, i. e., reduction of the size of the loss, and protection to denote loss prevention, i. e., reduction of the probability of loss, without differentiating between characteristics of private and public goods. In a more general setting without the distinction between private and public components, both reduce to shifting probability mass in the loss distribution function.[1] Unlike Varian [22], Grossklags *et al.* [7] assume both investment variables to have constant marginal impact across the range of investment opportunities (subject to interdependencies).

Computer security research has been effective in contributing to a better understanding of the uncertainties resulting from attackers' actions. However, this progress in measuring relevant parameters (e. g., attacker intent and attack probabilities) is only partially helpful to understand responders' actions. In particular, we need to have a better grasp of how these factors are perceived by defenders and translated into investment decisions. From behavioral research, it is well-understood that individuals exhibit different risk-coping mechanisms that may depend on a variety of factors (e. g., the amount at stake). Unfortunately, it is rarely the case that risk perception and resulting actions are perfectly in congruence (i. e., risk neutrality).

In fact, for a wide variety of risk scenarios individuals' actions demonstrate risk aversion [11]. Under this behavioral assumption and in the presence of uncertainty, the expected utility of wealth is less than the utility of expected wealth, where the expectations are taken over all possible outcomes of the random future state. To the best of our knowledge there exists no previous work that studies risk-averse agents' decision making in the presence of multiple security investment options (i. e., protection and self-insurance).

In the absence of regulation, institutional behavior is typically more aligned with risk-neutral decision-making, whereas individual decision makers' actions are typically consistent with risk-aversion. Risk aversion and contracts are the

---

[1] The term self-insurance in the sense of loss protection has also been used by Böhme and Kataria [4] in the context of cyber-insurance describing the option of a single decision maker who operates a large number of computing resources to achieve risk balancing within its own pool of resources rather than joining a risk pool on the insurance market.

only prerequisites for market insurance. More specifically, insurers are offering contracts to risk-averse agents, the insureds.[2] This risk-pooling should decrease the variance of losses and thereby increase overall welfare [16].

In practice, a number of obstacles have prevented the market for cyber-insurance from achieving maturity. Absence of reliable actuarial data to compute insurance premiums, lack of awareness among decision-makers contributing to too little demand, as well as legal and procedural hurdles have been identified in the "first generation" of cyber-insurance literature until about 2005 [3]. The latter aspect may cause frustration when claiming compensation for damages. Further, entities considering insurance must undergo a series of often invasive security evaluation procedures, revealing their IT infrastructures and policies [1,9]. Meanwhile, witnessing thousands of vulnerabilities, millions of attacks, and substantial improvement in defining security standards and computer forensics calls into question the validity of these factors to causally explain the lack of an insurance market. Consequently, a "second generation" of cyber-insurance literature emerged. Its authors link the market failure with fundamental properties of information technology, specifically correlated risk [2], information asymmetries between insurers and insureds [20], and interdependencies [18,20]. So far, these obstacles have been studied independent of the hybrid private–public good characteristic of network security. Our contribution in this paper is to marry both streams of research and characterize equilibria in a basic model of a security game with market insurance. To keep things tractable, we do not consider correlated risk and we remain in a regime of complete information except for the realization of future losses—see Böhme and Schwartz [5] for a discussion of the validity and implications of these conventions.

3   Model

We devise a stylized game-theoretic model with the intention to focus on the analysis of symmetric equilibria. Occam's razor was adjusted to emphasize the introduction of risk aversion and the option to obtain market insurance at endogenous but fair premiums. To that end, defenders act as players, attackers as nature, and insurers as mechanism, i.e., price-takers with perfect information about the players' actions.

3.1   Baseline Security Game

The baseline game includes neither risk aversion nor market insurance. Formally, the base model from which we develop our security games has the following payoff structure. Each of $N \in \mathbb{N}$ players has an initial wealth M_0. If a given player is

---

[2] There are situations where the purchase of insurance might serve as a strategic tool to achieve another purpose. For example, insureds can more credibly threaten with risky behaviors [17]. The purchase of insurance might also help to quell a stakeholder's fear, uncertainty, and doubt after a security breach. See, for example, banks' offers of identity theft insurance plans with a free trial period after large-scale data thefts.

attacked and compromised successfully she faces a loss L. Attacks arrive with an exogenous probability of p $(0 \leq p \leq 1)$. Players have two security actions at their disposition. Player i chooses a protection level $0 \leq e_i \leq 1$ and a self-insurance level $0 \leq s_i \leq 1$. Finally, $b \geq 0$ and $c \geq 0$ denote the unit cost of protection and self-insurance, respectively.

The post-event wealth function has the following structure:

$$M_1(s_i, e_i; b, c, M_0) = M_0(1 - q \cdot L \cdot (1 - s_i) - be_i - cs_i) \tag{1}$$

where $q \in \{0, 1\}$ is the realization of a random variable indicating loss $(q = 1)$ and no loss $(q = 0)$. The probability of loss is endogenous and depends on the probability of attack p scaled by the protection effort $H(e_i, e_{-i})$. $H : \mathbb{R}^N \mapsto [0, 1]$ is a contribution function aggregating the protection efforts of player i and all other players (denoted by suffix $-i$). H is monotonically increasing in all its parameters, thereby ensuring that protection generates positive externalities. For the analysis in this paper, we focus on the restricted case in which H describes a weakest link externality, i.e. $H(e_i, e_{-i}) = \min\{e_1, \ldots, e_N\}$.

The final utility is mapped to the utility domain by $u = U(M_1)$. As the players maximize expected utility, the combined payoff function of the baseline security game is

$$E(u_i) = p(1 - H(e_i, e_{-i})) \cdot U(M_0(1 - L \cdot (1 - s_i) - be_i - cs_i)) \\ + (1 - p(1 - H(e_i, e_{-i}))) \cdot U(M_0(1 - be_i - cs_i)). \tag{2}$$

Post-event wealth is divided into two cases depending on whether a loss occurs or not. In the bad case, new wealth is $M_1 = M_0(1 - L \cdot (1 - s_i) - be_i - cs_i)$. In the good case, new wealth is $M_1 = M_0(1 - be_i - cs_i)$.

3.2 Risk Aversion

Risk aversion is introduced by transforming wealth M_1 to utility $U(M_1)$ using a concave function of type CRRA[3],

$$U(M) = \begin{cases} \frac{M^{1-\sigma}}{1-\sigma} & \text{if } \sigma > 0, \sigma \neq 1 \\ \log(M) & \text{if } \sigma = 1, \end{cases} \tag{3}$$

so that $U'(x) = x^{-\sigma}$. $\sigma > 0$ is the degree of risk aversion, an exogenous parameter fixed to $\sigma = 1$ unless otherwise stated. The choice of the CRRA type is convenient because it allows us to derive conclusions that are independent of the initial wealth. This choice also follows established conventions in the cyber-insurance literature (e. g., [2,20]), although CARA-type[4] utility functions can be found as well [18]. Other researchers are agnostic about the shape of the utility function and just require concavity and twice differentiability [13].

---

[3] CRRA = constant relative risk aversion [19].
[4] CARA = constant absolute risk aversion.

3.3   Market Insurance

By augmenting the baseline security game with optional market insurance, players will receive an insurance payment, $0 \leq x_i \leq 1$, when a security compromise occurs and they have previously purchased insurance. We assume that agents cannot be overcompensated for losses through a combination of self-insurance and market insurance, i.e., $x_i + s_i \leq 1$. This reflects the principle of indemnity prevalent in the insurance industry. The cost of market insurance, π, is perfectly related to the loss probability and the potential loss in a market with a risk-neutral non-profit insurer who manages a pool of infinitely many homogeneous and independent risks, $\pi = Lp \cdot (1 - H(e_i, e_{-i}))$. However, every realistic (for-profit) insurer would require $\pi > Lp \cdot (1 - H(e_i, e_{-i}))$.

In the presence of market insurance Equation 2 becomes:

$$E(u_i) = p(1 - H(e_i, e_{-i})) \cdot U(M_0(1 - L \cdot (1 - s_i) - be_i - cs_i + x_i(1 - \pi)))$$
$$+ (1 - p(1 - H(e_i, e_{-i}))) \cdot U(M_0(1 - be_i - cs_i - \pi x_i)). \tag{4}$$

Now, in the bad case, new wealth is $M_1 = M_0(1 - L \cdot (1 - s_i) - be_i - cs_i - \pi x_i + x_i))$. In the good case, new wealth is $M_1 = M_0(1 - be_i - cs_i - \pi x_i)$.

3.4   Simplifications

To keep the number of parameters manageable, we assume that $b, c \leq L = 1$, and that, since decisions made on the basis of a CRRA utility function are invariant under multiplicative factors, M_0 can be eliminated. Table 1 in the appendix summarizes all symbols used in our model.

3.5   Payoff Dominance

Theorem 1. $E[u_i]$ *is bounded above by* $\max\{1 - b, 1 - c, 1 - \pi, 1 - p\}$. *Furthermore, the dominance is strict unless* $e_i \in \{0, 1\}$.

The theorem relies only on the affine structure of our wealth function, together with U being increasing and concave up; a full proof is in the appendix. We use this theorem to help isolate Nash equilibria. If the payoff of each player in a homogeneous strategy achieves the maximizing bound from the theorem, we may conclude that the strategy configuration is a Nash equilibrium.

Conversely, if a strategy configuration results in a utility for some player not conforming to one of the outcomes from the theorem, the only way this configuration can be an equilibrium is if at least one of the outcomes from the theorem is not possible to achieve. This observation can be strengthened by the following corollary.

Corollary 1. *In any hybrid equilibrium where there is a non-zero partial protection investment, the utility of each player is strictly less than* $\max\{U(1 - p), U(1 - b), U(1 - c), U(1 - \pi))\}$.

A proof of the corollary is also in the appendix.

4   Analysis

4.1   Base Model

We begin by briefly reviewing the equilibrium results from the base model (see [10]).

1. Protection equilibria
   If $b < p$ and $b < c$, then $(e_i, s_i) = (e_0, 0)$ (protection at level e_0) is a symmetric Nash equilibrium for any e_0 between $\frac{p-c}{p-b}$ and 1.
2. Self-insurance equilibria
   If $c < p$ then $(e_i, s_i) = (0, 1)$ (full self-insurance) is a symmetric Nash equilibrium.
3. Passivity equilibria
   If $p < c$, then $(e_i, s_i) = (0, 0)$ (passivity) is a symmetric Nash equilibrium.

The above are the only symmetric Nash equilibria for this game. Note that with the exception of partial protection equilibria, all equilibrium strategies are corner strategies. Among all protection equilibrium strategies, the strategy in which each player invests in full protection is Pareto-dominant.

4.2   Base Model with Risk Aversion

Incorporating risk aversion into the base model induces some changes. When the risk-aversion is positive, players have a strong aversion to very low wealth. In fact, for risk aversion coefficients $\sigma \geq 1$, the prospect of having zero wealth results in an infinitely negative utility.[5] The consequence is that players are no longer satisfied with any strategy in which there is the remote chance of obtaining a non-positive wealth.

We find four distinct types of symmetric Nash equilibrium in the base model supplemented by risk aversion, with $\sigma = 1$.

1. Full protection equilibria
   If $b < p$ and $b < c$, then $(e_i, s_i) = (1, 0)$ (full protection) is a symmetric Nash equilibrium.
2. Full self-insurance equilibria
   If $c \leq p$ then $(e_i, s_i) = (0, 1)$ (full self-insurance) is a symmetric Nash equilibrium.
3. Partial self-insurance equilibria
   If $p < c$, then $(e_i, s_i) = \left(0, \frac{p}{c}\right)$ (partial self-insurance at the indicated level) is a symmetric Nash equilibrium.

---

[5] For $0 < \sigma < 1$ players' utility at zero wealth is finite, but the derivative of utility tends to infinity as wealth approaches zero, so players still have an infinite aversion to retaining a non-positive wealth.

4. Combined protection and self-insurance equilibria
   If $p \leq c$, there exists a sufficiently small b such that for any choice of $e_0 < \frac{1-c}{b}$,
   $(e_i, s_i) = \left(e_0, \frac{p(1-e_0)}{c} + \frac{be_0(c-p(1-e_0))}{c(1-c)}\right)$ (partial protection with partial self-
   insurance) is a symmetric Nash equilibrium.

An algebraic expression for the maximum b to make this work is difficult to
produce (and in fact may not exist), but the existence of b itself follows from the
utility function U being differentiable on positive inputs. In the resulting hybrid
equilibrium, every player would prefer to invest in full protection because the
cost is cheap, but due to the interdependencies inherent in the weakest link game,
the maximum investment in protection cannot be set unilaterally, so players are
forced to make up for the resulting probability of loss by obtaining self-insurance.
If the incentives are such that full self-insurance is desirable, then the incentive to
protect will not remain. But if incentives are such that only partial self-insurance
is desirable (and if b is sufficiently small), then the configuration with both types
of investments is an equilibrium.

Note that passivity is never an equilibrium for any $\sigma \geq 0$, because in such cases
players have an infinitely-strong aversion to any non-zero chance of retaining zero
wealth.

4.3 Base Model with Risk Aversion and Market Insurance

When we incorporate market insurance, we arrive at more changes. The existence
of market insurance ensures that no partial self-insurance investment is optimal.
Such partial investments were only possible in the event $p < c$. But if $p < c$
and market insurance is available, then market insurance is always preferable
to self-insurance. Even if the reverse inequality holds, it is possible for market
insurance to be preferable to self-insurance if there is also a partial protection
investment.

1. Full market insurance
   If $p \leq c$ then $(e_i, s_i, x_i) = (0, 0, 1)$ (full market insurance) is a symmetric
   Nash equilibrium.
2. Full self-insurance
   If $c \leq p$, then $(e_i, s_i, x_i) = (0, 1, 0)$ (full self-insurance) is a symmetric Nash
   equilibrium.
3. Full protection
   If $b \leq \min\{c, p\}$, then $(e_i, s_i, x_i) = (1, 0, 0)$ (full protection) is a symmetric
   Nash equilibrium.
4. Partial market insurance and partial self-insurance
   If $c = p$, then for any x_0, s_0 with $s_0 + x_0 = 1$, $(e_i, s_i, x_i) = (0, s_0, x_0)$ is a
   symmetric Nash equilibrium.
5. Partial protection and full market insurance
   If $b \leq p$ and $be_0 + p(1 - e_0) < c$, then $(e_i, s_i, x_i) = (e_0, 0, 1)$ is a symmetric
   Nash equilibrium.

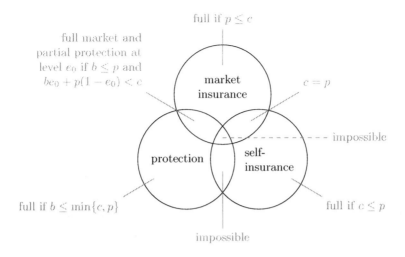

Fig. 1. Overview of feasible symmetric equilibria and corresponding conditions

The last case illustrates an instance in which the availability of market insurance has a positive effect on protection investment. In the same parameter configuration without availability of market insurance, individuals would instead be forced to turn to self-insurance to mitigate against the existing risk. If the additional (compatible) condition $c < p$ is added, then the incentive structure is such that players would prefer to defect to a full self-insurance strategy, and neglect any protection investment.

Figure 1 shows the equilibrium conditions for the case with risk aversion and market insurance.

5   Discussion

In the base model, we find that agents can only with difficulty coordinate on an equilibrium with full protection effort. In particular, the availability of alternative prevention equilibria at $e_0 < 1$ may function to disincentivize defenders to have faith in successful collective preventive actions. As a result, mitigation in the form of full self-insurance may appear more appealing. As risk-neutral decision makers, the agents refrain from security investments when the costs exceed potential losses (see passivity region in Figure 2.a).

Introducing risk aversion for the defender population serves to eliminate the inefficient partial protection equilibria. Further, complete inaction in the form of passivity equilibria disappears. Risk-averse decision makers are willing to invest in security measures costing more than expected losses (see equilibrium strategies for values larger than $pL = p = 0.5$ in Figure 2.b). For example, agents may select a partial self-insurance investment at a fixed level (i. e., $\frac{p}{c}$) when the cost of self-insurance exceeds expected losses.

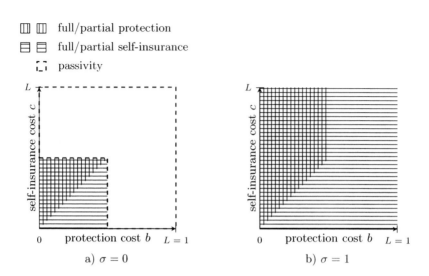

□□ □□ full/partial protection
目 目 full/partial self-insurance
⌐¬ passivity

Fig. 2. Symmetric equilibria in the (b,c)-plane for probabilities of attack $p = 0.5$ *without* (left) and *with* (right) risk aversion; no market insurance

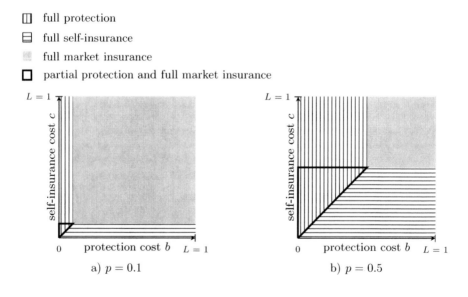

□ full protection
目 full self-insurance
▨ full market insurance
■ partial protection and full market insurance

Fig. 3. Symmetric equilibria in the (b,c)-plane for different probabilities of attack *with* risk aversion *and* market insurance

In contrast to the base model, we find that equilibria with a joint investment in protection and self-insurance may exist. These outcomes are a more adequate description of reality where a joint defense consisting of prevention and mitigation is common.

The equilibrium conditions including the market insurance option are depicted in Figures 3.a and 3.b. The presence of this third defense strategy serves to clarify the boundaries between the three different defense options. That is, for the most part specific parameter values directly dictate the optimal strategy. Full market insurance, full self-insurance and full protection split the parameter space. However, we observe a hybrid strategy with complementary full market insurance and partial protection investments competing with the full protection equilibrium. Our analysis finds that the hybrid option is payoff-inferior, but might nevertheless be chosen for managerial reasons or inherent unpredictabilities of protection options. Otherwise, full market insurance should only be selected when it is cheaper than both alternative options.

On a more abstract level, our analysis of security games with market insurance can be summarized in three key observations. First, market insurance equilibria exist, and all of them involve full insurance coverage. Second, market insurance is more prevalent for risks with small probability of occurrence. Third, (full) market insurance is a substitute for (expensive) self-insurance technologies, but complementary to (partial and cheap) protection mechanisms.

This leads us to the discussion of limitations of our model and possible extensions. The observation that market insurance responds in a complex manner to the relative cost of protection and self-insurance suggests further investigations are fruitful to account for non-linear cost functions. I.e., protection and self-insurance are likely to exhibit decreasing marginal returns in several scenarios — unlike market insurance which scales linearly as long as the risk is small (and uncorrelated) relative to the pool. The combined equilibrium of partial protection and full market insurance depends on the assumption that the insurer has perfect information about the insureds' protection efforts. If this assumption is relaxed, the arguments made by Shetty *et al.* for the case without self-insurance must be adapted to our security game [20].

Further investigations are needed for the case when insurers charge a strictly positive markup. This will introduce a "gap" of partial market and self-insurance and push the region with full market insurance further to the upper-right corner in Fig. 3. Strictly positive markups are more realistic for various reasons: insurance markets are not fully competitive, regulation requires insurers to be risk-averse, and network risks are often not only independent but correlated as well [5]. Risk correlation leads to longer right tails in the cumulative loss distribution and requires risk-averse insurers to set aside additional safety capital. The cost of this capital has to be added to the fair insurance premium.

To sum up, this paper closes a research gap by modeling network security investments that account for the choice between the hybrid goods of collective protection and individual mitigation and externally provided market insurance. To this end, we have characterized the equilibria of security games with risk aversion, and security games with risk aversion and market insurance. Overall, as several equilibria with full market insurance exist, market insurance has a place in security games. Moreover, it seems that the missing market problem for cyber-insurance is at least not exacerbated if the agents have a choice between protection and self-insurance.

References

1. Bandyopadhyay, T., Mookerjee, V., Rao, R.: Why IT managers don't go for cyber-insurance products. Communications of the ACM 52(11), 68–73 (2009)
2. Böhme, R.: Cyber-insurance revisited. In: Workshop on the Economics of Information Security (WEIS), Cambridge, MA (2005)
3. Böhme, R.: Towards insurable network architectures. it - Information Technology 52(5), 290–293 (2010)
4. Böhme, R., Kataria, G.: Models and measures for correlation in cyber-insurance. In: Workshop on the Economics of Information Security (WEIS). University of Cambridge, UK (2006)
5. Böhme, R., Schwartz, G.: Modeling cyber-insurance: Towards a unifying framework. In: Workshop on the Economics of Information Security (WEIS). Harvard University, Cambridge (2010)
6. Ehrlich, I., Becker, G.S.: Market insurance, self-insurance, and self-protection. Journal of Political Economy 80(4), 623–648 (1972)
7. Grossklags, J., Christin, N., Chuang, J.: Secure or insure? A game-theoretic analysis of information security games. In: Proceedings of the 2008 World Wide Web Conference (WWW 2008), Beijing, China, pp. 209–218 (April 2008)
8. Grossklags, J., Christin, N., Chuang, J.: Security and insurance management in networks with heterogeneous agents. In: Proceedings of the 9th ACM Conference on Electronic Commerce (EC 2008), Chicago, IL, pp. 160–169 (July 2008)
9. Grossklags, J., Radosavac, S., Cárdenas, A.A., Chuang, J.: Nudge: Intermediaries' Role in Interdependent Network Security. In: Acquisti, A., Smith, S.W., Sadeghi, A.-R. (eds.) TRUST 2010. LNCS, vol. 6101, pp. 323–336. Springer, Heidelberg (2010)
10. Grossklags, J.: Secure or Insure: An Economic Analysis of Security Interdependence and Investment Types. PhD thesis, University of California, Berkeley (2009)
11. Halek, M., Eisenhauer, J.: Demography of risk aversion. The Journal of Risk and Insurance 68(1), 1–24 (2001)
12. Hirshleifer, J.: From weakest-link to best-shot: The voluntary provision of public goods. Public Choice 41(3), 371–386 (1983)
13. Hofmann, A.: Internalizing externalities of loss prevention through insurance monopoly: An analysis of interdependent risks. Geneva Risk and Insurance Review 32(1), 91–111 (2007)
14. Holz, T., Engelberth, M., Freiling, F.: Learning More About the Underground Economy: A Case-Study of Keyloggers and Dropzones. In: Backes, M., Ning, P. (eds.) ESORICS 2009. LNCS, vol. 5789, pp. 1–18. Springer, Heidelberg (2009)
15. Kanich, C., Kreibich, C., Levchenko, K., Enright, B., Voelker, G., Paxson, V., Savage, S.: Spamalytics: An empirical analysis of spam marketing conversion. In: Proceedings of the Conference on Computer and Communications Security (CCS), Alexandria, VA (October 2008)

16. Kesan, J., Majuca, R., Yurcik, W.: The economic case for cyberinsurance. In: Proceedings of the Fourth Workshop on the Economics of Information Security (WEIS), Cambridge, MA (June 2005)
17. Kirstein, R.: Risk neutrality and strategic insurance. The Geneva Papers on Risk and Insurance 25, 251–261 (2000)
18. Ogut, H., Menon, N., Raghunathan, S.: Cyber insurance and IT security investment: Impact of interdependent risk. In: Fourth Workshop on the Economics of Information Security (WEIS), Cambridge, MA (June 2005)
19. Pratt, J.: Risk aversion in the small and in the large. Econometrica 32(1-2), 122–136 (1964)
20. Shetty, N., Schwartz, G., Felegyhazi, M., Walrand, J.: Competitive Cyber-Insurance and Internet Security. In: Workshop on Economics of Information Security 2009. University College London, England (2009)
21. Stone-Gross, B., Holz, T., Stringhini, G., Vigna, G.: The underground economy of spam: A botmaster's perspective of coordinating large-scale spam campaigns. In: Proceedings of the 4th USENIX Workshop on Large-Scale Exploits and Emergent Threats (LEET), Boston, MA (March 2011)
22. Varian, H.: System reliability and free riding. In: Camp, J., Lewis, S. (eds.) Economics of Information Security. Advances in Information Security, vol. 12, pp. 1–15. Kluwer Academic Publishers, Dordrecht (2004)

Appendix

Table 1. List of Symbols

Symbol	Type	Meaning	Constraints
b	parameter	cost of protection	$0 < b \leq 1$
c	parameter	cost of self-insurance	$0 < c \leq 1$
e_i	choice variable	level of player i's protection	
E	operator	expected value (over loss realization)	
H	function	protection contribution function	
L	constant	size of the loss	$L = 1$
M_0	constant	initial wealth	eliminated
M_1	variable	ex-post wealth	
N	parameter	number of players	$N > 1$
p	parameter	probability of loss	
π	variable	cost of market insurance	
q	random variable	realization of the loss	$q \in \{0, 1\}$
s_i	choice variable	level of player i's self-insurance	$s_i + x_i \leq 1$
σ	parameter	risk aversion	$\sigma \geq 0$
u_i	variable	player i's utility	
U	function	utility function	
x_i	choice variable	level of player i's market insurance	$s_i + x_i \leq 1$

A   Proof of Theorem 1

Proof. Assume that player strategies comprise a symmetric Nash equilibrium. Let e, s, x be the homogeneous protection, self-insurance, and market insurance investments, respectively. Then, we can write the expected utility of player i as

$$
\begin{aligned}
E[u_i] &= p(1-e) \cdot U(s+x-be-cs-\pi x)) + (1 - p(1-e)) \cdot U(1 - be - cs - \pi x) \\
&\leq U(p(1-e) \cdot (s+x-be-cs-\pi x)) + (1-p(1-e)) \cdot (1 - be - cs - \pi x)) \\
&= U(p(1-e)(s+x) + p(1-e)(-be-cs-\pi x) \\
&\quad + (1 - p(1-e) + (1 - p(1-e))(-be - cs - \pi x)) \\
&= U(p(1-e)(s+x) + (-be - cs - \pi x) + 1 - p(1-e)) \\
&= U(ps + px - pes - pex - be - cs - \pi x + 1 - p + pe) \\
&= U(1 - p + e(p - b) + x(p - \pi) + s(p - c) - ep(s + x)).
\end{aligned}
$$

Since U is increasing we can maximize the last formula in the derivation above by choosing e, s, x to maximize the quantity inside the U function.

Excluding the last term, that formula is linear; and the last term is strictly negative whenever at least one of e or $s + x$ is positive. So the choice if e, s, x to maximize the formula can be easily determined from $\min\{p, b, c, \pi\}$ – namely, we choose $(e, s, x) = (0, 0, 0)$ if p is the minimum, resulting in utility $U(1 - p)$; we choose $(e, s, x) = (1, 0, 0)$ if b is the smallest, obtaining utility $U(1-b)$; we choose $(e, s, x) = (0, 1, 0)$ if c is the least obtaining utility $U(1 - c)$; and if π is the min we choose $(e, s, x) = (0, 0, 1)$, obtaining utility $U(1 - \pi)$. If there are equalities among terms, then the proper choice of e, s, x to maximize the formula is not uniquely determined, but there is nothing about equality that would change the final utility. We conclude that for any choice of e, s, x, the expected utility of each player $E[u_i]$ cannot exceed $\max\{U(1 - p), U(1 - b), U(1 - c), U(1 - \pi)\}$.

For the strictness result, observe that the inequality in the second step follows from the fact that U is concave down. The only time that inequality is an equality is when one of the scaling factors $p(1 - e)$ or $1 - p(1 - e)$ is zero. Since we have assumed $p > 0$, the only way to have equality is if $e \in \{0, 1\}$.

B   Proof of Corollary 1

Proof. For the corollary, we first note that in any hybrid equilibrium in which there is a partial protection investment, we must necessarily have $\min\{b, p\} < \min\{c, \pi\}$. Otherwise, any player could unilaterally make an investment in full self-insurance or full market insurance and achieve the maximum bound from the theorem, which would necessarily be an improvement due to the strict inequality. Hence one of p or b minimizes $\{p, b, c, \pi\}$. If $p \leq b$, then setting $e > 1$ results in the final term in the derivation above being less than $U(1 - p)$. On the other hand, if $b \leq p$, then any investment $e < 1$ results in a final term in the above derivation being strictly less than $U(1 - b)$. [Investment in s or x (assuming $s + x \leq 1$) cannot make up for this, because the last term in the last formula subtracts the advantage gained from the protection investment].

Aegis
A Novel Cyber-Insurance Model

Ranjan Pal, Leana Golubchik, and Konstantinos Psounis

University of Southern California, USA
{rpal,leana,kpsounis}@usc.edu

Abstract. Recent works on Internet risk management have proposed the idea of cyber-insurance to eliminate risks due to security threats, which cannot be tackled through traditional means such as by using antivirus and antivirus softwares. In reality, an Internet user faces risks due to security attacks as well as risks due to non-security related failures (e.g., reliability faults in the form of hardware crash, buffer overflow, etc.). These risk types are often indistinguishable by a naive user. However, a cyber-insurance agency would most likely insure risks only due to security attacks. In this case, it becomes a challenge for an Internet user to choose the right type of cyber-insurance contract as traditional optimal contracts, i.e., contracts for security attacks only, might prove to be sub-optimal for himself.

In this paper, we address the problem of analyzing cyber-insurance solutions when a user faces risks due to both, security as well as non-security related failures. We propose *Aegis*, a simple and novel cyber-insurance model in which the user accepts a fraction *(strictly positive)* of loss recovery on himself and transfers rest of the loss recovery on the cyber-insurance agency. We mathematically show that only under conditions when buying cyber-insurance is mandatory, given an option, risk-averse Internet users would prefer Aegis contracts to traditional cyber-insurance contracts[1], under all premium types. This result firmly establishes the non-existence of traditional cyber-insurance markets when Aegis contracts are offered to users. We also derive an interesting counterintuitive result related to the Aegis framework: we show that an increase(decrease) in the premium of an Aegis contract *may not* always lead to decrease(increase) in its user demand. In the process, we also state the conditions under which the latter trend and its converse emerge. Our work proposes a new model of cyber-insurance for Internet security that extends all previous related models by accounting for the extra dimension of non-insurable risks. Aegis also incentivizes Internet users to take up more personal responsibility for protecting their systems.

Keywords: Aegis, risks, insurable, non-insurable.

---

[1] Traditional cyber-insurance contracts are those which do not operate on non-security related losses in addition to security related losses, and do not give a user that option of being liable for a fraction of insurer advertised loss coverage.

J.S. Baras, J. Katz, and E. Altman (Eds.): GameSec 2011, LNCS 7037, pp. 131–150, 2011.
© Springer-Verlag Berlin Heidelberg 2011

1   Introduction

The Internet has become a fundamental and an integral part of our daily lives. Billions of people nowadays are using the Internet for various types of applications. However, all these applications are running on a network, that was built under assumptions, some of which are no longer valid for today's applications, e,g., that all users on the Internet can be trusted and that there are no malicious elements propagating in the Internet. On the contrary, the infrastructure, the users, and the services offered on the Internet today are all subject to a wide variety of risks. These risks include denial of service attacks, intrusions of various kinds, hacking, phishing, worms, viruses, spams, etc. In order to counter the threats posed by the risks, Internet users[2] have traditionally resorted to antivirus and anti-spam softwares, firewalls, and other add-ons to reduce the likelihood of being affected by threats. In practice, a large industry (companies like *Symantec, McAfee,* etc.) as well as considerable research efforts are centered around developing and deploying tools and techniques to detect threats and anomalies in order to protect the Internet infrastructure and its users from the resulting negative impact.

In the past one and half decade, protection techniques from a variety of computer science fields such as cryptography, hardware engineering, and software engineering have continually made improvements. Inspite of such improvements, recent articles by Schneier [1] and Anderson [2][3] have stated that it is impossible to achieve a 100% Internet security protection. The authors attribute this impossibility primarily to four reasons:

– New viruses, worms, spams, and botnets evolve periodically at a rapid pace and as a result it is extremely difficult and expensive to design a security solution that is a panacea for all risks.
– The Internet is a distributed system, where the system users have divergent security interests and incentives, leading to the problem of 'misaligned incentives' amongst users. For example, a rational Internet user might well spend $20 to stop a virus trashing its hard disk, but would hardly have any incentive to invest sufficient amounts in security solutions to prevent a service-denial attack on a wealthy corporation like an Amazon or a Microsoft [4]. Thus, the problem of misaligned incentives can be resolved only if liabilities are assigned to parties (users) that can best manage risk.
– The risks faced by Internet users are often correlated and interdependent. A user taking protective action in an Internet like distributed system creates positive externalities [5] for other networked users that in turn may discourage them from making appropriate security investments, leading to the 'free-riding' problem [6][7][8][9].
– Network externalities affect the adoption of technology. Katz and Shapiro [10] have determined that externalities lead to the classic S-shaped adoption curve, according to which slow early adoption gives way to rapid deployment once the number of users reaches a critical mass. The initial deployment is

---
[2] The term 'users' may refer to both, individuals and organizations.

subject to user benefits exceeding adoption costs, which occurs only if a minimum number of users adopt a technology; so everyone might wait for others to go first, and the technology never gets deployed. For example DNSSEC, and S-BGP are secure protocols that have been developed to better DNS and BGP in terms of security performance. However, the challenge is getting them deployed by providing sufficient internal benefits to adopting entities.

In view of the above mentioned inevitable barriers to 100% risk mitigation, the need arises for alternative methods of risk management in the Internet. Anderson and Moore [3] state that microeconomics, game theory, and psychology will play as vital a role in effective risk management in the modern and future Internet, as did the mathematics of cryptography a quarter century ago. In this regard, *cyber-insurance* is a psycho-economic-driven risk-management technique, where risks are transferred to a third party, i.e., an insurance company, in return for a fee, i.e., the *insurance premium*. The concept of cyber-insurance is growing in importance amongst security engineers. The reason for this is three fold: (i) ideally, cyber-insurance increases Internet safety because the insured increases self-defense as a rational response to the reduction in insurance premium [11][12][13][14], a fact that has also been mathematically proven by the authors in [15][16], (ii) in the IT industry, the mindset of 'absolute protection' is slowly changing with the realization that absolute security is impossible and too expensive to even approach while adequate security is good enough to enable normal functions - the rest of the risk that cannot be mitigated can be transferred to a third party [17], and (iii) cyber-insurance will lead to a market solution that will be aligned with economic incentives of cyber-insurers and users (individuals/organizations) - the cyber-insurers will earn profit from appropriately pricing premiums, whereas users will seek to hedge potential losses. In practice, users generally employ a simultaneous combination of retaining, mitigating, and insuring risks [18].

Research Motivation: The concept of cyber-insurance as proposed in the security literature covers losses only due to security attacks. However, in reality, security losses are not the only form of losses. Non-security losses (e.g., reliability losses) form a major loss type, where a user suffers, either because of hardware malfunction due to a manufacturing defect or a software failure (e.g., buffer overflow caused by non-malicious programming or operational errors[3])[19]. A naive Internet user would not be able to distinguish between a security or a non-security failure and might be at a disadvantage w.r.t. buying traditional cyber-insurance contracts. That is, on facing a risk, the user would not know whether the cause of the risk is a security attack or a non-security related failure[4]. The disadvantage is due to the fact that traditional cyber-insurance would only cover those losses due to security attacks, whereas an Internet user may incur a loss that occurs due to a non-security problem and not get covered for

---

[3] A buffer overflow can also be caused by a malicious attack by hackers. Example of such attacks include the Morris worm, Slapper worm, and Blaster worm attacks on Windows PCs.

[4] Irrespective of whether a loss due to a risk is because of a security attack or a non-security failure, the effects felt by a user are the same in both cases.

it[5]. In such cases, it is an interesting problem to investigate the demand for traditional cyber-insurance as it seems logical to believe that an Internet user might not be in favor of transferring complete loss recovery liability to a cyber-insurer as the former would would have to pay the premium and at the same time bear the valuation of the loss on being affected by non-security related losses. In this paper, we analyze the situation of Internet users buying cyber-insurance when they face risks that may arise due to non-security failures or security attacks. We propose an alternative model of cyber-insurance, i.e., Aegis, in this regard and show that given an option between cyber-insurance and Aegis contracts, an Internet user would *always* prefer the latter. We make the following contributions in the paper.

1. We propose a novel[6] model of cyber-insurance, Aegis, in which Internet users need not transfer the total loss recovery liability to a cyber-insurer, and may keep some liability to themselves, i.e., an Internet user may not transfer the entire risk to an insurance company. Thus, as an example, an Internet user may rest 80% of his loss recovery liability to a cyber-insurer and may want to bear the remaining 20% on his own. Our model captures the realistic scenario that Internet users could face risks from security attacks as well as from non-security related failures. It is based on the concept of co-insurance in the traditional insurance domain. (See Section 2.)

2. We mathematically show that when Internet users are risk-averse, Aegis contracts are *always* the user preferred policies when compared to traditional cyber-insurance contracts. In this regard, the latter result de-establishes a market for traditional cyber-insurance. The availability of Aegis contracts also *incentivizes* risk-averse Internet users to rest some loss coverage liability upon themselves rather than shifting it all to a cyber-insurer. (See Section 3.)

---

[5] We assume here that the loss covering agency can distinguish between both types of losses and it does not find it suitable to cover losses due to hardware or software malfunctions, as it feels that they should be the responsibility of the hardware and software vendors (e.g., some computer service agencies in India employ experts who could distinguish between the two loss types, and these experts may be hired by the loss recovery agency also.).

[6] Our cyber-insurance model is novel because we model partial insurance, whereas existing works related to traditional cyber-insurance model full and partial insurance coverage but not partial insurance. The notion of partial insurance can be explained as follows: in traditional cyber-insurance models, only the cyber-insurer has the say on the amount of coverage it would provide to its clients and in turn the premiums it would charge, whereas in the Aegis model, the clients get to decide on the fraction of the total amount of advertised insurance coverage it wants and in turn the proportional premiums it would pay, given an advertised contract. Thus, in traditional cyber-insurance, it is mandatory for users to accept the insurance policy in full, whereas in the Aegis model users have the option of accepting the insurance policy in partial.

3. We mathematically show that a risk-averse Internet user would prefer cyber-insurance of some type (Aegis or traditional) *only* if it is mandatory for him to buy some kind of insurance, given that he faces risks due to both, security as well as non-security failures. (See Section 3.)

4. We mathematically show the following counterintuitive results: (i) an increase in the premium of an Aegis contract *may not* always lead to a decrease in its user demand and (ii) a decrease in the premium of an Aegis contract may not always lead to an increase in its user demand. In the process, we also state the conditions under which these trends emerge. The conditions give a guideline to cyber-insurers on how to increase or decrease their premiums in order to increase user demands for cyber-insurance. (See Section 4.)

2   The Aegis Cyber-Insurance Model

We consider the scenario where an Internet user faces risks[7] that arise due to security attacks from worms, viruses, etc., as well as due to non-security related failures. One example of non-security related problems arises due to reliability faults. In a seminal paper [19], the authors identified operational and programming errors, manufacturing problems of software and hardware vendors, and buffer overflow as some examples of system reliability faults, which have effects on Internet users that are identical to the effects when they are affected by certain security threats (e.g., buffer overflow due to a malicious attack). On facing the negative effects, an Internet user in general cannot distinguish between the loss type. In this paper, we assume that a loss occurs either due to a security attack or a non-security related failure and not both, i.e., a unit of damage cannot occur simultaneously due to a security and a non-security failure. For example, a file or a part of it that has been damaged by a security attack cannot be damaged by a non-security fault at the same time.

We assume that cyber-insurers[8] offer Aegis contracts to their clients, Aegis contracts unlike traditional insurance contracts allow the user to rest some fraction of loss recovery liability upon itself. For example, if the value[9] of a loss incurred by an Internet user equals L, and the insurance coverage advertised by an insurer equals $L - d$, where $d \geq 0$, an Aegis contract would allow its client to rest a fraction, $1 - \theta$, of the coverage on itself and the remaining θ part on the cyber-insurer, whereas a traditional contract would fix the value of θ to 1. Our concept of Aegis contracts are based on the theory of *co-insurance* in general insurance literature. It is logical to believe that a user will not prefer a $1 - \theta$ value

---

[7] A risk is defined as the chance that a user faces a certain amount of loss.

[8] A cyber-insurer could be an ISP, a third-party agency, or the government.

[9] In this paper, like in all of existing cyber-insurance literature, we assume that loss and coverage have the same scalar unit. In reality, this may not be true. As an example, losing a valuable file may not be compensated by replacing the same file. In return, monetary compensation may result. Considering appropriate units of loss and coverage is an area of future work.

that is large as it would mean that it wants to rest a substantial loss recovery liability on itself, thereby diminishing the importance of buying cyber-insurance. We assume that the value of θ is fixed between the user and the cyber-insurer prior to contract operation.

Most of our analysis in the paper will revolve around the final wealth of a risk-averse Internet user who may be subject to risks due to both security attacks and non-security related failures. We have the following equation regarding its final wealth according to the Aegis concept:

$$W = w_0 + v - L_S - L_{NS} + \theta(I(L_S) - P), \tag{1}$$

where W is a random variable representing the final wealth of a user, $w_0 + v$ is his constant initial wealth, with v^{10} being the constant total value of the object subject to loss as a result of a security attack or a non-security attack, L_S is a random variable denoting loss due to security attacks, L_{NS} is the random variable denoting loss due to non security related failures, and $I(L_S)$ is the cyber-insurance function that decides the amount of coverage to be provided in the event of a security-related loss, where $0 \le I(L_S) \le L_S$. We assume that both L_S and L_{NS} lie in the interval $[0, v]$. As mentioned previously, a given amount of loss can be caused either by a security attack or a by a non-security fault and not by both. In this sense the loss types are *not independent* but are *negatively correlated*. P is the premium[11] charged to users in insurable losses and is defined as $P = (1 + \lambda)E(I(L_S))$. λ is the loading factor and is zero for fair premiums and greater than zero for unfair premiums. We define $\theta \epsilon [0, 1]$ as the *level of cyber-insurance liability* opted for by a user. For example, a value of $\theta = 0.6$, implies that the user transfers 60% of its insurance coverage liability to the cyber-insurer and keeps the rest 40% of the coverage liability on himself, where the insurance coverage could be either full or partial. We observe from Equation (1) that depending on the liability level, a user pays proportional premiums to the cyber-insurer.

We define the expected utility of final wealth[12] of an Internet user as

$$E(W) = A + B + C + D, \tag{2}$$

where

$$A = \int \int_{0 < L_S \le v, L_{NS} = 0} u(w_0 + v - L_S - L_{NS} + \theta(I(L_S) - P)) \cdot g(L_S, L_{NS}) dL_1 \cdot dL_{NS},$$

$$B = \int \int_{0 < L_{NS} \le V, L_S = 0} u(w_0 + v - L_S - L_{NS} + \theta(I(L_S) - P)) \cdot g(L_S, L_{NS}) dL_S \cdot dL_{NS},$$

---

[10] We divide the fixed initial wealth of a user into two parts for modeling simplicity.

[11] P is the premium corresponding to a θ value of 1, where θ is the level of cyber-insurance liability opted by a user.

[12] In economic and risk analyses, dealing with the expected utility of final wealth is a standard approach and it arises from the von Neumann-Morgenstern model of expected utility [20].

$$C = \int \int_{0 < L_S, 0 < L_{NS}} u(w_0 + v - L_S - L_{NS} + \theta(I(L_S) - P)) \cdot g(L_S, L_{NS}) dL_S \cdot dL_{NS},$$

and

$$D = \beta \cdot u(w_0 + v - \theta \cdot P),$$

with A, B, C, and D being the components of expected utility of final wealth when there is a loss due to a security attack only, a non-security related failure only, a security attack as well as a non-security related failure, and no failure of any kind, respectively. u is a twice continuously differentiable risk-averse concave utility function of wealth of a user.

We define the joint probability density function, $g()$, of L_S and L_{NS} as

$$g(L_S, L_{NS}) = \begin{cases} \alpha \cdot f_S(L_S) \ 0 < L_S \le v, L_{NS} = 0 \\ (1 - \alpha - \beta) \cdot f_{NS}(L_{NS}) \ 0 < L_{NS} \le v, L_S = 0, \\ 0 \ 0 < L_S \le v, 0 < L_{NS} \le v \end{cases} \tag{3}$$

where α is the probability[13] of loss occurring due to a security attack, and β is the probability of no attack due to either a security or a non security attack. $f_S(L_S)$ and $f_{NS}(L_{NS})$ are the univariate density functions of losses due to a security attack and non security attack respectively. The joint probability density function has three components: 1) the case where there is a loss only due to a security attack, 2) the case when there is a loss only due to a non-security related failure, and 3) the case when a loss occurs due to both types of risks.

Based on $g()$, Equation (1) can be re-written as

$$E(W) = A1 + B1 + C1, \tag{4}$$

where

$$A1 = \int_0^v u(w_0 + v - L_S + \theta(I(L_S) - P))\alpha \cdot f_S(L_S) dL_S,$$

$$B1 = \int_0^v u(v_0 + v - L_{NS} - \theta(P))(1 - \alpha - \beta) \cdot f_{NS}(L_{NS}) dL_{NS},$$

and

$$C1 = \beta \cdot u(w_0 + v - \theta \cdot P),$$

with $A1$, $B1$, and $C1$ being the components of expected utility of final wealth when there is a loss due to a security attack only, a non-security related failure only, and no failure of any kind, respectively.

In the following sections, we adopt the Aegis model of cyber-insurance and derive results in the form of theorems and propositions.

3   Efficacy of Aegis Contracts

In this section, we investigate whether Aegis contracts are preferred by Internet users over traditional cyber-insurance contracts, and if yes, then under what

---

[13] We plan to estimate α using correlation models.

conditions. In this regard, we state the following theorems that establish results regarding the user demand for Aegis contracts when compared to traditional cyber-insurance contracts.

Theorem 1. *Risk-averse Internet users always prefer Aegis contracts to traditional cyber-insurance contracts when non-insurable losses exist, irrespective of whether the cyber-insurance premium charged in an Aegis contract is fair ($\lambda = 0$) or unfair ($\lambda > 0$)*[14].

Proof: Taking the first derivative of $E(W)$ w.r.t. θ, and equating it to zero, we get the first order condition as

$$\frac{dE(W)}{d\theta} = A2 + B2 + C2 = 0, \tag{5}$$

where

$$A2 = \int_0^v u'(w_0 + v - L_S + \theta(I_{L_S} - P))(I(L_S) - P)\alpha \cdot f_S(L_S)dL_S,$$

$$B2 = \int_0^v u'(w_0 + v - L_{NS} - \theta(P))(-P)(1 - \alpha - \beta) \cdot f_{NS}(L_{NS})dL_{NS},$$

and

$$C2 = \beta \cdot u'(w_0 + v - \theta \cdot P)(-P).$$

Now substituting $I(L_S) = L_S$ (indicating full coverage) and $\theta = 1$ (indicating no co-insurance) into the first order condition, we get

$$\frac{dE(W)}{d\theta} = A3 + B3 + C3 = 0, \tag{6}$$

where

$$A3 = \int_0^v u'(w_0 + v - P)(L_S - P)\alpha \cdot f_S(L_S)dL_S,$$

$$B3 = \int_0^v u'(w_0 + v - L_S - P)(-P)(1 - \alpha - \beta) \cdot f_{NS}(L_{NS})dL_{NS},$$

and

$$C3 = \beta \cdot u'(w_0 + v - P)(-P).$$

Re-arranging the integrals we get

$$A3 = u'(w_0 + v - P) \cdot \alpha \int_0^v (L_S - P)f_S(L_S)dL_S,$$

and

$$B3 = (-P)(1 - \alpha - \beta) \int_0^v u'(w_0 + v - L_{NS} - P)f_{NS}(L_{NS})dL_{NS},$$

---

[14] The comparison is based on *equal* degrees of fairness or unfairness between an Aegis contract and a traditional cyber-insurance contract.

Now using the fact that $E(I(L_S)) = \alpha \cdot \int_0^v L_S \cdot f_S(L_S)dL_S = P$ (fair premiums), we have the following equation

$$\frac{dE(W)}{d\theta} = A4 + B4, \tag{7}$$

where

$$A4 = u'(w_0 + v - P)(1 - \alpha - \beta)P$$

and

$$B4 = (-P)(1 - \alpha - \beta)\int_0^v u'(w_0 + v - L_{NS} - P)f_{NS}(L_{NS})dL_{NS}.$$

Since a user has a risk-averse utility function, we have $u'(w_0 + v - L_{NS} - P) > u'(w_0 + v - P) \forall L_{NS} > 0$. Thus, $\frac{dE(W)}{d\theta} < 0$ at $\theta = 1$. This indicates that the optimal value of θ is less than 1 for fair insurance premiums. On the other hand, even if we consider unfair premiums with a load factor $\lambda > 0$, we get $\frac{dE(W)}{d\theta} < 0$. Therefore in this case also the optimal value of θ is less than 1.   **Q.E.D.**

Implications of Theorem 1. The theorem implies that risk-averse users would always choose Aegis cyber-insurance contracts over traditional cyber-insurance contracts, when given an option.
Intuition Behind Theorem 1. In situations where a risk-averse user cannot distinguish between losses due to a security attack or a non-security failure, he would be conservative in his investments in insurance (as he could pay premiums and still not get covered due to a non-insurable loss) and would prefer to invest more in self-effort for taking care of his own system so as to minimize the chances of a loss. *Thus, in a sense the Aegis model incentivizes risk-averse Internet users to invest more in taking care of their own systems than simply rest the entire coverage liability upon a cyber-insurer.*

Theorem 2. *When risks due to non-insurable losses are increased in a first order stochastic dominant[15] sense, the demand for traditional cyber-insurance amongst all risk-averse Internet users decreases.*

Proof. Again consider the first order condition

$$\frac{dE(W)}{d\theta} = A2 + B2 + C2 = 0, \tag{8}$$

---

[15] Let X and Y be two random variables representing risks. Then X is said to be smaller than Y in first order stochastic dominance, denoted as $X \leq_{ST} Y$ if the inequality $VaR[X;p] \leq VaR[Y;p]$ is satisfied for all $p \in [0,1]$, where $VaR[X;p]$ is the value at risk and is equal to $F_X^{-1}(p)$. First order stochastic dominance implies dominance of higher orders. We adopt the stochastic dominant approach to comparing risks because a simple comparison between various moments of two distributions may not be enough for a correct prediction about the dominance of one distribution over another.

where

$$A2 = \int_0^v u'(w_0 + v - L_S + \theta(I_{L_S} - P))(I(L_S) - P)\alpha \cdot f_S(L_S)dL_S,$$

$$B2 = \int_0^v u'(w_0 + v - L_{NS} - \theta(P))(-P)(1 - \alpha - \beta) \cdot f_{NS}(L_{NS})dL_{NS},$$

and

$$C2 = \beta \cdot u'(w_0 + v - \theta \cdot P)(-P).$$

We observe that when L_{NS} is increased in a first order stochastic dominant sense[16] and $f_S(L_S)$ and β remain unchanged, the premium for insurance does not change. An increase in L_{NS} in the first order stochastic dominant sense increases the magnitude of $\int_0^v u'(w_0 + v - L_{NS} - \theta(P))(-P)(1 - \alpha - \beta) \cdot f_{NS}(L_{NS})dL_{NS}$, whenever $u'(w_0 + v - L_{NS} - \theta(P))$ is increasing in L_{NS}. This happens when $u(W)$ is concave, which is the exactly the case in our definition of u. Thus, an increase in L_{NS} in a first order stochastic dominant sense leads to the first order expression, $\frac{dE(W)}{d\theta}$, to become increasingly negative and results in reductions in θ, implying the lowering of demand for cyber-insurance.   **Q.E.D.**

Implications of Theorem 2. The theorem simply implies the intuitive fact that an increase in the risk due to non-insurable losses leads to a decrease in the demand of traditional cyber-insurance contracts, irrespective of the degree of risk-averseness of a user.

Intuition Behind Theorem 2. The implications of Theorem 2 hold as the user would think that there are greater chances of it being affected by a loss and not being covered at the same time. An increase in the risk due to non-insurable losses also decreases the demand for Aegis contracts. However, according to Theorem 1, for the same amount of risk, Aegis contracts are preferred to traditional cyber-insurance contracts.

Theorem 3. *When the risk due to non-insurable losses increases in the first order stochastic dominant sense, the expected utility of final wealth for any cyber-insurance contract (Aegis and traditional) falls when compared to the alternative of no cyber-insurance, for risk averse Internet users.*

Proof. The expected utility of any cyber-insurance contract is given by the following

$$E(W) = A1 + B1 + C1, \tag{9}$$

---

[16] Let X and Y be two random variables representing risks. Then X is said to be smaller than Y in first order stochastic dominance, denoted as $X \leq_{ST} Y$ if the inequality $VaR[X; p] \leq VaR[Y; p]$ is satisfied for all $p \in [0, 1]$, where $VaR[X; p]$ is the value at risk and is equal to $F_X^{-1}(p)$. First order stochastic dominance implies dominance of higher orders. We adopt the stochastic dominant approach to comparing risks because a simple comparison between various moments of two distributions may not be enough for a correct prediction about the dominance of one distribution over another.

where

$$A1 = \int_0^v u(w_0 + v - L_S + \theta(I(L_S) - P))\alpha \cdot f_S(L_S)dL_S,$$

$$B1 = \int_0^v u(w_0 + v - L_{NS} - \theta(P))(1 - \alpha - \beta) \cdot f_{NS}(L_{NS})dL_{NS},$$

and

$$C1 = \beta \cdot u(w_0 + v - \theta \cdot P).$$

When $\theta = 0$ (the case for no cyber-insurance), $E(W)$ reduces to

$$E(W) = A1' + B1' + C1', \tag{10}$$

where

$$A1' = \int_0^v u(w_0 + v - L_S)\alpha \cdot f_S(L_S)dL_S,$$

$$B1' = \int_0^v u(w_0 + v - L_{NS})(1 - \alpha - \beta) \cdot f_2(L_{NS})dL_{NS},$$

and

$$C1' = \beta \cdot u(w_0 + v).$$

Increases in L_{NS} affect only the second terms in each of these utility expressions. Thus, we need to consider the change in the second order terms in the two utility expressions to observe the impact of the increase in L_{NS}. The difference in the second order terms is given as

$$R1 - R2,$$

where

$$R1 = \int_0^v u(w_0 + v - L_{NS} - \theta(P))(1 - \alpha - \beta) \cdot f_{NS}(L_{NS})dL_{NS}$$

and

$$R2 = \int_0^v u(w_0 + v - L_{NS})(1 - \alpha - \beta) \cdot f_{NS}(L_{NS})dL_{NS}.$$

Thus, $R1 - R2$ evaluates to

$$\int_0^v [u(w_0 + v - L_{NS} - \theta(P)) - u(w_0 + v - L_{NS})](1 - \alpha - \beta) \cdot f_{NS}(L_{NS})dL_{NS},$$

where $[u(w_0 + v - L_{NS} - \theta(P)) - u(w_0 + v - L_{NS})]$ is decreasing in L_{NS} under risk aversion and concave under user prudence. Thus, increases in L_{NS} in the first order stochastic dominant sense reduces the expected utility of cyber-insurance relative to no cyber-insurance.   **Q.E.D.**

Implications of Theorem 3. Theorem 3 provides us with an explanation of why risk-averse Internet users would be reluctant to buy cyber-insurance of any kind given an option between choosing and not choosing insurance, when risks due to non-security related losses are present along with risks due to security attacks.

Intuition Behind Theorem 3. Theorem 3 holds because the expected utility to a risk-averse Internet user opting for a zero level of cyber-insurance liability is greater than that obtained when he opts for a positive level of cyber-insurance liability.

Combining the results in Theorems 1, 2, and 3, we conclude the following:

– In the presence of non-insurable losses, the market for traditional cyber-insurance may not exist.
– When risk-averse Internet users have an option between traditional cyber-insurance, Aegis contracts, and no cyber-insurance, they may prefer the last option. Thus, Aegis contracts might be preferred by Internet users over traditional cyber-insurance contracts *only* if it is mandatory for them to buy some kind of insurance. In general, Internet service providers (ISPs) or cyber-insurance agencies might force its clients on regulatory grounds to sign up for some positive amount of cyber-insurance to ensure a more secure and robust Internet.

4   Sensitivity Analysis of User Demands

In this section we conduct a sensitivity analysis of user demands for Aegis contracts. We investigate whether an increase in the premium charged by a contract results in an increase/decrease in user demand for the contract. The user demand is reflected in the θ value, i.e., user demand indicates the fraction of loss coverage liability a risk-averse user is willing to rest on the cyber-insurance agency. In an Aegis contract, to avoid insurance costs not related to a security attack, a risk-averse user takes up a fraction of loss coverage liability on himself as it does not know beforehand whether he is affected by a security or a non security threat. Thus, intuitively, a decrease in a contract premium may not always lead to a user increasing his demand and analogously an increase in the premium may not always lead to a decrease in the user demands. The exact nature of the relationship between the premiums and user demand in this case depends on the degree of risk averseness of a user. To make the latter statement clear, consider an Internet user who is very risk averse. It would not matter to that user if there is a slight decrease in the premium amount because he might still not transfer additional loss coverage liability to the cyber-insurer, given that he is unsure about whether the risk he faces is due to a security attack or a non security related issue. On the other hand a not so risk averse user may not decrease the amount of loss coverage liability rested upon a cyber-insurer, even if there is a slight increase in the cyber-insurance premiums. In this section we study the conditions under which there is an increase/decrease of user demand for Aegis contracts with change in contract premiums. We first provide the basic setup for sensitivity analysis, which is then followed by the study of the analysis results.

4.1   Analysis Setup

Let a user's realized final wealth be represented as

$$W = w - L + \theta(L - P). \tag{11}$$

Substituting $P = \lambda'E(L)$, we get

$$W = w - L + \theta(L - \lambda'E(L)), \tag{12}$$

where λ' equals $(1+\lambda)$, w is equal to w_0+v, θ lies in the interval $[0,1]$, $\lambda \geq 1$ is the gross loading factor of insurance, $L = L_S+L_{NS}$, and $\lambda E(L) = \alpha \int_0^v L \cdot f_S(L)dL$ is the premium payment for full insurance[17] with E being the expectation operator. The user is interested in maximizing his expected utility of final wealth in the von Neumann-Morgenstern expected utility sense and chooses a corresponding θ to achieve the purpose. Thus, we have the following optimization problem.

$$argmax_\theta E(U(W)) = E[U(w - L + \theta(L - \lambda'E(L)))],$$

where $0 \leq \theta \leq 1$. The first order condition for an optimum θ is given by

$$E'_\theta(U(W)) = E[U'(W)(L - \lambda'E(L))] = 0, \tag{13}$$

which occurs at an optimal $\theta = \theta^*$. Integrating by parts the LHS of the first order condition and equating it to zero, we get

$$M1 + M2 = 0, \tag{14}$$

where

$$M1 = U'(W(0)) \int_0^v (L - \lambda'E(L))dF_S(L)$$

and

$$M2 = \int_0^v U''(W(L))W'(L) \left(\int_L^v (t - \lambda'E(L))dF_S(t) \right) dL.$$

Here $W(x)$ is the value of W at $L = x$ and $W'(L) = -(1 - \theta) \leq 0$. The second order condition is given by

$$E''_\theta(U(W)) = E[U''(W)(L - \lambda'E(L))^2] < 0, \tag{16}$$

which is always satisfied for $U'' < 0$. We now consider the following condition C, which we assume to hold for the rest of the paper.

Condition C - *The utility function U for a user is twice continuously differentiable, thrice piecewise continuously differentiable[18] and exhibits $U' > 0$, $U'' < 0$ with the coefficient of risk aversion, A, being bounded from above.*

The condition states the nature of the user utility function U, which is in accordance with the standard user utility function used in the insurance literature, with the additional restriction of thrice piecewise continuous differentiability of U to make the coefficient of risk aversion well-defined for all W. We adopt the

---

[17] By the term 'full insurance', we imply a user resting its complete loss liability on the cyber-insurer, i.e., $\theta = 1$. Full insurance here does not indicate full insurance coverage.

[18] We consider the thrice piecewise continuously differentiable property of U so that $A'(W)$ becomes piecewise continuous and is thus defined for all W.

standard *Arrow-Pratt* risk aversion measure [21], according to which the coefficient of risk aversion is expressed as (i) $A = A(W) = -\frac{U''(W)}{U'(W)}$ for an *absolute*
risk averse measure and (ii) $R = R(W) = -\frac{WU''(W)}{U'(W)}$ for a *relative* risk averse
measure.

4.2   Sensitivity Analysis Study

In this section we study the change in user demands for Aegis contracts with variations in cyber-insurance premiums, under two standard risk-averse measures:
(1) the decreasing absolute risk averse measure and (2) the decreasing relative
risk averse measure. The term 'decreasing' in both the risk measures implies
that the risk averse mentality of users decrease with the increase in their wealth,
which is intuitive from a user perspective. We are interested in investigating the
sign of the quantity, $\frac{d\theta^*}{d\lambda'}$. The nature of the sign drives the conditions for an
Aegis contract to be either more or less preferred by Internet users when there
is an increase in the premiums, i.e., if $\frac{d\theta^*}{d\lambda'} \leq 0$, an increase in cyber-insurance
premium implies decrease in user demand, and $\frac{d\theta^*}{d\lambda'} \geq 0$ implies an increase in
user demand with increase in premiums.

We have the following theorem and its corresponding proposition related to
the conditions under which Internet users increase or decrease their demands for
Aegis contracts, when the users are risk-averse in an *absolute* sense.

Theorem 4. *For any arbitrary* w, λ', F, *and any* U *satisfying condition* C, *(i)*
$\frac{d\theta^*}{d\lambda'} \geq 0$ *if and only if there exists* $\rho \in \mathbb{R}$ *such that*

$$\int_L^w [A(W(x))\theta^*(x - \lambda'E(L)) - 1]dF(x) \geq \rho \int_L^w \theta^*(x - \lambda'E(L))dF(x), \quad (17)$$

and (ii) $\frac{d\theta^*}{d\lambda'} < 0$ *if and only if there exists* $\rho \in \mathbb{R}$ *such that*

$$\int_L^w [A(W(x))\theta^*(x - \lambda'E(L)) - 1]dF(x) < \rho \int_L^w \theta^*(x - \lambda'E(L))dF(x), \quad (18)$$

where $L \in [0, w]$ *and* $F(\cdot)$ *is the distribution function of loss* L.

Proof. We know that $\frac{d\theta^*}{d\lambda'} = -\frac{E'_{\theta\lambda'}}{E''_\theta}$. Now $\frac{d\theta^*}{d\lambda} \leq 0$ if and only if the following
relationship holds because $E''_\theta < 0$.

$$E'_{\theta\lambda'}(U(W(L))) = E[-U''(W(L))\theta^* E(L)(L - \lambda'E(L)) - U'(W(L))E(L)] \leq 0 \tag{19}$$

or

$$E\left\{\left(A(W(L)) - \frac{1}{\theta(L - \lambda'E(L))}\right)U'(W(L))(L - \lambda'E(L))\right\} \leq 0 \tag{20}$$

The LHS of Equation 19 can be expressed via integration by parts as

$$\int_0^w [A(W(L))\theta^*(L - \lambda'E(L)) - 1]U'(W(L))dF(L)$$

which evaluates to

$$X + Y,$$

where

$$X = U'(W(0)) \int_0^w [A(W(L))\theta^*(L - \lambda'E(L)) - 1]dF(L)$$

and

$$Y = \int_0^w U''(W(L))W'(L) \left\{ \int_L^w [A(W(t))\theta^*(x - \lambda'E(L)) - 1]dF(x) \right\} dL.$$

Now $X + Y \geq M + N$, where

$$M = U'(W(0)) \int_0^w \rho(L - \lambda'E(L))dF(L)$$

and

$$N = \int_0^w U''(W(L))W'(L) \cdot \left(\rho \int_L^w (x - \lambda'E(L))dF(x) \right) dL.$$

Thus, $\frac{d\theta^*}{d\lambda'} \geq 0$, and the sufficient condition is proved. The proof of the necessary condition follows from Proposition 1' in [22]. Reversing Equation 16 we get the necessary and sufficient conditions for $\frac{d\theta^*}{d\lambda'} \leq 0$, which is condition (ii) in Theorem 4.   **Q.E.D.**

Proposition 1. *There exists a $\rho \epsilon \mathbb{R} - \{0\}$ such that Theorem 4 holds if the following two conditions are satisfied.*

$$\frac{(1 - \theta^*)A'}{A} \leq \theta^* A \tag{21}$$

and

$$\int_0^w A(W(L)) \left\{ L - \lambda'E(L) - \frac{1}{\theta^* A(W(L))} \right\} dF(L) > 0 \tag{22}$$

Proof. We observe the following relation

$$E1 < E2, \tag{23}$$

where

$$E1 = \int_0^w A(W(L)) \left\{ L - \lambda'E(L) - \frac{1}{\theta^* A(W(L))} \right\} dF(L)$$

and

$$E2 = \int_0^w A(W(L))(L - \lambda'E(L))dF(L).$$

Thus, from Equation 21 in the theorem statement, we have

$$\int_0^w A(W(L))(L - \lambda'E(L))dF(L) > 0 \tag{25}$$

According to Lemma 2 in [22], there exists $\rho^* \in \mathbb{R} - \{0\}$ such that the following relation holds for all $L \in [0, w]$.

$$\int_L^w A(W(x))(x - \lambda'E(L))dF(x) \geq \rho^* \int_L^w (x - \lambda'E(L))dF(x) \qquad (26)$$

Now, $\frac{d[L-\lambda'E(L)-(\theta^* A(W(L)))^{-1}]}{dL} \geq 0$ if and only if $(1 - \theta^* \frac{A'}{A} \leq \theta^* A$. Since both $L - \lambda'E(L))d(F(x)$ and $L - \lambda'E(L) - (\theta^* A(W(L)))^{-1}$ are increasing and become negative when L is sufficiently small, and using arguments presented in Lemma 2 in [22], we can show that Equations 21 and 22 imply

$$\int_L^w A(W(t))(t - \lambda'E(L))d(F(t) > 0 \qquad (27)$$

and

$$\int_L^w A(W(t))[t - \lambda'E(L) - (\theta^* A(W(t)))^{-1}]dF(t) > 0. \qquad (28)$$

for all $L \in [0, w]$. Now choosing a $\delta > 0$ sufficiently small gives

$$P1 > P2 > 0, \qquad (29)$$

where

$$P1 = \int_L^w A(W(t))[t - \lambda'E(L) - (\theta^* A(W(t)))^{-1}]dF(t)$$

and

$$P2 = \delta \int_L^w A(W(t))(t - \lambda'E(L))d(F(t),$$

for all $L \in [0, w]$. Now, setting $\rho = \rho^* \delta$, we get for all $L \in [0, w]$, the following relation

$$\int_L^w A(W(t))[t - \lambda'E(L) - (\theta^* A(W(t)))^{-1}]dF(t) \geq \rho \int_L^w (t - \lambda'E(L))dF(t) \quad (30)$$

This proves the proposition   **Q.E.D.**

Notes on Theorem 4 and Proposition 1. Theorem 4 and Proposition 1 are related to each other in the sense that Theorem 4 provides the necessary and sufficient conditions under which Internet users increase/decrease demands of Aegis contracts. The intuition behind the result in Theorem 4 is based on expected utility comparisons. For an increase in the λ value, the expected utilities of a user are compared with and without a corresponding increase in θ value. We say that user demands for Aegis contracts increase (decrease) if there is an increase (decrease) in expected utility with an increase in the θ value, and we find the conditions for such situations to arise. Proposition 1 states that Theorem 4 always holds provided certain conditions are met.

We have the following theorem that states the conditions under which Internet users increase or decrease their demands for Aegis contracts, when the users are risk averse in a *relative* sense.

Theorem 5. *For any arbitrary* w, λ', F, *and any* U *satisfying condition* C, *(i)* $\frac{d\theta^*}{d\lambda'} \geq 0$ *only if* $R(W) > 1$ *and (ii)* $\frac{d\theta^*}{d\lambda'} < 0$ *only if* $R(W) \leq 1$, *where* $W \in [W(w), W(0)]$.

Proof. We can rewrite Equation 16 as follows

$$\int_L^w \{\theta^*[A(W(x)) - \rho](x - \lambda' E(L)) - 1\}dF(x) \geq 0, \qquad (31)$$

which can be further rewritten as

$$\int_L^w \{(\mathring{R}(W(x)) - 1) - A(W(x))(w_0 - x) - \rho(x - \lambda' E(L))\}dF(x) \geq 0. \qquad (32)$$

The integral in Equation 23 is non-negative for all $L \in [0, w]$ only if $R(W) > 1$ for some W. To see this it suffices to realize that $-A(W(L))(w_0 - L) < 0$ for all $L \in [0, w)$ as $L \leq w_0$ and there exists $L \in [0, w]$ at which $-\int_L^w \rho(L - \lambda' E(L))dF(x) < 0$ as $\int_L^w (x - \lambda' E(L))dF(x))$ alternates in sign on $(0, w)$. Now suppose by contradiction that $R(W) \leq 1$ for all W. Substituting this into Equation 23 violates the condition stated in Equation 22 for some $L \in [0, w]$. Again by Theorem 4, we have that there exists utility function U satisfying condition C such that $\frac{d\theta^*}{d\lambda'} \geq 0$ - a contradiction. Since F is arbitrary, the result (i) in the theorem follows. By reversing the sign of the condition on $R(W)$ the result (ii) in the theorem follows.   **Q.E.D.**

Implications of Theorem 5. The theorem implies that above a certain level of the degree of relative risk averseness, a user prefers Aegis contracts even if there is an increase in contract premiums.

Intuition Behind Theorem 5. The coefficient of relative risk aversion is measured relative to the wealth of a user and thus more his wealth, lesser would be his concerns about losing money due to paying more cyber-insurance premiums, and not getting coverage on being affected by a non-security failure. The intuition is similar for the case when below a certain threshold of relative risk averseness, users reduce their demand for Aegis contracts.

5   Related Work

The field of cyber-insurance in networked environments has been fueled by recent results on the amount of individual user self-defense investments in the presence of network externalities[19]. The authors in [6][7][23][24][8][9] mathematically show that Internet users invest too little in self-defense mechanisms relative to the socially efficient level, due to the presence of network externalities. These works highlight the role of positive externalities in preventing users from investing optimally in self-defense. Thus, one challenge to improving overall network

---

[19]   An externality is a positive (external benefits) or negative (external costs) impact on an user not directly involved in an economic transaction.

security lies in incentivizing end-users to invest in a sufficient amount of self-defense in spite of the positive externalities they experience from other users in the network. In response to this challenge, the works in [23][24] modeled network externalities and showed that a tipping phenomenon is possible, i.e., in a situation where the level of self-defense is low, if a certain fraction of population decides to invest in self-defense mechanisms, then a large cascade of adoption in security features could be triggered, thereby strengthening the overall Internet security. However, these works did not state how the tipping phenomenon could be realized in practice. In a series of recent works [15][16], Lelarge and Bolot have stated that under conditions of no *information asymmetry* [25][26] between the insurer and the insured, cyber-insurance *incentivizes* Internet user investments in self-defense mechanisms, thereby paving the path to triggering a cascade of adoption. They also showed that investments in both self-defense mechanisms and insurance schemes are quite inter-related in maintaining a socially efficient level of security on the Internet. In a follow up work on joint self-defense and cyber-insurance investments, the authors in [27] show that Internet users invest more efficiently in self-defense investments in a cooperative environment when compared to a non-cooperative one, in relation to achieving a socially efficient level of security on the Internet.

In spite of Lelarge and Bolot highlighting the role of cyber-insurance for networked environments in incentivizing increasing of user security investments, it is common knowledge that the market for cyber-insurance has not yet blossomed with respect to its promised potential. Most recent works [28] [29] have attributed this to (1) *interdependent security* (i.e., the effects of security investments of a user on the security of other network users connected to it), (2) *correlated risk* (i.e., the risk faced by a user due to risks faced by other network users), and (3) *information asymmetries* (i.e., the asymmetry between the insurer and the insured due to one having some specific information about its risks that the other does not have). In a recent work [30], the authors have designed mechanisms to overcome the market existence problem due to information asymmetry, and show that a market for cyber-insurance exists in a single cyber-insurer setting.

However, none of the above mentioned works related to cyber-insurance address the scenario where a user faces risks due to security attacks as well as due to non-security related failures. The works consider attacks that occur due to security lapses only. In reality, an Internet user faces both types of risks and cannot distinguish between the types that caused a loss. Under such scenarios, it is not obvious that users would want to rest the full loss recovery liability to a cyber-insurer. We address the case when an Internet user faces risks due to both security as well as non-security problems, and show that users always prefer to rest some liability upon themselves, thus de-establishing the market for traditional cyber-insurance. However, the Aegis framework being a type of a cyber-insurance framework also faces problems identical to the traditional cyber-insurance framework, viz., that of interdependent security, correlated risk, and information asymmetry.

6   Conclusion

In this paper we proposed Aegis, a novel cyber-insurance model in which an Internet user accepts a fraction (strictly positive) of loss recovery on himself and transfers the rest of the loss recovery on the cyber-insurance agency. Our model is specifically suited to situations when a user cannot distinguish between similar types of losses that arise due to either a security attack or a non-security related failure. We showed that given an option, Internet users would prefer Aegis contracts to traditional cyber-insurance contracts, under all premium types. The latter result firmly establishes the non-existence of traditional cyber-insurance markets when Aegis contracts are offered to users. Furthermore, the Aegis model incentivizes risk-averse Internet users to invest more in taking care of their own systems than simply rest the entire coverage liability upon a cyber-insurer. We also derived two interesting counterintuitive results related to the Aegis framework, i.e., we showed that an increase (decrease) in the premium of an Aegis contract *may not* always leads to a decrease (increase) in its user demand. *Finally, through a simple model of cyber-insurance we show that only under conditions when buying some type of cyber-insurance is made mandatory, does a market exist, and that too for idealistic situations when information asymmetry is absent.* Thus, it is important that (i) the insuring agency (if it is not the ISP or the government) partners with the regulators to make cyber-insurance mandatory for Internet users, and (ii) information asymmetry be taken into equation to check whether a market for cyber-insurance can be made to exist in its presence. As part of future work, we plan to investigate the efficacy of Aegis contracts under information asymmetry.

References

1. Schneier, B.: Secrets and Lies: Digital Security in a Networked World. John Wiley and Sons (2001)
2. Anderson, R.: Why Information Security is Hard - An Economic Perspective. In: Annual Computer Security Applications Conference (2001)
3. Anderson, R., Moore, T.: Information Security Economics and Beyond. Information Security Summit (2008)
4. Varian, H.: Managing Online Security Risks. The New York Times (June 1, 2000)
5. Kunreuther, H., Heal, G.: Interdependent Security. Journal of Risk and Uncertainty 26 (2002)
6. Grossklags, J., Christin, G., Chuang, J.: Security and Insurance Management in Networks with Heterogenous Agents. In: ACM EC (2008)
7. Jiang, L., Ananthram, V., Walrand, J.: How Bad are Selfish Investments in Network Security. IEEE Transactions On Networking (2010)
8. Ko-Miura, A.R., Yolken, B., Bambos, N., Mitchell, J.: Security Investment Games of Interdependent Organizations. Allerton (2008)
9. Omic, J., Orda, A., Mieghem, V.P.: Protecting Against Network Infections: A Game-Theoretic Perspective. In: IEEE INFOCOM (2009)
10. Katz, M., Shapiro, C.: Network Externalities, Competition, and Compatibility. The American Economic Review 75(3) (1985)

11. Kesan, J., Majuca, R., Yurcik, W.: The Economic Case for Cyber-Insurance: In Securing Privacy in the Internet Age. Stanford University Press (2005)
12. Kesan, J., Majuca, R., Yurcik, W.: Cyberinsurance As A Market-Based Solution To The Problem of Cyber-Security: A Case Study. In: WEIS (2005)
13. Scheier, B.: Its The Economics Stupid. In: WEIS (2002)
14. Yurcik, W., Doss, D.: Cyberinsurance: A Market Solution To The Internet Security Market Failure. In: WEIS (2002)
15. Lelarge, M., Bolot, J.: Cyberinsurance As An Incentive for Internet Security. In: WEIS (2008)
16. Lelarge, M., Bolot, J.: Economic Incentives to Increase Security in the Internet: The Case for Insurance. In: IEEE INFOCOM (2009)
17. Majuca, R.P., Yurcik, W., Kesan, J.P.: The Evolution of Cyberinsurance. Information Systems Frontier (2005)
18. Schneier, B.: Insurance and the Computer Industry. Communications of the ACM 44(3) (2001)
19. Honeyman, P., Schwarz, G.: Interdependence of Reliability and Security. In: WEIS (2007)
20. Neumann, J.V., Morgenstern, O.: Theory of Games and Economic Behavior. Princeton University Press (2009)
21. Mascollel, A., Winston, M.D., Green, J.R.: Microeconomic Theory. Oxford University Press (1985)
22. Hau, A.: When is A Coinsurance-Type Insurance Policy Inferior or Even Giffen. Journal of Risk and Insurance 75(2) (2008)
23. Lelarge, M., Bolot, J.: A Local Mean Field Analysis of Security Investments in Networks. In: ACM NetEcon (2008)
24. Lelarge, M., Bolot, J.: Network Externalities and The Deployment of Security Features and Protocols in the Internet. In: ACM SIGMETRICS (2008)
25. Internet Wikipedia Source. Information Asymmetry
26. Varian, H.R.: Microeconomic Analysis. Norton (1992)
27. Pal, R., Golubchik, L.: Analyzing Self-Defense Investments In The Internet Under Cyberinsurance Coverage. In: IEEE ICDCS (2010)
28. Bohme, R., Schwartz, G.: Modeling Cyberinsurance: Towards A Unifying Framework. In: WEIS (2010)
29. Shetty, N., Schwarz, G., Feleghyazi, M., Walrand, J.: Competitive Cyberinsurance and Internet Security. In: WEIS (2009)
30. Pal, R., Golubchik, L.: Pricing and Investments in Internet Security. Arxiv (2011)

Maximizing Influence in Competitive Environments: A Game-Theoretic Approach

Andrew Clark and Radha Poovendran

Network Security Lab, Department of Electrical Engineering,
University of Washington, Seattle, WA 98195
{awclark,rp3}@uw.edu

Abstract. Ideas, ranging from product preferences to political views, spread through social interactions. These interactions may determine how ideas are adopted within a market and which, if any, become dominant. In this paper, we introduce a model for Dynamic Influence in Competitive Environments (DICE). We show that existing models of influence propagation, including linear threshold and independent cascade models, can be derived as special cases of DICE. Using DICE, we explore two scenarios of competing ideas, including the case where a newcomer competes with a leader with an already-established idea, as well as the case where multiple competing ideas are introduced simultaneously. We formulate the former as a Stackelberg game and the latter as a simultaneous-move game of complete information. Moreover, we show that, in both cases, the payoff functions for both players are submodular, leading to efficient algorithms for each player to approximate his optimal strategy. We illustrate our approach using the Wiki-vote social network dataset.

Keywords: Social network, influence propagation, noncooperative game.

1 Introduction

Ideas spread rapidly through human social interactions, especially when enabled by modern technology, including blogs, online social networking sites, and mobile and pervasive computing. Such interactions can be used to convey information to the public at little direct cost. In politics, both traditional social networks (such as groups of politically like-minded people) and new, online social networks (such as Facebook and Twitter) have been instrumental in spreading revolutionary sentiment [12]. Commercial marketing campaigns have also leveraged social media, with companies using online social networks to enhance word-of-mouth effects in advertising [13].

In the applications listed above, multiple, competing ideas, potentially interfering with one another, may propagate simultaneously through the network. This competition may take different forms [9]. In the *leader-follower* (or *Stackelberg*) case, a well-established idea, such as a market-leading product or prevailing political belief, is challenged by newcomers, denoted as *followers*. Alternatively,

J.S. Baras, J. Katz, and E. Altman (Eds.): GameSec 2011, LNCS 7037, pp. 151–162, 2011.
© Springer-Verlag Berlin Heidelberg 2011

two or more ideas may be introduced simultaneously, as in a political election with two or more parties.

In all of these competitive scenarios, the success or failure of each idea may depend on how the idea is introduced into the network, in particular who the initial holders of the idea (denoted as *seeds*) are. For instance, an idea espoused by the owner of a popular blog may reach a large number of people, while an idea held by a handful of isolated individuals may not spread at all [11]. In addition, there may be scenarios where no idea is able to completely dominate the other ideas. Effective introduction of an idea into a social network therefore requires an understanding of how competing ideas will propagate through the network, as well as a tractable framework for choosing the set of nodes in the network that must initially hold the idea. At present, however, while there are formulations of the competitive influence maximization problem [2,7], they do not lead to computationally tractable solution algorithms for the three classes of players listed above.

Our contributions in this paper are two-fold. As our first contribution, we introduce a model for Dynamic Influence in Competitive Environments (DICE). Under DICE, each individual adopts an idea based on observations of his neighbors' current beliefs, leading to a Markov model of idea propagation. Unlike existing approaches, DICE allows nodes to switch between adopted ideas over time. This allows modeling of the case where a new idea is able to overtake or replace an existing, well-established idea. We further show how to leverage the Markovian properties of our proposed model to compute the expected number of individuals holding each idea, as well as the probability that a certain individual holds a given idea, in steady-state.

As our second contribution, we develop game-theoretic formulations for competition between two ideas within the Stackelberg and simultaneous competitive environments described above. Our influence model leads to an average-case optimization problem for each player. We show that, for the case of a social network with strongly connected components (as in [1]), the objective function for each player is *submodular*. As a result, solution algorithms can be developed for each player that approximate the optimal strategy up to a provable bound.

The rest of this paper is organized as follows. In Section 2, we review the related work on influence propagation. In Section 3, we define and analyze our proposed influence propagation models. In Sections 4 and 5, we introduce non-cooperative game formulations for the Stackelberg and simultaneous competition models, respectively. Simulations are presented in Section 6, and Section 7 concludes the paper and gives directions for future work.

2   Related Work and Preliminaries

2.1   Influence Propagation Models

The first mathematical model of influence propagation was the threshold model of [5]. This model assumes that an individual will adopt a given idea if a threshold number of its acquaintances adopt the idea first. The threshold model can be motivated both by sociological observation and as the equilibrium of a non-cooperative game between the individuals comprising the network. Another class of propagation models is based on *cascading* phenomena, in which each individual attempts to convince his or her neighbor of the idea, succeeds with a certain probability, and otherwise fails and does not try again [4].

Markov models for propagation of belief through a network have been proposed in the context of gossip and rumor spreading [3,1]. In such models, each individual's belief is represented by a real number. The individuals reach consensus on a global belief by taking randomized weighted averages of their neighbors' beliefs. While our approach uses a Markov model of influence propagation, we study the case where ideas are competing and mutually exclusive, thus ruling out averaging and consensus.

2.2   Maximizing Spread of Influence

The problem of influence maximization in social networks was first proposed in [11], in the context of marketing. In [6], the authors analyzed the problem of choosing an optimal set of k seed nodes in order to maximize the spread of influence. It was shown that, for a generalized influence model taking both the cascade and threshold models as special cases, the influence maximization problem is *submodular*, enabling the use of greedy approximation heuristics.

Extensions to the case of multiple competing ideas have been explored recently. In [2], the cascade model is extended to competing ideas, and it is shown that, for a follower, the influence maximization problem is submodular. Strategies for leaders, however, are only computable under specific assumptions about the network topology. In [7], the connection between influence maximization and competitve facility location was explored, with the observation that, under a basic diffusion model related to [4], approximating the optimal strategy for the leader is NP-hard.

2.3   Background on Submodular Functions

The notion of *submodularity* will be used to derive solution algorithms for the formulations of Section 4. The notion of submodularity is defined as follows.

Definition 1. *Let V be a finite set. A function $f : 2^V \rightarrow \mathbf{R}$ is submodular if, for any $S \subseteq T \subseteq V$ and any $v \in V \setminus T$,*

$$f(S + v) - f(S) \geq f(T + v) - f(T) \tag{1}$$

Definition 1 can be understood as a "diminishing returns" property, in which the incremental utility of adding an element to a set decreases as the set grows. For additional background on submodular functions, including the following lemma, see [10].

Lemma 1. *A nonnegative weighted sum of submodular functions is submodular.*

3   Proposed Influence Propagation Model: DICE

In this section, the Dynamic Influence in Competitive Environments (DICE) model is defined. The social network is defined by a graph $G = (V, E)$, where V is the set of n individuals (also referred to as *nodes*) and E is the set of social relationships, with $(i, j) \in E$ if individual i has influence over individual j. $N(i)$ is the set of individuals who have influence over i.

A set of m ideas, indexed $\mathcal{I} = \{I_1, \ldots, I_m\}$, is present in the network. I_k is assumed to have an originator $\mathcal{O}_k$. At each time t, each individual $i \in V$ has a state $x_i(t) \in \{0, \ldots, m\}$, where $x_i(t) = k$ if i holds idea I_k at time t and $x_i(t) = 0$ if i has not adopted an idea at time t. Further, it is assumed that each node may be aware of multiple ideas, even if it only holds one of them. Let $\mathcal{I}_i(t) \subseteq \mathcal{I}$ be the set of ideas that i is aware of at time t.

The propagation of ideas under DICE proceeds as follows. At time $t = 0$, let V_k be the set of individuals with $x_i(0) = k$. All nodes $i \in V \setminus \cup_{k=1}^{m} V_k$ have $x_i(0) = 0$. At each subsequent time step t, each individual i chooses a node $j \in N(i) \cup \{i\}$ with probability $d_{ij} > 0$. If $j = i$, then i chooses an idea $I_l \in \mathcal{I}_i(t)$ with probability $P_i(k, l)$, where $k = x_i(t)$. Individual i then updates its state to $x_i(t + 1) = l$. If $j \neq i$, then i sets $x_i(t + 1) = x_j(t)$ with probability $P_{ij}(x_i(t), x_j(t))$, and sets $x_i(t + 1) = x_i(t)$ otherwise. In either case, i updates $\mathcal{I}_i$ according to $\mathcal{I}_i(t + 1) = \mathcal{I}_i(t) \cup \{x_j(t)\}$.

This approach can be generalized to include probabilistic social network models, such as those in [6,4], as follows. Suppose that there is a base topology $G = (V, E_0)$, and let $\mathcal{P}$ be a probability distribution, where $\mathcal{P}(E)$ is the probability that a given edge set $E \subseteq E_0$ occurs. Then for a given realization $E \subseteq E_0$, let

$$d_{ij}^E = \frac{d_{ij}}{\sum_{(i,j) \in E} d_{ij}} \tag{2}$$

while the values of $P_i(k, l)$ and $P_{ij}(x_i(t), x_j(t))$ remain unchanged.

DICE contains several existing influence models as special cases. These connections are summarized in Table 1.

3.1   Distribution of Ideas in Steady-State

The eventual popularity of a given idea can be studied by examining the asymptotic distribution of ideas. This steady-state distribution determines the probability $\pi_i(k)$ that individual i holds idea I_k for large values of t, given an initial distribution $V_1, \ldots, V_m$. The total expected number of nodes holding ideas

Table 1. Existing influence models as realizations of DICE. The triggering model is equivalent in steady-state, while the remaining models have the same dynamics.

Existing model	Parameters
Triggering model [6]	Number of ideas $m = 1$
Generalized linear threshold model [5]	$P_i(k, l) \equiv \frac{1}{|\mathcal{I}_i(t)|}$, choice of $P_{ij}(k, l)$
Independent cascade model [2]	$\mathcal{P}(E) = \prod_{(u,v) \in E} p_{U,V}$

$I_1, \ldots, I_m$ can then be used to evaluate the effectiveness of using $V_1, \ldots, V_m$ as seed nodes. The following theorem gives necessary conditions for this distribution to exist.

Theorem 1. *Suppose that, for each $i \in V$ and $k, l \in \mathcal{I}_i(t)$, $d_{ii} > 0$ and $P_i(k, l) > 0$. Then for a given collection of seed nodes $V_1, \ldots, V_m$, the proposed influence propagation model converges to a unique stationary distribution for $(x_1, \ldots, x_n)$, where $\pi_i(k|V_1, \ldots, V_m)$ denotes the stationary probability of node i holding idea I_k. Furthermore, let*

$$\mathcal{I}_i(V_1, \ldots, V_m) \triangleq \{ I_k \in \mathcal{I} : \exists \ path \ from \ V_k \ to \ i \ \} \tag{3}$$

If G can be decomposed into strongly connected components, then

$$\pi_i(k|V_1, \ldots, V_m) = \pi_i(k|V_1', \ldots, V_m') \tag{4}$$

for any $V_1, \ldots, V_m$ and $V_1', \ldots, V_m'$ satisfying $\mathcal{I}_i(V_1, \ldots, V_m) = \mathcal{I}_i(V_1', \ldots, V_m')$.

Due to space constraints, a full proof is not given here. A sketch of the proof is as follows. By the definition of DICE, every individual becomes aware of the ideas held by its neighbors, even if it does not immediately adopt the ideas. This, together with the assumptions that $d_{ii} > 0$ and $P_i(k, l) > 0$, implies that i becomes aware of every idea in $\mathcal{I}_i$ within finite time. Hence for sufficiently large t, the vector $(x_1(t), \ldots, x_n(t))$ can take any value in the space $\mathcal{S} = \mathcal{I}_1 \times \cdots \times \mathcal{I}_n$. Moreover, since $d_{ii} > 0$ and $P_i(k, l) > 0$, there is a nonzero probability of transitioning into any state $s \in \mathcal{S}$ at each time step. This implies that the chain, restricted to $\mathcal{S}$, is irreducible and aperiodic, and hence has a unique stationary distribution over $\mathcal{S}$. Finally, if G has strongly connected components, then for every component G_l and $i, j \in G_l$, the relationship $\mathcal{I}_i = \mathcal{I}_i(t) = \mathcal{I}_j(t) = \mathcal{I}_j$ holds for sufficiently large values of t, regardless of which specific nodes are in the sets $V_k \cap G_l$.

4   Problem Formulation: Leader-Follower Model

In this section, we formulate the competition between leader and follower ideas as a Stackelberg game. Analysis and algorithms for both the leader and follower strategies are provided as well. Although DICE can be applied to an arbitrary finite number of ideas, the case of two ideas is considered in the following sections in order to ensure simplicity of the solution algorithms.

4.1   Game Definition

The leader-follower game is defined as follows.

Definition 2. *The Stackelberg competing ideas game consists of two players $\mathcal{O}_k$, $k = 1, 2$, each of which owns a competing idea I_k. One of the players (without loss of generality, assume it is $\mathcal{O}_1$) selects a set of individuals V_1 to implant with*

I_1 at time 0. The second player, $\mathcal{O}_2$, observes V_1 and then chooses a set of individuals $V_2 \subseteq V \setminus V_1$ in which to implant I_2. Player $\mathcal{O}_k$'s payoff U_k is given by

$$U_k(V_1, V_2) = \sum_{E \subseteq E_0} \left[\mathcal{P}(E) \left(\sum_{i \in V} \pi_i(k|V_1, V_2, E) \right) \right] - c|V_k| \qquad (5)$$

where $c > 0$ is the cost associated with implanting an idea and $\pi_i(k|V_1, V_2, E)$ is the steady-state probability that individual i will hold idea I_k given initial sets V_1 and V_2 and edge set E.

Under this formulation, the goal of player $\mathcal{O}_2$ is to find the set $V_2^*(V_1)$ satisfying

$$V_2^*(V_1) = \arg \max_{V_2 : V_2 \cap V_1 = \emptyset} U_2(V_1, V_2) \qquad (6)$$

Meanwhile, the goal of player $\mathcal{O}_1$ is to find the set V_1^* satisfying

$$V_1^* = \arg \max_{V_1} U_1(V_1, V_2^*(V_1)) \qquad (7)$$

4.2   Solving Stackelberg Game for Follower

The goal of the follower is to find the set of seed nodes that maximizes the number of individuals holding I_2 in steady-state, given knowledge of the leader's seed nodes V_1. The follower's optimal strategy when G is deterministic is given by Proposition 1. When the topology is probabilistic, the follower's strategy can be found by using submodular optimization techniques, as described in Theorem 2.

Proposition 1. *Suppose that the underlying interaction topology G can be divided into strongly connected components $G_1, \ldots, G_L$. Then the follower's best response consists of a single node v_l from each connected component G_l satisfying*

$$\sum_{i \in V} \pi_i(2|V_1, v_l) > c \qquad (8)$$

Proof. First, note that, for each component G_l and each $i \in G_l$, $\pi_i(2|V_1, V_2)$ depends only on whether $G_l \cap V_2$ is nonempty by Theorem 1. This, coupled with the fact that

$$\sum_i \pi_2(2|V_1, V_2) - c|G_l \cap V_2| \leq \sum_i \pi_2(2|V_1, V_2) - c \qquad (9)$$

implies that any V_2 with $|G_l \cap V_2| > 1$ is suboptimal.

Now, $U_2(V_1, V_2|E)$ can be rewritten as

$$U_2(V_1, V_2|E) = \sum_{l=1}^{L} \left(\sum_{i \in G_l} (\pi_i(2|V_1, V_2)) - c|G_l \cap V_2| \right) \qquad (10)$$

Since propagation of ideas in disconnected components is independent, each term of the outer sum of (10) can be considered independently. Thus $|G_l \cap V_2| = 1$ is optimal iff the corresponding term of (10) is positive, which occurs iff (8) holds. □

Theorem 2. *When the interaction topology is stochastic and V_1 is fixed, $U_2(V_1, V_2)$ is a submodular function of V_2.*

Proof. By Proposition 1, the incremental gain from adding $v \in G_l$ to V_2 is equal to

$$U_2(V_1, V_2 + v) - U_2(V_1, V_2) = \begin{cases} \sum_{i \in G_l} \pi_i(2|V_1, v) - c, & V_2 \cap G_l = \emptyset \\ -c, & \text{else} \end{cases} \quad (11)$$

Now, if $S \subseteq T$ and $T \cap G_l = \emptyset$, then $S \cap G_l = \emptyset$. Hence

$$U_2(V_1, S + v) - U_2(V_1, S) \geq U_2(V_1, T + v) - U_2(V_1, T) \quad (12)$$

proving that $U_2(V_1, \cdot)$ is submodular. In general, U_2 is given by

$$U_2(V_1, V_2) = \sum_{E \subseteq E_0} U_2(V_1, V_2|E)\mathcal{P}(E) \quad (13)$$

which is a nonnegative weighted sum of submodular functions, and hence is submodular. □

The submodularity of $U_2(V_1, \cdot)$ implies that (6) is a *submodular maximization problem*. Although the submodular maximization problem is NP-hard, algorithms have been proposed that are guaranteed to achieve a provable approximation bound in polynomial time [10].

An algorithm for solving (6) is as follows. Initialize $V_2^0 = \emptyset$. At each subsequent iteration t, find a node v^* satisfying

$$v^* = \arg \max_{v \in V \setminus (V_1 \cup V_2^t)} U_2(V_1, V_2 + \{v\}) - U_2(V_1, V_2) \quad (14)$$

If $U_2(V_1, V_2^t \cup \{v^*\}) - U_2(V_1, V_2^t) > 0$, then set $V_2^{t+1} = V_2^t \cup \{v^*\}$, increment t, and continue. Otherwise the algorithm terminates. A pseudo-code description of the algorithm is given in Figure 1.

Submodularity of $U_2(V_1, \cdot)$ and Proposition 4.1 of [10], yields the following proposition on the optimality of the algorithm.

Proposition 2. *The algorithm of Figure 1 returns a set $\tilde{V}_2$ such that*

$$U_2(V_1, V_2^*) - U_2(V_1, \tilde{V}_2) \leq c|\tilde{V}_2| \quad (15)$$

4.3 Solving Stackelberg Game for Leader

As a preliminary, the following lemma describes the leader's payoff for a given set of seed nodes V_1.

Lemma 2. *For fixed topology $G = (V, E)$, the payoff for the leader is given by*

$$U_1(V_1, V_2^*(V_1)|E) = \sum_{l: G_l \cap V_1 \neq \emptyset} \sum_{i \in G_l} w_i - |V_1|c \quad (16)$$

```
FOLLOWER_SEED_NODE_SELECTION
Input: Set V₁, topology G = (V, E), distribution 𝒫(E)
Output: Set V₂
V₂⁰ ← ∅
t ← 0
while(1)
    v ← arg max_{v∈V\V₁} U₂(V₁, V₂ᵗ + v) − U₂(V₁, V₂ᵗ)
    if U₂(V₁, V₂ᵗ + v) > U₂(V₁, V₂ᵗ)
        V₂^{t+1} ← V₂ᵗ ∪ {v}
        t ← t + 1
    else
        break
return V₂ᵗ
```

Fig. 1. Pseudo-code description for submodular maximization of follower payoff

where w_i is given by

$$w_i = \begin{cases} \pi_i(1|I_1, I_2), & \sum_{i\in G_l} \pi_i(2|I_1, I_2) > c \\ \pi_i(1|I_1) & else \end{cases} \qquad (17)$$

and $\pi_i(1|I_1, I_2)$ and $\pi_i(1|I_1)$ are the stationary probability that i holds idea I_1 when $\mathcal{I}_i = \{I_1, I_2\}$ and $\mathcal{I}_i = \{I_1\}$, respectively.

Proof. By Proposition 1, the follower's best response is to add a node $v_l \in G_l$ iff $\sum_{i\in G_l} \pi_i(2|I_1, I_2) > c$. In this case, the leader's payoff is $\sum_{i\in G_l} \pi_i(1|I_1, I_2)$ by Definition 2. Otherwise, the nodes in G_l only become aware of I_1, and so the leader's payoff is $\sum_{i\in G_l} \pi_i(1|I_1)$.   □

This leads to the following theorem, analogous to Theorem 2.

Theorem 3. *$U_1(V_1, V_2^*(V_1))$ is submodular as a function of V_1.*

Proof. By Lemma 2, the incremental gain from adding v to V_1 when G is deterministic is given by

$$U_1(V_1 + v, V_2^*(V_1 + v)) - U_1(V_1, V_2^*(V_1)) = \begin{cases} \sum_{i\in G_l} w_i - c, & V_1 \cap G_l = \emptyset \\ -c, & else \end{cases} \qquad (18)$$

Hence the incremental gain is positive iff $V_1 \cap G_l = \emptyset$. Given $S \subseteq T \subseteq V$, $T \cap G_l = \emptyset$ implies that $S \cap G_l = \emptyset$. Thus

$$U_1(S+v, V_2^*(S+v)) - U_1(S, V_2^*(S)) \geq U_1(T+v, V_2^*(T+v)) - U_1(T, V_2^*(T)) \qquad (19)$$

as desired. As in Theorem 2, when G is stochastic, $U1$ is a nonnegative weighted sum of submodular functions, and is therefore submodular.   □

Theorem 3 implies that an algorithm analogous to that in Figure 1 can be used to solve the leader's optimization problem (7). A pseudo-code description of the algorithm is contained in Figure 2.

```
LEADER_SEED_NODE_SELECTION
Input: Topology G = (V, E), distribution P(E)
Output: Set V₁
V₁⁰ ← ∅
t ← 0
while(1)
    v ← arg max_{v∈V} U₁(V₁ᵗ + v, V₂*(V₁ᵗ + v))
    if U₁(V₁ᵗ + v, V₂*(V₁ᵗ + v)) > U₁(V₁ᵗ, V₂*(V₁ᵗ))
        V₁^{t+1} ← V₁ᵗ ∪ {v}
        t ← t + 1
    else
        break
return V₁ᵗ
```

Fig. 2. Pseudo-code description for submodular maximization of leader payoff

5   Problem Formulation: Simultaneous Model

Under the simultaneous-move game, the originators of competing ideas simultaneously choose sets of seed nodes. This models the case where two ideas are introduced at the same time, or, more generally, when neither player is able to observe the other's choice of seed nodes before introducing his idea.

Definition 3. *The simultaneous-move game consists of two players $\mathcal{O}_k$, $k = 1, 2$, each of which owns a competing idea I_k. The players simultaneously select sets V_k of individuals to implant with idea I_k at time 0. (If the players choose the same individual, then that individual adopts one of the ideas but is aware of both of them). Player $\mathcal{O}_k$'s payoff is given by*

$$U_k(V_1, V_2) = \sum_{i \in V} \pi_i(k|V_1, V_2) - c|V_k| \tag{20}$$

where c and π_i are defined as in Definition 2.

In what follows, analysis of the simultaneous-move game under DICE is provided. The first observation is that, when the topology is deterministic, the game can be decomposed into a set of L games, each played on a different connected component G_l of the social network G. For a given component G_l, each player chooses whether or not to choose V_k such that $G_l \cap V_k \neq \emptyset$. The resulting payoff matrix is given by Table 2, where $E_k = \sum_{i \in G_l} \pi_i(k)$, H denotes the case where $G_l \cap V_k \neq \emptyset$, and H' denotes the case where $G_l \cap V_k = \emptyset$.

The following theorem describes the Nash equilibria of the game.

Theorem 4. *For the simultaneous-move game with a single component G_l,*

   (i) *If $|G_l| < c$, then the game has a unique Nash equilibrium of (H', H').*
   (ii) *If $E_1 > c$ (resp. E_2) and $E_2 < c$ (resp. E_1), then the game has a unique Nash equilibrium of (H, H') (resp. (H', H)).*

Table 2. Payoffs of $\mathcal{O}_1$ and $\mathcal{O}_2$ for simultaneous-move game

	H	H'
H	$(E_1 - c, E_2 - c)$	$(|G_l| - c, 0)$
H'	$(0, |G_l| - c)$	$(0, 0)$

(iii) *If $E_k > c$ for $k = 1, 2$, there are two pure strategy Nash equilibria and one Nash equilibrium in mixed strategies.*

Proof. Points (i) and (ii) follow by inspection of the payoff matrix, noting that the equilibria Pareto dominate the other possible strategies. When the conditions of (iii) hold, (H, H') and (H', H) are Nash equilibria by inspection. To find the mixed Nash equilibrium, note that it occurs when both parties are indifferent between playing H and H'. Let p_k denote the probability that player k plays H. Then player 1's payoff from playing H is $p_2(E_1 - c) + (1 - p_2)(|G_l| - c)$ while the payoff from playing H' is 0. Setting these equal yields $p_2 = \frac{|G_l| - c}{|G_l| - E_1}$. Thus the mixed strategy equilibrium is given by

$$p_1 = \frac{|G_l| - c}{|G_l| - E_2}, \qquad p_2 = \frac{|G_l| - c}{|G_l| - E_1} \qquad (21)$$

as desired.   □

6   Simulation Study

In this section, a simulation study of the leader-follower game of Section 4 is presented. Simulations were performed using Matlab on the Wiki-vote dataset [8]. A link (i, j) exists if user i voted in favor of user j becoming an administrator. The original data set had $|V| = 7115$; in order to reduce runtime, randomly chosen subsets of V were used for simulation. It was assumed that each edge in E_0 was added to E with probability chosen uniformly at random from $[0, 0.5]$. For each edge (i, j), the probability that individual i changes from I_1 to I_2 (or vice versa) based on j's input was chosen uniformly at random from $[0, 1]$. The probability of an individual i spontaneously switching between ideas was chosen uniformly at random from $[0, 0.05]$. The remaining simulation parameters are summarized in Table 3.

Table 3. Simulation parameters

Parameter	Values Used
Number of nodes, n	$n = 100, 200, 300, 400, 500, 600, 700$
Number of ideas, m	2
Probability of self-determination d_{ii}	0.5
Probability of i choosing j, d_{ij}	$\frac{0.5}{|N(i)|}$
Cost of adding an individual to V_k, c	5 (low cost) and 15 (high cost)

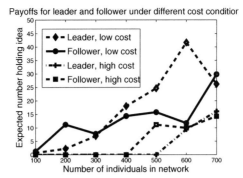

Fig. 3. Simulation of the leader-follower game using the Wiki-vote dataset. The payoff for each player increases with network size; the rate of increase depends on the cost c. The leader's payoff typically exceeds the follower's.

Figure 3 shows the number of individuals holding ideas I_1 and I_2 in steady-state for different values of n and c. The payoffs of both players increase with network size. However, in most cases, the payoff of the leader exceeds the payoff of the follower. This is because the follower must choose whether to compete with the leader for influence over highly-connected clusters of individuals. When the follower has no incentive to do so, the leader may automatically gain control of these clusters at minimal cost.

Increasing the cost c gives each player less incentive to target individual nodes. Figure 3 suggests that there is a cutoff $n(c)$ on the network size in order for idea originators to be willing to introduce their ideas. This may be interpreted as a "barrier to entry" for ideas to enter the marketplace [9].

7   Conclusions and Future Work

In this paper, the problem of maximizing influence of competing ideas was studied. The Dynamic Influence in Competitive Environments (DICE) model, which uses Markov processes to model the propagation of ideas through a social network, was introduced. Based on DICE, game-theoretic models of competition between ideas were developed, including a Stackelberg game modeling the interaction between a leader and a follower in a marketplace, as well as a model for simultaneous introduction of ideas. It was shown that computationally tractable algorithms can be used to approximate the solution for both players.

In our proposed formulation it was assumed that both players have complete, full knowledge of the network topology and each other's attributes. Our plan of future work is to develop models of competition with incomplete information. We will also extend the static games analyzed in this paper to dynamic games, in which the owner of each idea adapts his strategy in response to the actions of his competitors. Another direction of future work is to improve on the speed and accuracy of the solution algorithms of Section 4.

References

1. Acemoglu, D., Ozdaglar, A., ParandehGheibi, A.: Spread of (mis) information in social networks. Games and Economic Behavior (2010)
2. Bharathi, S., Kempe, D., Salek, M.: Competitive Influence Maximization in Social Networks. In: Deng, X., Graham, F.C. (eds.) WINE 2007. LNCS, vol. 4858, pp. 306–311. Springer, Heidelberg (2007)
3. Boyd, S., Ghosh, A., Prabhakar, B., Shah, D.: Gossip algorithms: Design, analysis and applications. In: INFOCOM 2005, vol. 3, pp. 1653–1664. IEEE (2005)
4. Goldenberg, J., Libai, B., Muller, E.: Talk of the network: A complex systems look at the underlying process of word-of-mouth. Marketing Letters 12(3), 211–223 (2001)
5. Granovetter, M.: Threshold models of collective behavior. The American Journal of Sociology 83(6), 1420–1443 (1978)
6. Kempe, D., Kleinberg, J., Tardos, É.: Maximizing the spread of influence through a social network. In: Proceedings of the Ninth ACM SIGKDD International Conference on Knowledge Discovery and Data Mining, pp. 137–146. ACM (2003)
7. Kostka, J., Oswald, Y.A., Wattenhofer, R.: Word of Mouth: Rumor Dissemination in Social Networks. In: Shvartsman, A.A., Felber, P. (eds.) SIROCCO 2008. LNCS, vol. 5058, pp. 185–196. Springer, Heidelberg (2008)
8. Leskovec, J., Huttenlocher, D., Kleinberg, J.: Signed networks in social media. In: Proceedings of the 28th International Conference on Human Factors in Computing Systems, pp. 1361–1370. ACM (2010)
9. Myerson, R.: Game theory: analysis of conflict. Harvard Univ. Pr. (1997)
10. Nemhauser, G., Wolsey, L., Fisher, M.: An analysis of approximations for maximizing submodular set functionsi. Mathematical Programming 14(1), 265–294 (1978)
11. Richardson, M., Domingos, P.: Mining knowledge-sharing sites for viral marketing. In: Proceedings of the Eighth ACM SIGKDD International Conference on Knowledge Discovery and Data Mining, pp. 61–70. ACM (2002)
12. Shirky, C.: The political power of social media. Foreign Affairs 90(1), 28–41 (2011)
13. Tuten, T.: Advertising 2.0: social media marketing in a Web 2.0 world. Praeger Publishers (2008)

Collaborative Location Privacy with Rational Users

Francisco Santos, Mathias Humbert, Reza Shokri, and Jean-Pierre Hubaux

School of Computer and Communication Sciences, EPFL, Switzerland
`firstname.lastname@epfl.ch`

Abstract. Recent smartphones incorporate embedded GPS devices that enable users to obtain geographic information about their surroundings by providing a location-based service (LBS) with their current coordinates. However, LBS providers collect a significant amount of data from mobile users and could be tempted to misuse it, by compromising a customer's location privacy (her ability to control the information about her past and present location). Many solutions to mitigate this privacy threat focus on changing both the architecture of location-based systems and the business models of LBS providers. MobiCrowd does not introduce changes to the existing business practices of LBS providers, rather it requires mobile devices to communicate wirelessly in a peer-to-peer fashion. To lessen the privacy loss, users seeking geographic information try to obtain this data by querying neighboring nodes, instead of connecting to the LBS. However, such a solution will only function if users are willing to share regional data obtained from the LBS provider. We model this collaborative location-data sharing problem with rational agents following threshold strategies. Initially, we study agent cooperation by using pure game theory and then by combining game theory with an epidemic model that is enhanced to support threshold strategies to address a complex multi-agent scenario. From our game-theoretic analysis, we derive cooperative and non-cooperative Nash equilibria and the optimal threshold that maximizes agents' expected utility.

1 Introduction

Today's smartphones are often equipped with GPS devices that enable their users to obtain contextual information about their surroundings, such as the location of the nearest supermarket, without needing to ask directions from other people. To obtain such contextual information, users normally query a **location-based service** (LBS), such as Google Maps (`http://maps.google.com`) that, given a current position, can provide detailed information about points-of-interest in the region and stepwise instructions to reach a particular destination. The downside of using an LBS system is the possible loss of **location privacy** [5][10][12], defined as the ability for a user to control how, where, and when information about her current and past location is used and by whom [1][4].

MobiCrowd [14] mitigates the loss in location privacy by assuming users carry location-aware wireless devices capable of peer-to-peer communication, through

J.S. Baras, J. Katz, and E. Altman (Eds.): GameSec 2011, LNCS 7037, pp. 163–181, 2011.

which they can share regional data and, in this way, reduce the fraction of queries dispatched to the LBS. MobiCrowd devices are equipped with a **mobile proxy** that stores the results of LBS queries in a buffer. When a user issues a new query, the mobile proxy scans the buffer for the information. If the query cannot be answered by the local cache, it is broadcast to peer devices within range. Should these peers be unable or unwilling to answer the query, the device finally prompts the LBS server. The main advantage of this new scheme is that it helps protect users' location privacy and requires no changes to the current business practices of LBS providers and only minimal changes to the architecture of conventional location-based services.

Our goal is to learn whether the overall level of cooperation amongst users is sufficiently high for a solution such as MobiCrowd to work. For this purpose, we model the problem by using game theory [2][13]. This discipline provides a rich set of analytical tools through which researchers study the interaction between agents as decision-makers, notably in the context of cooperation in wireless communications [7] and in location privacy [1][4][8][9][11]. Combining game theory with epidemic models, such as the susceptible-infected-removed (SIR) model [6], is a practical approach to studying strategic behavior for large populations of agents. This method is used in [3] to explore why rational individuals might preclude the eradication of a vaccine-preventable disease by weighing the risks of vaccination and infection; and in [15] to study how the investment in security by self-interested agents affects the propagation of a computer network infection.

In our game-theoretic model, we represent users as agents who follow threshold strategies. We assume that users are **rational**, meaning that they have knowledge of their actions, reason about uncertainty, have clear preferences expressed through a utility function, and choose actions in their own self-interest by maximizing this utility function [13]. Using this model, we define two infinitely repeated games of imperfect information. In the first game, we study an elementary, two-agent interaction and use pure game theory to derive Nash and Pareto optimal equilibria. In the second, we analyze a complex, multi-agent interaction by using a modified version of the epidemic model in [14] to support threshold strategies and derive the optimal threshold that maximizes agents' payoffs.

2 System Model

In this section we define a game-theoretic framework from which we scrutinize the MobiCrowd architecture. Consider an agent i confined to a single region. As it interacts with other agents in the region and the LBS, it switches between three distinct roles: seeker (K), informed (I), and removed (R) [14]. **Seekers** try to obtain regional data by querying other peers or, ultimately, the LBS server. **Informed** agents have data on the region and accept to spread this information according to a threshold strategy; they become removed once the data on their mobile proxy expires. Finally, **removed** agents are not interested in obtaining regional data but can become information seekers later in the game. The set of roles that each agent may have is defined as follows: $\Pi_R = \{K, I, R\}$. We illustrate the interaction between the three agent roles in Figure 1.

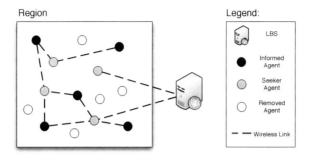

Fig. 1. Operating principle of MobiCrowd. Seekers attempt to obtain data on this region through informed agents or the LBS using wireless communication links. Removed agents do not have any regional data and do not want to obtain such information, abstaining from any interaction with either seekers or informed agents.

As agent i communicates with its peers, it records (i) the number of times it received data from other agents, $\mathrm{rc}_i(t)$, and (ii) the number of times it answered queries from its peers, $\mathrm{tr}_i(t)$, up to and excluding time t, where $\mathrm{rc}_i(0) = \mathrm{tr}_i(0) = 0$. The goal is to define an agent's **cumulative cooperation effort**: the total number of times an agent receives help in excess of the amount of times it shared data with a peer. An agent uses this value to decide whether to cooperate or defect when receiving queries from its peers. Let $\phi_i(t)$ denote the cumulative cooperation effort of agent i, defined as follows:

$$\phi_i(t) = \mathrm{rc}_i(t) - \mathrm{tr}_i(t) \ . \tag{1}$$

Further, let the set Π_ϕ equal the range of the cumulative cooperation effort: $\Pi_\phi = \{\phi^{\min}, \ldots, 0, \ldots, \phi^{\max}\}$, where $\phi^{\min}$ and $\phi^{\max}$ are, respectively, the minimum and maximum bounds for $\phi_i(t)$.

There are two variables that define the state of an agent in the system: (i) its current role and (ii) its cumulative cooperation effort. Hence, by combining these two variables, we can define the full set of agent states. Let $\mathcal{S}$ denote the set of agent states, computed as the cartesian product of Π_R and Π_ϕ: $\mathcal{S} = \Pi_R \times \Pi_\phi$. We adopt the following notation to refer to each state: $\mathcal{X}_\phi$, where $\mathcal{X}$ represents one of the three possible roles and ϕ is the cumulative cooperation effort. Additionally, we express the fraction of agents in each state at time t as $X_\phi(t)$.

An agent i decides to share regional facts with seekers according to a threshold strategy $s_i(t)$. It cooperates with a seeker if (i) it is informed and (ii) its cumulative cooperation effort, $\phi_i(t)$, is above a common threshold α that expresses the amount of sharing agents accept to perform before expecting some aid in return; lowering α improves an agent's capacity to collaborate. If either condition is false, the agent defects:

$$s_i(t) = \begin{cases} \text{cooperate} & \text{if } \phi_i(t) > \alpha \wedge i \in \mathcal{I}_{\phi_i(t)} \ , \\ \text{defect} & \text{otherwise} \ . \end{cases} \tag{2}$$

As in [14], we assume that agents share a single kind of contextual information. To consider multiple types of regional data, we can apply the same system model to each type of data separately; hence, no loss of generality results from this simplifying assumption.

In our work, we consider a payoff model that reflects the costs and benefits experienced by MobiCrowd users as they interchange data with peers and connect to the LBS. Consider a simple interaction between a seeker i, an informed agent j, and the LBS. When i prompts j and j cooperates, i receives a benefit of b_{inf}, expressing the agent's information gain, and j incurs a communication cost of c_{com} for sharing data. If j defects it receives no payoff and i must seek help from the LBS, earning a lesser payoff of $b_{inf} - c_{srv}$, where c_{srv} denotes the privacy loss implied when communicating with the location-based service. The utility of agent i at time t is defined as the difference between the benefit obtained, $b_i(t)$, and the cost incurred, $c_i(t)$, at time t:

$$u_i(t) = b_i(t) - c_i(t) \ . \tag{3}$$

According to the operating principle of MobiCrowd [14], users that seek geographic data query nearby informed peers to avoid the privacy threat posed by the LBS; hence, we assume that the LBS privacy cost (c_{srv}) exceeds the peer-to-peer communication cost (c_{com}). As users ultimately endeavor to become informed, we further assume that the benefit of acquiring data (b_{inf}) tops the privacy and communication costs: $0 \leq c_{com} < c_{srv} < b_{inf} \leq 1$.

In infinitely repeated games, agents lack knowledge of the precise game duration, thus, they value present payoffs more than future rewards. To capture this fact, we define the **aggregate discounted reward** of an agent for two distinct cases: discrete time and continuous time games. In the discrete case, the payoff of an agent i is observed at several discrete time instances or game stages and the aggregate discounted reward is the sum of its payoff in the immediate stage game, plus the sum of aggregate rewards discounted by a constant $\delta \in]0, 1[$.

Definition 1. *Given an infinite sequence of payoffs $u_i(0), u_i(1), \ldots$ for agent i, and a discount factor δ with $0 < \delta < 1$, the **aggregate discounted reward** of i in a discrete time game is: $\sum_{t=0}^{\infty} \delta^t u_i(t) = \sum_{t=0}^{\infty} \delta^t (b_i(t) - c_i(t))$.*

Additionally, we consider a continuous time game, in which case the aggregate discounted reward is given by the integral of an agent's payoff for the duration of the game, discounted by a constant $\delta \in]0, 1[$.

Definition 2. *Given the utility function $u_i(t)$ for agent i and a discount factor δ with $0 < \delta < 1$, the **aggregate discounted reward** of i in a continuous time game is: $\int_0^{\infty} \delta^t u_i(t) \, dt = \int_0^{\infty} \delta^t (b_i(t) - c_i(t)) \, dt$.*

3 Two-Agent Game

We first model the problem as an infinitely repeated two-agent game of imperfect information and we constrain both agents to a single map region. To avoid this

Table 1. List of Symbols

Symbol	Description
n	size of the agent population
$\phi_i(t)$	cumulative cooperation effort of agent i
$\mathrm{rc}_i(t)$	number of times agent i received data up to and excluding time t
$\mathrm{tr}_i(t)$	number of times agent i shared data up to and excluding time t
$s_i(t)$	pure strategy followed by agent i at time t
α	threshold value used in $s_i(t)$ to trigger cooperative and non--cooperative agent behavior
α^{opt}	common threshold value that maximizes agents' expected payoff
$u_i(t)$	agent i's payoff at time t
$b_i(t)$	agent i's benefit (if any) at time t
$c_i(t)$	agent i's cost (if any) at time t
b_{inf}	payoff for obtaining regional data from a peer or the LBS
c_{com}	communication cost of sharing regional data with a peer
c_{srv}	privacy cost incurred when calling the LBS server
δ	discount factor used to calculate agents' aggregate payoffs
U_i	agent i's expected aggregate discounted payoff
$\widehat{U}$	estimate of an agent's expected aggregate discounted payoff
U_{total}	total discounted payoff of the game
$\widehat{U}_{\mathrm{total}}$	estimate of the game's total discounted payoff
$\mathrm{rc}_{\mathrm{total}}$	total rate at which seekers receive data from informed peers
$\widehat{\mathrm{rc}}_{\mathrm{total}}$	estimate of the total rate at which seekers receive data from informed peers
$\mathrm{tr}_{\mathrm{total}}$	total rate at which informed agents share data with seekers
$\widehat{\mathrm{tr}}_{\mathrm{total}}$	estimate of the total rate at which informed agents share data with seekers
$\widehat{\mathrm{sv}}_{\mathrm{total}}$	estimate of the total rate at which seekers receive data from the LBS
$\mathcal{S}$	set of game states
$\mathit{\Pi}_\phi$	range of the cumulative cooperation effort: $\{\phi^{\mathrm{min}}, \ldots, 0, \ldots, \phi^{\mathrm{max}}\}$
$\mathit{\Pi}_R$	set of possible agent roles: $\{K, I, R\}$
$\mathcal{K}_\phi$	an agent i is in this game state at time t if it wants to obtain regional data and $\phi_i(t) = \phi$
$K_\phi(t)$	fraction of agents in state $\mathcal{K}_\phi$ at time t
$\mathcal{I}_\phi$	an agent i is in this game state at time t if it has regional data and $\phi_i(t) = \phi$
$I_\phi(t)$	fraction of agents in state $\mathcal{I}_\phi$ at time t
$\mathcal{R}_\phi$	an agent i is in this game state at time t if it has no regional data, does not wish to obtain this data, and $\phi_i(t) = \phi$
$R_\phi(t)$	fraction of agents in state $\mathcal{R}_\phi$ at time t

privacy threat, agents 1 and 2 want to obtain information about this region by querying their partner, instead of calling the LBS. In the beginning, no agent has the desired data, consequently one agent must call the LBS. The data retrieved from the LBS expires after one game epoch, at which time it is deleted from the agents' mobile proxy. At the start of each stage, exactly one agent — chosen uniformly at random — will be responsible for querying the server. Upon querying the LBS, an agent i becomes informed and follows a threshold strategy to decide whether to cooperate and share data with its peer j or defect by refusing to share, according to the strategy function $s_i(t)$ (2).

When agent i cooperates at time t, i.e. $s_i(t) =$ cooperate, i incurs a communication cost of c_{com} and j gains a benefit of b_{inf}. If i defects, $s_i(t) =$ defect, it obtains no payoff and j must call the LBS, earning $b_{\text{inf}} - c_{\text{srv}}$. The utility function, shared by both agents, is defined succinctly as follows:

$$
u_i(t) = \begin{cases}
b_{\text{inf}} & \text{if } s_j(t) = \text{cooperate} , \\
b_{\text{inf}} - c_{\text{srv}} & \text{if } s_j(t) = \text{defect} , \\
-c_{\text{com}} & \text{if } s_i(t) = \text{cooperate} , \\
0 & \text{if } s_i(t) = \text{defect} .
\end{cases}
\tag{4}
$$

Agent i uses the same utility and strategy functions as its peer j, knows the action chosen by j in the current game stage, but it does not record j's past moves. The ability to memorize the history of actions played by an agent is unrealistic due to the memory constraints of existing smartphone devices. Hence, i is unable to compute j's cumulative cooperation effort for all game stages. For this reason, agents i and j have imperfect information about the game state.

4   Two-Agent Game Analysis

The goal of this section is to determine under which conditions the threshold strategy introduced previously will yield a cooperative equilibrium. Let U_i denote the expected aggregate discounted reward for agent i. Agents acting rationally will only cooperate if the expected discounted benefit is greater than the expected discounted cost incurred, as defined next:

$$
U_i = \sum_{t=0}^{\infty} \delta^t \left(b_i(t) - c_i(t) \right) > 0 .
\tag{5}
$$

We now calculate the expected discounted aggregate reward, assuming that agents are chosen randomly at each turn to query the LBS.

Lemma 1. *If exactly one agent is chosen at each turn to query the LBS and the probability of being chosen equals $1/2$, then agent i's expected aggregate discounted reward, as a function of the threshold α, is given by*

$$
U_i(\alpha) = \begin{cases}
\dfrac{1}{1-\delta} \cdot \dfrac{b_{inf} - c_{com}}{2} & \text{if } \alpha \leq -1 , \\
\dfrac{1}{1-\delta} \cdot \dfrac{b_{inf} - c_{srv}}{2} & \text{otherwise} .
\end{cases}
\tag{6}
$$

Proof. The probability that an agent i is chosen to query the LBS is $1/2$. If agent i is chosen to query the LBS in turn k, it will cooperate with the second agent if $\phi_i(k) > \alpha$, incurring a cost of c_{com}, and defect otherwise, receiving no payoff. If agent j is selected to query the LBS in turn k, it will cooperate with agent i if $\phi_j(k) > \alpha$, resulting in a benefit of b_{inf} for agent i, and defect if $\phi_j(k) \leq \alpha$, causing agent i to call the LBS server, yielding a lesser benefit of $b_{\text{inf}} - c_{\text{srv}}$. Let $U_i(\alpha)$ denote the expected aggregate discounted reward for agent i as a function of the threshold value α. Then $U_i(\alpha)$ can be calculated as follows:

$$U_i(\alpha) = \frac{1}{2} \sum_{t=0}^{\infty} \delta^t \left[-c_{\text{com}} \cdot 1_{\{\phi_i(t)>\alpha\}} + 0 \cdot 1_{\{\phi_i(t)\leq\alpha\}} \right] +$$
$$\delta^t \left[b_{\text{inf}} \cdot 1_{\{\phi_j(t)>\alpha\}} + (b_{\text{inf}} - c_{\text{srv}}) \cdot 1_{\{\phi_j(t)\leq\alpha\}} \right] , \qquad (7)$$

where $1_{predicate}$ is an indicator function yielding 1 each time the predicate is true and zero when the predicate is false.

As we do not know the value of $\phi_i(t)$ for the duration of the game, we must calculate the average cumulative cooperation effort ($\overline{\phi}$) based on the average number of times an agent receives data ($\overline{rc}$) and shares data ($\overline{tr}$), as follows:

$$\overline{\phi} = \overline{rc} - \overline{tr} . \qquad (8)$$

As the two agents have equal probabilities of being chosen to query the LBS at each turn, we expect that each agent is chosen for $m/2$ game stages, where m is the length of the game. Additionally, agent i will share data with its peer j with an estimated probability of: $\Pr\{\overline{\phi} > \alpha\}$. Hence, $\overline{rc}$ and $\overline{tr}$ are given by

$$\overline{rc} = (m/2) \cdot \Pr\{\overline{\phi} > \alpha\} , \qquad (9)$$

$$\overline{tr} = (m/2) \cdot \Pr\{\overline{\phi} > \alpha\} , \qquad (10)$$

implying that the average cumulative cooperation effort is zero: $\overline{\phi} = 0$. Using the previous results, we can now estimate the value of $U_i(\alpha)$ as follows:

$$U_i(\alpha) = \frac{1}{2} \sum_{t=0}^{\infty} \delta^t \left[(b_{\text{inf}} - c_{\text{com}}) \cdot 1_{\{\alpha<0\}} + (b_{\text{inf}} - c_{\text{srv}}) \cdot 1_{\{\alpha\geq0\}} \right] =$$
$$= \begin{cases} \dfrac{1}{1-\delta} \cdot \dfrac{b_{\text{inf}} - c_{\text{com}}}{2} & \text{if } \alpha \leq -1 , \\ \dfrac{1}{1-\delta} \cdot \dfrac{b_{\text{inf}} - c_{\text{srv}}}{2} & \text{otherwise} . \end{cases}$$

$\square$

In the following theorem we characterize the possible equilibria resulting from the game described in Lemma 1.

Theorem 1. *The game described in Lemma 1 has two Nash equilibria: (i) both agents choose a threshold $\alpha = -1$, ensuring the minimal level of cooperation* *necessary to achieve the maximal payoff and (ii) both agents opt for a threshold of $\alpha \geq 0$, thus defecting throughout the game. Further, the choice of $\alpha = -1$ results in a Pareto optimal equilibrium.*

Proof. Agents i and j can opt for a threshold of $\alpha > -1$ or, equivalently, $\alpha \geq 0$, leading to a non-cooperative Nash equilibrium. As the function $\phi_i(t)$ is initially zero, it will never be greater than zero, and agents will never cooperate: $\forall_{i,t} \phi_i(t) \not> 0$. If agent j independently chooses a threshold α' smaller than zero, it will cooperate, at most, $-\alpha'$ times, incurring a cost proportional to c_{com}.

The choice of $\alpha = -1$ is also a Nash equilibrium as it is the maximal threshold that still allows mutual cooperation between both agents, minimizes the number of times they share data, and maximizes their expected discounted payoffs:

$$\max U_{i,j}(\alpha) = \frac{1}{1 - \delta} \cdot \frac{b_{\text{inf}} - c_{\text{com}}}{2} \qquad \text{for } \alpha \leq -1. \tag{11}$$

If agent j independently decreases its threshold to $\alpha' < \alpha$, it will increase its capacity to cooperate with i, potentially incurring a higher sharing cost, proportional to c_{com}, and consequently obtaining a lower payoff than (11). Assuming j chooses, instead, $\alpha' \geq 0$, it will always defect and agent i will cooperate, at most, once with j, at which time j earns b_{inf} and i has a cost of c_{com}. As this event occurs, at most, once for an infinitely repeated game, we do not consider this an advantage for player j. Instead, we assume that the game outcome is equivalent to when both agents defect throughout the game.

We now compare the two possible choices of threshold values: $\alpha = -1$ and $\alpha \geq 0$. Clearly, $U_i(-1) > U_i(\alpha')$, for $\alpha' \geq 0$ and $0 \leq c_{\text{com}} < c_{\text{srv}} < b_{\text{inf}} \leq 1$. Hence, the strategical choice of $\alpha = -1$ Pareto dominates the choice $\alpha \geq 0$. As there is no other strategy that Pareto dominates the choice of $\alpha = -1$, we conclude that this choice by both agents is Pareto optimal. □

5   Multiple-agent Game

In an effort to match more closely a real interaction between MobiCrowd devices, we now consider a game with multiple agents and represent time as a continuous measure. To analyze the behavior of a large population of agents ($n >> 2$), we alter the epidemic model in [14] to support threshold strategies and we define the rate at which agents switch between states, as shown in Figure 2. The letters $\mathcal{K}$, $\mathcal{I}$, and $\mathcal{R}$ in the state-diagram stand for the three agent roles, seeker ($\mathcal{K}$), informed ($\mathcal{I}$), and removed ($\mathcal{R}$) and the integer suffix next to each letter denotes the cumulative cooperation effort. The Cartesian product of the set of roles and the range of the cumulative cooperation effort generates the full set of agent states: $\mathcal{S} = \Pi_R \times \Pi_\phi$.

Assume that only seekers are present at the start of the game; so all agents are at state $\mathcal{K}_0$ (the cumulative cooperation effort is initially zero). With no informed agents in the vicinity, seekers must acquire regional data via the LBS, following transition $\mathcal{K}_0 \rightarrow \mathcal{I}_0$, in which case their cumulative cooperation effort remains constant at zero. These newly informed agents can now cooperate with seekers at state $\mathcal{K}_0$. A seeker receiving regional data through an informed peer improves its cumulative cooperation effort by one and then jumps to a higher layer in the state diagram. Whereas, informed agents sharing data suffer a decrement in their cumulative cooperation effort and fall to a lower level in the state diagram.

An informed agent i with the minimum cumulative cooperation level (i.e. that has reached the bottom level in the state chart) does not cooperate with seekers as the condition to cooperate, $\phi_i(t) > \phi^{\min} = \alpha$, no longer holds (2). To reach the maximum cumulative cooperation effort, an agent i must act solely as a recipient of information and all its $(n-1)$ peers act as providers of information; agent i's peers descend to the bottom layer of the diagram and reach the minimum cumulative cooperation effort.

Theorem 2. *The cumulative cooperation effort of an agent i, $\phi_i(t)$, has a maximum of ϕ^{max} and minimum of ϕ^{min}, where*

$$\phi^{max} = (1-n)\alpha \ , \tag{12}$$
$$\phi^{min} = \alpha \ , \tag{13}$$

for an n-agent game and a common threshold value of α.

In other words, there is an upper limit to the amount of cooperation an agent i is forced to provide by following the threshold strategy $s_i(t)$, which is proportional to the number of agents n in the game and the threshold value α. If agent i's cooperation effort reaches the maximum value, $\phi_i(t) = \phi^{max}$, i will cooperate, at most, $\phi^{max} - \phi^{min} = -n\alpha$ times, for $\phi_i(t) = 0$, i will cooperate, at most, $-\alpha$ times, and if $\phi_i(t) = \phi^{min}$, i will defect.

Proof. Consider an n-agent game and let $\mathcal{G}$ be the group of agents consisting of all but agent p_1. Hence, the cardinality of $\mathcal{G}$ is given by: $|\mathcal{G}| = n - 1$. The cumulative cooperation effort of all agents is initially equal to zero. Each time t the regional data in p_1's cache expires, an agent p, satisfying $\{p \in \mathcal{G} | \phi_p(t) > \alpha\}$, contacts the LBS server and shares the retrieved data with p_1, upon p_1's request. This benevolent sharing activity ceases at time t' when all agents $p \in \mathcal{G}$ have a cumulative cooperation effort of: $\phi_p(t') = \alpha$. If there was an agent $p^* \in \mathcal{G}$ such that $\phi_{p^*}(t') = \alpha + 1 > \alpha$, then p^* could share data with p_1 once more before defecting. Each agent in $\mathcal{G}$ can, thus, share at most $-\alpha$ data items with p_1 before defecting. Hence, p_1's cumulative cooperation effort reaches a maximum at time t', given by:

$$\phi_{p_1}(t') = |\mathcal{G}|(-\alpha) = (1-n)\alpha \ ,$$

and those of $p \in \mathcal{G}$ reach a minimum of

$$\phi_p(t') = \alpha \ . \qquad \Box$$

The arrows in the state chart denote the rates at which agents switch between states, where r_{con} refers to the contact rate between game agents in a region, $1/r_{\mathrm{srv}}$ is the average waiting time before contacting the LBS server, $1/r_{\mathrm{inf}}$ is the average information lifetime, and finally, r_{req} is the rate at which MobiCrowd users request information. Here $A(t)$ is proportional to the fraction of informed agents with a cumulative cooperation effort above the minimum (ϕ^{min}), i.e. the pool of informed agents that can cooperate, and $B(t)$ is proportional to the fraction of seekers with a cumulative cooperation effort below the maximal value (ϕ^{max}), i.e. the group of seekers capable of querying informed peers. In Table 2 we define the payoffs associated with each state transition.

Table 2. Transition Rates and Payoffs

Transition	Pre-condition	Rate	Utility
1) $\mathcal{K}_\phi \rightarrow \mathcal{I}_{\phi+1}$	$\phi < \phi^{\mathrm{max}}$	$A(t) \cdot K_\phi(t)$	b_{inf}
2) $\mathcal{K}_\phi \rightarrow \mathcal{I}_\phi$	–	$r_{\mathrm{srv}} \cdot K_\phi(t)$	$b_{\mathrm{inf}} - c_{\mathrm{srv}}$
3) $\mathcal{I}_{\phi+1} \rightarrow \mathcal{I}_\phi$	$\phi < \phi^{\mathrm{max}}$	$B(t) \cdot I_{\phi+1}(t)$	$-c_{\mathrm{com}}$
4) $\mathcal{I}_\phi \rightarrow \mathcal{R}_\phi$	–	$r_{\mathrm{inf}} \cdot I_\phi(t)$	0
5) $\mathcal{R}_\phi \rightarrow \mathcal{K}_\phi$	–	$r_{\mathrm{req}} \cdot R_\phi(t)$	0

where:

$$A(t) = r_{\mathrm{con}} \sum_{l=\phi^{\mathrm{min}}+1}^{\phi^{\mathrm{max}}} I_l(t) \text{ and } B(t) = r_{\mathrm{con}} \sum_{l=\phi^{\mathrm{min}}}^{\phi^{\mathrm{max}}-1} K_l(t) \ .$$

When a seeker successfully queries an informed agent, it becomes informed, improves its cumulative cooperation effort by 1 and obtains a benefit of b_{inf} as defined by Transition 1. Note that only informed agents with a cumulative cooperation effort greater than ϕ^{min} (or α) will cooperate with seekers. By following Transition 2, seekers can become informed by connecting to the LBS instead of communicating with a peer. However, in this case, they receive a payoff of just $b_{\mathrm{inf}} - c_{\mathrm{srv}}$ and their cumulative cooperation efforts remain constant. When a seeker i reaches the highest cumulative cooperation effort, $\phi_i(t) = \phi^{\mathrm{max}}$, it can only become informed through the LBS server, by following Transition 2, as there is no other agent j available such that $\phi_j(t) > \alpha$ (see proof of Theorem 2). Each time informed agents share data with a seeker, their cumulative cooperation effort drops by 1 unit and they incur a communication cost of c_{com}, according to Transition 3. Finally, the remaining two transitions, 4 and 5, define the rates at which agents switch from the informed to the removed state and from the removed to the seeker state, respectively.

As we define the transition rates between agent states for the n-agent game, we ensure that the total information shared is equal to the total amount of information received by seekers. Even though agents are constantly updating the regional data stored in their mobile proxies, this property must still hold because each fresh data item shared by an informed agent must be received by a seeker. This property implies that there is no loss of information during the data sharing process between informed agents and seekers; we assume lost packets are retransmitted. This concept is expressed formally in the following theorem.

Theorem 3. *The total amount of information shared by informed agents must equal the total amount of information received by seekers, assuming the information sharing process incurs no loss of data.*

Proof. The total rate at which seekers receive data from informed agents equals

$$\text{rc}_{\text{total}}(\alpha, t) = \sum_{\phi=\alpha}^{(1-n)\alpha - 1} K_\phi(t) \cdot r_{\text{con}} \left(I_{\alpha+1}(t) + \ldots + I_{(1-n)\alpha}(t) \right) . \quad (14)$$

A seeker agent i at state $\mathcal{K}_{\phi\text{max}}$ cannot receive data from an informed agent because state $\mathcal{K}_{\phi\text{max}}$ is only reached when all other agents j have the lowest possible cumulative cooperation effort, $\forall_j \phi_j(t) = \alpha$, as explained in the proof of Theorem 2, hence any informed agent $j \in \mathcal{I}_{\phi\text{min}}$ following threshold strategy $s_j(t)$ will defect.

Similarly, the total rate at which informed agents share data with seekers is given by the following:

$$\text{tr}_{\text{total}}(\alpha, t) = \sum_{\phi=\alpha+1}^{(1-n)\alpha} I_\phi(t) \cdot r_{\text{con}} \left(K_\alpha(t) + \ldots + K_{(1-n)\alpha-1}(t) \right) . \quad (15)$$

Expanding $\text{rc}_{\text{total}}(\alpha, t)$ gives

$$\text{rc}_{\text{total}}(\alpha, t) = K_\alpha(t) \cdot r_{\text{con}} \left(I_{\alpha+1}(t) + \ldots + I_{(1-n)\alpha}(t) \right) + \ldots$$
$$+ K_{(1-n)\alpha-1}(t) \cdot r_{\text{con}} \left(I_{\alpha+1}(t) + \ldots + I_{(1-n)\alpha}(t) \right) . \quad (16)$$

By factorizing (16) in terms of $I_\phi(t)$, for $\alpha < \phi \leq (1-n)\alpha$, we obtain

$$I_{\alpha+1}(t) \cdot r_{\text{con}} \left(K_\alpha(t) + \ldots + K_{(1-n)\alpha-1}(t) \right) + \ldots$$
$$+ I_{(1-n)\alpha}(t) \cdot r_{\text{con}} \left(K_\alpha(t) + \ldots + K_{(1-n)\alpha-1}(t) \right) = \text{tr}_{\text{total}}(\alpha, t) . \quad (17)$$

As both rates are equal, $\text{rc}_{\text{total}}(\alpha, t) = \text{tr}_{\text{total}}(\alpha, t)$, the total amount of information shared must equal the total amount of information received:

$$\int_0^\infty \text{rc}_{\text{total}}(\alpha, t) \, \mathrm{d}t = \int_0^\infty \text{tr}_{\text{total}}(\alpha, t) \, \mathrm{d}t . \quad (18)$$

□

To calculate the total discounted game payoff, we map each transition to a utility value (see Table 2). We assign the highest payoff of b_{inf} to transition $\mathcal{K}_\phi \to \mathcal{I}_{\phi+1}$, through which seekers acquire regional data from an informed peer. Seekers earn a lesser payoff of $b_{\text{inf}} - c_{\text{srv}}$ per query when they acquire data from the LBS using transition $\mathcal{K}_\phi \to \mathcal{I}_\phi$ due to the privacy loss incurred. Informed agents suffer a penalty of c_{com} whenever they share data with seekers using transition $\mathcal{I}_{\phi+1} \to \mathcal{I}_\phi$. By multiplying the transition rates and the transition utility values we obtain the game's payoff rate $u(\alpha, t)$ (see Appendix A). Applying the discount factor δ to the payoff rate $u(\alpha, t)$ and integrating the result over the whole duration of the game gives the total game discounted reward; in other words, the payoff accumulated by the whole agent population:

$$U_{\text{total}}(\alpha) = \int_0^\infty \delta^t u(\alpha, t)\, \mathrm{d}t \ .$$

(19)

For a finite population of agents n, the expected total discounted payoff of each agent is calculated simply as

$$U(\alpha) = \frac{1}{n} \cdot U_{\text{total}}(\alpha) \ .$$

(20)

6   Multiple-agent Game Analysis

The quantity of agents in each state is controlled by a system of non-linear differential equations (see Appendix B), derived from the agent state-chart in Figure 2. Ideally, the exact expressions for $K_\phi(t)$, $I_\phi(t)$, and $R_\phi(t)$ could be found by solving the system of equations. Unfortunately, the basic W. Kermack and A. McKendrick SIR model, on which our work relies, cannot be solved analytically [6][1]. The additional complexity of our own system of differential equations only lessens the chances of finding an analytical solution. However, we can still solve the system of differential equations by using numerical methods [6]. We resort to a numerical ODE solver in Mathematica to compute functions $K_\phi(t)$, $I_\phi(t)$, and $R_\phi(t)$, to analyze the game's evolution and its steady-state equilibrium.

Our first concern in the game analysis is to study the evolution of the rates at which (i) seekers become informed by contacting informed peers ($\mathcal{K}_\phi \rightarrow \mathcal{I}_{\phi+1}$), (ii) informed agents share data with seekers ($\mathcal{I}_{\phi+1} \rightarrow \mathcal{I}_\phi$), and (iii) the rate at which seekers become informed by contacting the LBS ($\mathcal{K}_\phi \rightarrow \mathcal{I}_\phi$). We do this because these are the only three transitions that have utilities different from zero (see Table 2). From the plots in Figure 3, it is clear that the rate at which informed agents share data equals the rate at which seekers obtain data from informed peers (plots of $\mathcal{K}_\phi \rightarrow \mathcal{I}_{\phi+1}$ overlap those of $\mathcal{I}_{\phi+1} \rightarrow \mathcal{I}_\phi$ in 3a and 3d). Varying the contact rate, r_{con}, average information lifetime, $1/r_{\text{inf}}$, average time before calling the server, $1/r_{\text{srv}}$, and the request rate, r_{req}, affects the proportion of agents occupying each of the three roles (i.e. seeker, informed, and removed) at equilibrium. By balancing the transition rates (Figure 3d), it is possible to achieve a uniform proportion of agents in each of the three roles (Figure 3e).

Decreasing the common threshold α increases the fraction of informed agents willing to cooperate, i.e. all informed agents i whose cumulative cooperation effort is above the threshold $\phi_i(t) > \phi^{\min} = \alpha$. Consequently, seekers begin querying informed peers ($\mathcal{K}_\phi \rightarrow \mathcal{I}_{\phi+1}$) more often than the LBS ($\mathcal{K}_\phi \rightarrow \mathcal{I}_\phi$), thus raising their payoffs from $b_{\text{inf}} - c_{\text{srv}}$ to b_{inf} per query. The total discounted payoff stabilizes when the cumulative cooperation effort of most informed agents is above the threshold and the rate of transition $\mathcal{K}_\phi \rightarrow \mathcal{I}_{\phi+1}$ no longer increases.

---

[1] See Section 1.5 of [6].

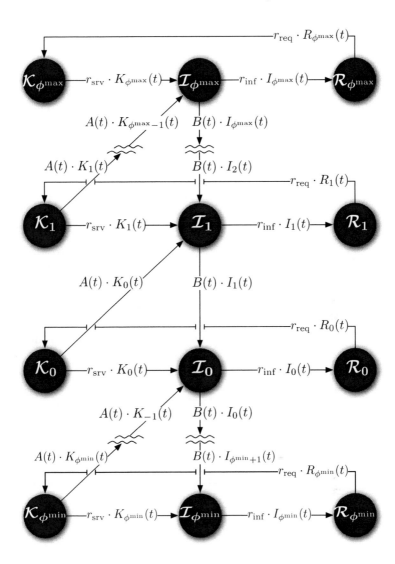

Fig. 2. State diagram of the n-agent game. Initially all agents have a null cumulative cooperation effort and are spread out across the three states: $\mathcal{K}_0$, $\mathcal{I}_0$, and $\mathcal{R}_0$. A seeker at state $\mathcal{K}_0$ can become informed by contacting an informed agent at states $\mathcal{I}_{\phi^{\min}+1}, \dots, \mathcal{I}_0, \dots, \mathcal{I}_{\phi^{\max}}$ and switch to state $\mathcal{I}_1$. If this fails, the same seeker can obtain regional data through the LBS, maintaining its cumulative cooperation effort, and switch to state $\mathcal{I}_0$. Informed agents at states $\mathcal{I}_{\phi^{\min}+1}, \dots, \mathcal{I}_0, \dots, \mathcal{I}_{\phi^{\max}}$ can share data with seekers, in which case their cumulative cooperation effort drops by one unit, and they fall to a lower level in the diagram. An informed agent at state $\mathcal{I}_{\phi^{\min}}$ does not cooperate with seekers and seekers at state $\mathcal{K}_{\phi^{\max}}$ can only become informed via the LBS. Once the information expires, informed agents become removed and can later become seekers. The transitions represent the rates at which agents change state.

This phenomenon is illustrated in Figures 3c and 3f, where the total discounted payoff function $U_{\text{total}}(\alpha)$ is almost level for $\alpha \leq -4$ in 3c and $\alpha \leq -5$ in 3f.

We proceed to estimate the distribution of agents for the whole range of cumulative cooperation values, i.e. $\phi \in [\phi^{\min}, \phi^{\max}] = [\alpha, (1-n)\alpha]$, assuming a uniform distribution of agents for the three possible roles at the game's steady state equilibrium. As we represent the population of agents in each state as a fraction of unity and given that agents can only be in a single state at a time, the sum $K_\phi(t) + I_\phi(t) + R_\phi(t)$ can be interpreted as the probability that a given agent i has a cumulative cooperation effort equal to ϕ at time t:

$$\Pr\{\exists_i : \phi_i(t) = \phi\} = K_\phi(t) + I_\phi(t) + R_\phi(t) \ . \tag{21}$$

For a large population of agents $(n >> 2)$ and at the game's steady-state equilibrium $(t \to \infty$ or $t_\infty)$, we estimate this probability (21) as

$$\Pr\{\exists_i : \phi_i(t_\infty) = \phi\} \approx \frac{1}{1-\alpha} \cdot \frac{1}{(1-1/\alpha)^{\phi-\alpha}} \ , \tag{22}$$

where $\sum_{\phi=\alpha}^{(1-n)\alpha} \frac{1}{1-\alpha} \cdot \frac{1}{(1-1/\alpha)^{\phi-\alpha}} \to 1$, for $\alpha \leq -1$ and $n \to \infty$. In Figure 4 we plot (21) and (22) for a population of $n = 40$ agents and a threshold of $\alpha = -2$. The estimate (22) is particularly relevant to characterize agents' rational behavior at equilibrium, which allows them to quantify the rates of all transitions with utilities different from zero, and to be able to choose the optimal threshold α^{opt} to maximize their payoff.

Assuming we reach a uniform distribution of agents occupying each of the three roles at the game's steady-state equilibrium, as with the case illustrated in Figure 3e, we can estimate the probability that a given agent i is in, for example, state $\mathcal{K}_\phi$ as: $1/3 \cdot \Pr\{\exists_i : \phi_i(t_\infty) = \phi\}$ (the same holds for $\mathcal{I}_\phi$ and $\mathcal{R}_\phi$). This enables us to estimate the total rate at which information is shared by informed agents, $\widehat{\text{tr}}_{\text{total}}(\alpha) \approx \text{tr}_{\text{total}}(\alpha, t_\infty)$, received by seekers contacting informed peers, $\widehat{\text{rc}}_{\text{total}}(\alpha) \approx \text{rc}_{\text{total}}(\alpha, t_\infty)$, and received by seekers contacting the LBS, $\widehat{\text{sv}}_{\text{total}}(\alpha)$, at the game's steady-state equilibrium, as shown below:

$$\widehat{\text{tr}}(\alpha) = \frac{r_{\text{con}}}{3^2} \cdot \frac{\alpha}{\alpha-1} \left[\left(\frac{\alpha-1}{\alpha} \right)^{n\alpha} - 1 \right]^2 \ , \tag{23}$$

$$\widehat{\text{rc}}(\alpha) = \widehat{\text{tr}}(\alpha) \ , \tag{24}$$

$$\widehat{\text{sv}}(\alpha) = \frac{r_{\text{srv}}}{3} \left[1 - \left(\frac{\alpha-1}{\alpha} \right)^{n\alpha-1} \right] \ . \tag{25}$$

Using (23)-(25) we estimate the game's total payoff at equilibrium as

$$\widehat{U}_{\text{total}}(\alpha) = b_{\text{inf}} \cdot \widehat{\text{rc}}(\alpha) - c_{\text{com}} \cdot \widehat{\text{tr}}(\alpha) + (b_{\text{inf}} - c_{\text{srv}}) \cdot \widehat{\text{sv}}(\alpha) \tag{26}$$

and an agent i's total payoff at equilibrium as

$$\widehat{U}(\alpha) = \frac{1}{n} \cdot \widehat{U}_{\text{total}}(\alpha) \ , \tag{27}$$

for a large, but finite agent population of size n $(n >> 2)$.

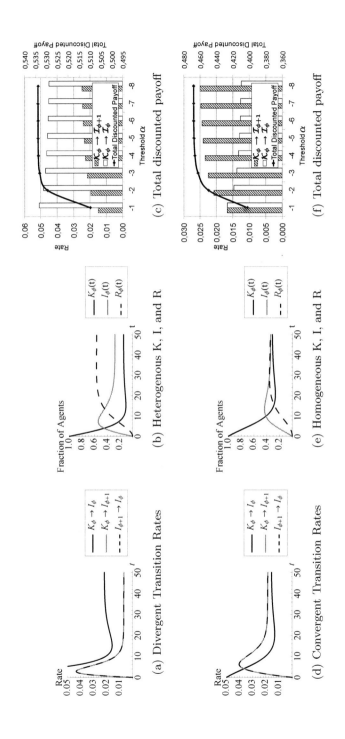

Fig. 3. Transition rates, distribution of seeker (K), informed (I), and removed (R) agents, and the total discounted payoff at the game's steady-state equilibrium, for a population of $n = 40$ agents, of which all are initially seekers ($K_0(0) = 1$), a discount factor of $\delta = 0.9$, and the following benefits and costs: $b_{inf} = 1$, $c_{srv} = 0.4$, and $c_{com} = 0.1$. Plots (a), (b), and (c) use: $r_{con} = 0.25$, $r_{inf} = 0.1$, $r_{req} = 0.05$, and $r_{srv} = 0.15$. Plots (d), (e), and (f) use: $r_{con} = 0.3$, $r_{inf} = 0.1$, $r_{req} = 0.1$, and $r_{srv} = 0.05$. With divergent transition rates and for $\alpha = -1$ (a) we obtain a heterogeneous distribution of K, I, and R populations at equilibrium (b). With convergent transition rates and for $\alpha = -1$ (d), we obtain a homogeneous distribution of K, I, and R populations at equilibrium (e). The total discounted payoff function $U_{total}(\alpha)$ stabilizes when almost all informed agents have a cumulative cooperation effort superior to the threshold α; in (c) this occurs at $\alpha \leq -4$ and in (f) at $\alpha \leq -5$.

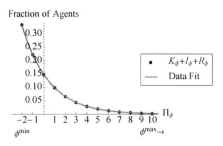

Fig. 4. Distribution of the cumulative cooperation effort. The plot shows the fraction of agents with a specific cumulative cooperation effort, for $\Pi_\phi = [\phi^{\min}, \phi^{\max}]$, assuming we have $1/3$ of each type of agent at the game's steady-state equilibrium.

Agents acting to maximize their payoffs will benefit by decreasing their threshold. A lower threshold implies that more agents are willing to cooperate when informed. Seekers can obtain regional data from this larger pool of cooperative informed agents, instead of using the LBS and gathering a higher payoff. Clearly, the benefit of acquiring data from an informed peer and sharing it with another agent is greater than the payoff achieved when connecting to the LBS: $b_{\inf} - c_{\mathrm{com}} > b_{\inf} - c_{\mathrm{srv}}$, provided $0 \le c_{\mathrm{com}} < c_{\mathrm{srv}} < b_{\inf} \le 1$. In order to consider an optimal threshold (α^{opt}), as illustrated in Figures 3c and 3f, we define a tolerance $\epsilon > 0$ such that an agent i will not ponder reducing α to maximize its payoff if the added benefit is less than ϵ:

$$\alpha^{\mathrm{opt}} = \max\{\alpha \le -1 : \widehat{U}(\alpha - 1) - \widehat{U}(\alpha) < \epsilon\} \ . \tag{28}$$

7   Conclusion

The MobiCrowd system architecture introduces only minor changes to the way traditional location-based schemes operate, thus enabling users to obtain geographic data from other peers and to potentially improve their location privacy by reducing the amount of queries that participants send to the LBS server [14]. However, the improvement in privacy is only possible if users are willing to cooperate. The goal of our work is to learn if users adopting the MobiCrowd system would be willing to collaborate by sharing regional data between themselves in order to avoid the privacy threat of connecting directly to the LBS provider. To study this problem we first developed a game-theoretic framework, from which we defined two infinitely repeated games of imperfect information.

In the first game, we model two agents that are chosen to query the LBS with equal probability. We derived two Nash equilibria in this game, one favors mutual cooperation that is Pareto optimal and the second favors mutual defection. In the second game, by modifying the original MobiCrowd SIR model [14] to support agent rational behavior using threshold strategies we represent an interaction between multiple agents confined to a single region. Due to the complexity of modeling peer-to-peer interactions in this case, we use transition rates to define the pace at which agents change state over time and we assign payoffs to each transition. From the analysis of this game, we derive the optimal threshold that maximizes an agent's expected payoff, for a large population of agents, assuming a uniform distribution of agents at the steady-state equilibrium.

Our results show that rational agents attempting to maximize their payoffs will benefit by sharing data with their peers, both in the controlled environment of the two-agent game and in the more realistic n-agent scenario. As a future development of this work, we plan to analyze the same contextual-data sharing problem considering threshold strategies with an independent (per agent) threshold, experiment with different types of reactive strategies, and introduce statistical processes describing the rate at which agents enter and leave a region.

References

1. Alpcan, T.: Network Security: A Decision and Game-Theoretic Approach. Cambridge University Press (2010)
2. Başar, T., Olsder, G.: Dynamic noncooperative game theory. Society for Industrial Mathematics (1999)
3. Bauch, C., Earn, D.: Vaccination and the theory of games. Proceedings of the National Academy of Sciences of the United States of America 101(36), 13391 (2004)
4. Buttyan, L., Hubaux, J.P.: Security and Cooperation in Wireless Networks: Thwarting Malicious and Selfish Behavior in the Age of Ubiquitous Computing. Cambridge University Press, New York (2007)
5. Dhar, S., Varshney, U.: Challenges and business models for mobile location-based services and advertising. Commun. ACM 54, 121–128 (2011), http://doi.acm.org/10.1145/1941487.1941515
6. Earn, D.: A Light Introduction to Modelling Recurrent Epidemics. In: Brauer, F., van den Driessche, P., Wu, J. (eds.) Mathematical Epidemiology. Lecture Notes in Mathematics, vol. 1945, pp. 3–17. Springer, Heidelberg (2008)
7. Felegyhazi, M., Hubaux, J., Buttyan, L.: Nash equilibria of packet forwarding strategies in wireless ad hoc networks. IEEE Transactions on Mobile Computing 5(5), 463–476 (2006)
8. Freudiger, J., Manshaei, M.H., Hubaux, J.P., Parkes, D.C.: On Non-cooperative Location Privacy: A Game-theoretic Analysis. In: ACM Conference on Computer and Communications Security (CCS) (2009), http://www.sigsac.org/ccs/CCS2009/
9. Humbert, M., Manshaei, M.H., Freudiger, J., Hubaux, J.-P.: Tracking Games in Mobile Networks. In: Alpcan, T., Buttyán, L., Baras, J.S. (eds.) GameSec 2010. LNCS, vol. 6442, pp. 38–57. Springer, Heidelberg (2010)
10. Krumm, J.: A survey of computational location privacy. Personal and Ubiquitous Computing 13(6), 391–399 (2009)
11. Manshaei, M., Zhu, Q., Alpcan, T., Basar, T., Hubaux, J.P.: Game Theory Meets Network Security and Privacy. Tech. rep., École Polytechnique Fédérale de Lausanne, Lausanne (2010)

12. Olumofin, F., Tysowski, P.K., Goldberg, I., Hengartner, U.: Achieving Efficient Query Privacy for Location Based Services. In: Atallah, M.J., Hopper, N.J. (eds.) PETS 2010. LNCS, vol. 6205, pp. 93–110. Springer, Heidelberg (2010)
13. Osborne, M., Rubinstein, A.: A Course in Game Theory. The MIT press (1994)
14. Shokri, R., Papadimitratos, P., Theodorakopoulos, G., Hubaux, J.P.: Collaborative Location Privacy. In: 2011 IEEE 8th International Conference on Mobile Adhoc and Sensor Systems, MASS (2011)
15. Theodorakopoulos, G., Le Boudec, J.Y., Baras, J.S.: Selfish Response to Epidemic Propagation. IEEE Transactions on Automatic Control (2010)

Appendix A   Total Discounted Payoff

Using the transitions, rates and payoffs (defined in Table 2), we can calculate the total discounted game payoff by integrating over the duration of the game the product of the payoff rate, $u(\alpha, t)$, and the discount factor, $\delta \in]0, 1[$. The payoff rate, $u(\alpha, t)$, is defined as the difference between the benefit rate $b(\alpha, t)$, and the cost rate, $c(\alpha, t)$:

$$u(\alpha, t) = b(\alpha, t) - c(\alpha, t) \ , \qquad (29)$$

where

$$b(\alpha, t) = b_{\inf} \sum_{l=\alpha}^{(1-n)\alpha-1} A(t) \cdot K_l(t) + (b_{\inf} - c_{\mathrm{srv}}) \sum_{l=\alpha}^{(1-n)\alpha} r_{\mathrm{srv}} \cdot K_l(t) \ , \qquad (30)$$

and

$$c(\alpha, t) = c_{\mathrm{com}} \sum_{l=\alpha+1}^{(1-n)\alpha} B(t) \cdot K_l(t) \ . \qquad (31)$$

We now apply Definition 2 to calculate the total discounted reward as a function of the threshold α:

$$U_{\mathrm{total}}(\alpha) = \int_0^\infty \delta^t u(\alpha, t) \, \mathrm{d}t \ . \qquad (32)$$

Appendix B   Game Dynamics

The system of non-linear differential equations governing the fraction of agents in each state, $K_\phi(t)$, $I_\phi(t)$, $R_\phi(t)$, for $\phi \in [\phi^{\min}, \phi^{\max}]$, is

$$K'_{\phi^{\max}}(t) = r_{\mathrm{req}} \cdot R_{\phi^{\max}}(t) - r_{\mathrm{srv}} \cdot K_{\phi^{\max}}(t) \ , \tag{33}$$

$$\begin{aligned} I'_{\phi^{\max}}(t) = {} & A(t) \cdot K_{\phi^{\max}-1}(t) + r_{\mathrm{srv}} \cdot K_{\phi^{\max}}(t) \\ & - (B(t) + r_{\mathrm{inf}}) I_{\phi^{\max}}(t) \ , \end{aligned} \tag{34}$$

$$R'_{\phi^{\max}}(t) = r_{\mathrm{inf}} \cdot I_{\phi^{\max}}(t) - r_{\mathrm{req}} \cdot R_{\phi^{\max}}(t) \ , \tag{35}$$

$$K'_{\phi^{\min}<\phi<\phi^{\max}}(t) = r_{\mathrm{req}} \cdot R_\phi(t) - (A(t) + r_{\mathrm{srv}}) K_\phi(t) \ , \tag{36}$$

$$\begin{aligned} I'_{\phi^{\min}<\phi<\phi^{\max}}(t) = {} & A(t) \cdot K_{\phi-1}(t) + r_{\mathrm{srv}} \cdot K_\phi(t) - (B(t) + r_{\mathrm{inf}}) I_\phi(t) \\ & + B(t) \cdot I_{\phi+1}(t) \ , \end{aligned} \tag{37}$$

$$R'_{\phi^{\min}<\phi<\phi^{\max}}(t) = r_{\mathrm{inf}} \cdot I_\phi(t) - r_{\mathrm{req}} \cdot R_\phi(t) \ , \tag{38}$$

$$K'_{\phi^{\min}}(t) = r_{\mathrm{req}} \cdot R_{\phi^{\min}}(t) - (A(t) + r_{\mathrm{srv}}) K_{\phi^{\min}}(t) \ , \tag{39}$$

$$I'_{\phi^{\min}}(t) = r_{\mathrm{srv}} \cdot K_{\phi^{\min}}(t) + B(t) \cdot I_{\phi^{\min}+1}(t) - r_{\mathrm{inf}} \cdot I_{\phi^{\min}}(t) \ , \tag{40}$$

$$R'_{\phi^{\min}}(t) = r_{\mathrm{inf}} \cdot I_{\phi^{\min}}(t) - r_{\mathrm{req}} \cdot R_{\phi^{\min}}(t) \ . \tag{41}$$

As agents can only be in a single state at a time, the following relationship must also hold:

$$\sum_{l=\phi^{\min}}^{\phi^{\max}} K_l(t) + \sum_{l=\phi^{\min}}^{\phi^{\max}} I_l(t) + \sum_{l=\phi^{\min}}^{\phi^{\max}} R_l(t) = 1 \ . \tag{42}$$

Digital Trust Games: An Experimental Study[*]

Tansu Alpcan[1], Albert Levi[2], and Erkay Savaş[2]

[1] Dept. of Electrical and Electronic Engineering,
The University of Melbourne, Australia
tansualpcan@gmail.com
http://www.tansu.alpcan.org
[2] Sabanci University, Istanbul, Turkey
{levi,erkays}@sabanciuniv.edu

Abstract. An experimental study of the digital trust game in [2] is presented. The study consists of an initial survey followed by a four-part dynamic experiment investigating various aspects of digital trust decisions. Digital trust in online environments differs from its offline variants due to its unique characteristics such as near instantaneous communication, transient and impersonal nature of interactions, immediate access to opinions of others, and availability of high amount of (but often low quality) information. It is observed that while the game theory provides a suitable analytical framework for quantitative analysis of digital trust decisions, the model in [2] has its shortcomings. Firstly, the subjects do not seem to adopt an iterative best or gradient response strategy. They exhibit significant (mental) inertia and only respond to new information or significant situation changes. Secondly, they take into account signals from their social circle much more than aggregate signals such as average scores. Both of these results and additional insights gained have important implications for future game theoretic modeling efforts of digital trust.

Keywords: Digital trust, decision making, game theory, social influences, survey, dynamic experiment.

1 Introduction

Trust relationships in online environments differ from those in offline media due to fundamental differences between the two. Traditionally, trust is established through personal interactions and their history, based on which the involved parties make their trust decisions. The situation is quite different in an **online medium**. However, *digital trust* and reputation still play a fundamental role in many types of online interactions ranging from communication and entertainment services to e-commerce [17, 5]. This new paradigm of digital trust has naturally attracted the attention of the research community as a result of its increasing importance [10, 1].

Trust and related interactions, whether they occur in digital or offline environments, naturally involve **multi-person decision making**. Therefore, it is not surprising to see

---

[*] This work is supported by Deutsche Telekom Laboratories. T. Alpcan was with Technical Univ. of Berlin and Deutsche Telekom Laboratories during this research.

J.S. Baras, J. Katz, and E. Altman (Eds.): GameSec 2011, LNCS 7037, pp. 182–200, 2011.

that the topic has been studied by the **game theory** community [7, 8, 3, 11] often with a focus on economical aspects or offline repeated games. While some these game models are also (partially) applicable to digital trust in online environments, the change of medium from offline to online necessitates a fresh approach due to fundamental differences in their characteristics [14]. Near instantaneous communication, transient and impersonal nature of interactions, immediate access to opinions of others, availability of high amount of (but often low quality) information, online social networking effects are among the properties of online media that distinguishes digital trust from its offline versions [6, 16, 15].

This experimental study builds upon the earlier game theoretic model introduced in [2], which focuses on community effects and other factors in digital trust decisions. Quantitative analysis of digital trust decisions and factors influencing them are at the center of the model considered. Thus, it differs from a recent work [12] where no analytical modeling effort is present. The developed model in [2] constitutes a starting point for a quantitative and analytical approach to digital trust from a decision making perspective. However, that model has only been analyzed numerically until this paper. **The goal of this paper** is to take the next natural step in the scientific process: an experimental study of the digital trust game model. Although the experiments conducted have limitations and are of modest scale, they still provide invaluable insights to where the model successfully captures reality and at which points the inherent modeling assumptions fail.

Experimental studies investigating the gap between game theoretic models and real human behavior have a long history. The book "behavioral game theory" [4] covers the field extensively reporting the results of numerous past experiments. The experimental results presented in this paper support a conclusion of that book stating that game theoretic models capture some aspects of the human decision making processes surprisingly well while failing to match reality in others.

The main contributions of this work are summarized as follows:

– Experimental study of a digital trust game model which provides valuable insights to how well the model captures real trust decisions.
– Presentation of experimental results indicating
  • actual decisions of subjects do not necessarily follow gradient or best response dynamics in parallel update schemes that are often used in game models.
  • many social factors such as peer-pressure or influence, social networking effects, and anonymity can be captured using appropriate and tractable cost structures in digital trust games.
– A discussion various interpretations of the results obtained and potential evolution of the model in light of the observations.

The next section gives an overview of the digital game model that inspired the survey and experiments for completeness. Section 3 presents the experiment setup which is followed by the results in Section 4. Interpretation and discussion of the results obtained are in Section 5. The paper ends with concluding remarks of Section 6. All of the survey and experiment results are in Appendix A. Appendix B gives a brief overview of the web-based software used along with screenshots.

2   Digital Trust Game Model

In this section the digital trust game model of [2] is summarized for completeness as it constitutes the basis of the experimental study in this paper. For further details on analysis and simulations we refer to that paper.

Consider the set of N users or agents, $\mathcal{A}$, in a standard online environment. A *digital trust game* is played among the users of the set $\mathcal{A}$, who make trust decisions on, for example, an e-commerce website or seller. Each user i has an initial trust score (level), $e_i \in \mathbb{R}$, of the website. Using this initial score e_i as a starting point, the user decides on her/his trust level, $x_i \in \mathbb{R}$, of the seller after exchanging information with the rest of the community. The individual opinion or trust score, x_i, is influenced by various community effects as well as the properties of the user. The opinions of all the users represented by the vector

$$\mathbf{x} = [x_1, \ldots, x_N] \in \mathcal{X} \subset \mathbb{R}^N$$

define the decision space of the digital trust game.

In the game, $x_i = 0$ corresponds to a neutral or default opinion of user i on the seller. Consequently, the positive values, $x_i > 0$ represent a positive opinion and negative ones, $x_i < 0$, a negative opinion. The users' opinions are not only a function of the initial reputation and image but also of factors capturing community influences. The decision process of a user i can be modeled by the minimization of a well-defined cost function that quantifies the factors affecting the opinion of the agent. As one possibility, the following cost function is adopted here:

$$J_i(x_i, \mathbf{x}_{-i}) := \frac{\alpha_i}{2} x_i^2 + \frac{\beta_i}{2} \left(x_i - \frac{1}{N-1} \sum_{j \neq i} x_j \right)^2 + \frac{\gamma_i}{2} (x_i - e_i)^2, \qquad (1)$$

where $0 \leq \alpha_i, \beta_i, \gamma_i \leq 1, \alpha_i + \beta_i + \gamma_i = 1 \; \forall i$, and $\mathbf{x}_{-i} := [x_1, \ldots, x_{i-1}, x_{i+1}, \ldots, x_N]$. It is naturally possible to consider different types of cost functions. This particular one is chosen for its nice analytical properties as a first order approximation.

The **first term**, $\alpha_i x_i^2$, in the cost function (1) quantifies the timidness of user i, especially in the case when the trust decisions are publicized. The term quadratically penalizes any positive or negative opinion of the user forcing it to the neutral or zero opinion. Users with different properties can be represented by choosing the weighting parameter α appropriately. A *timid* user, who is reluctant to pass judgment, is expected to have a high α whereas a *self-assertive* or opinionated one is captured by a small α parameter value. The **second term** in the cost function quantifies the influence of *peer pressure* on the user. Here, peer pressure is modeled using a quadratic cost on any opinion deviating from the mean value of others. An individualistic or independent user is represented with a small β value. On the other hand, a user who follows the crowd is expected to have a high-valued β parameter. The **third term**, $\gamma_i(x_i - e_i)^2$, captures the effect of the initial trust e_i of the user on the final opinion x_i. A *steadfast* user who does not change its own opinion as a result of community interactions or sharing is represented by a high γ value. On the other hand, a user who updates its opinion easily has a small γ parameter in the respective cost function. Notice that the weighting parameters α, β, γ are normalized in such a way that the factors discussed above are balanced with each other. Hence, the inherent trade-offs between the factors are captured by the cost function and the game.

The set of players or users $\mathcal{A}$, the decision space $\mathcal{X}$, and the cost functions J_i $\forall i$ define together the digital trust game, $\mathcal{G}_1(\mathcal{A}, \mathcal{X}, J)$. In this noncooperative game each individual user i minimizes own cost J_i by choosing own opinion (trust decision), $x_i \in \mathbb{R}$, given the opinions (trust decisions) of others, $\mathbf{x}_{-i}$, i.e.

$$x_i = \arg \min_{x_i} J_i(x_i, \mathbf{x}_{-i}). \tag{2}$$

It has been shown in [2] that the static digital trust game $\mathcal{G}_1$ defined admits a unique Nash equilibrium solution, which can be even analytically described in the case of symmetric user costs. It has been also shown that the parallel update algorithm (PUA) summarized in Algorithm 1 and its asynchronous variant converge to this unique Nash equilibrium very fast (geometrically). In PUA, each user i updates its own opinion $x_i(t)$ together (in parallel) with all other agents at the same discrete time instances $t = 1, 2, \ldots$ according to own best response function:

$$x_i(t+1) = \frac{\beta_i}{N-1} \sum_{j \neq i} x_j(t) + \gamma_i e_i, \quad \forall i, \tag{3}$$

which is directly computed from the convex cost function (1).

Algorithm 1. Parallel Update Algorithm (PUA)

Input: Individual trust values e, convergence threshold ε.
Initialize trust values $x_i(0) = e_i$ $\forall i$ and time step $t = 0$.
while $\|x(t+1) - x(t)\| > \varepsilon$ **do**
   $t = t + 1$
   Compute $s(t) := \sum_i x_i(t)$
   for $i = 1$ to N **do**
      Compute $x_i(t+1) = \dfrac{\beta_i}{N-1} (s(t) - x_i(t)) + \gamma_i e_i.$
   end for
end while

---

3   Experiment Setup and Execution

The experiment is conducted using a Web-based program over the Internet. It consists of two parts: a standard survey and a dynamic experiment, which investigates the iterative decision process in digital trust games. Although there are adequate survey software which can be used for the survey part, there is no suitable software for conducting the dynamic experiment within the context of the model in Section 2. Hence, an all new custom software has been written using the Django Web framework[1] in Python[2] programming language. Technical information about the software is provided in Appendix B.

The survey and experiment are completed by a total of 28 volunteered participants who are senior and graduate level Computer Science students of Sabanci University on May 23, 2011. Participants have come together in a classroom environment and used their computers during the entire experiment. All stages of the experiment have been supervised by a single instructor who has acted as the administrator.

First the supervisor has asked the participants to create an online account for the experiment system. Once the accounts have been created, the supervisor has delivered an overview speech to explain what will be performed during the experiments.

After that, the supervisor has asked the volunteers to complete the survey part. It is important that participants start the second part (dynamic experiments) synchronously. Therefore, the supervisor has made an announcement in which the participants are requested to wait for each other in order to continue with the second part (dynamic experiment) after finishing the first part (survey). The supervisor has not made extra verbal explanations for the survey questions with the exception that he clarified the distinction between the questions 5-7 (anonymous evaluation) and 8-10 (non-anonymous evaluation). The completion of survey questions has taken 12 minutes for all participants.

Once the survey part is finished, the supervisor has instructed the participants about the dynamic experiments part. He has specifically mentioned that there are four different parts in this experiment and each of them has five rounds. He has also specified that the participants should act synchronously such that all participants should finish entering scores in a round in order to proceed with the next round. The reason is that the participants should see the updated average scores at the beginning of each round, except the initial round.

At the beginning of each experiment part, the scenario displayed on the screen has also been explained verbally by the administrator. Each participant enters his/her initial score in the first round of each experiment segment. As mentioned before, all of the participants should complete a particular round in order to continue with the next round. Actually the online system shows a "please wait" screen to a particular participant while other participants record their scores for a round; however, the participants have been specifically instructed by the supervisor not to close their windows, restart, or press back button of the browser. Round transitions have been performed without intervention of the supervisor since the participant screens are automatically refreshed and next round information is shown once all participants finish their entry. It has taken approximately 30 minutes to complete all four parts of the dynamic experiment.

4   Experiment Results

4.1   Survey Results

The survey consists of ten single or multiple choice questions answered by all participants before moving on to the dynamic part of the experiment. The questions and their answers in graphical (histogram) representation are shown in **Appendix A**.

---

[1] www.djangoproject.com/

[2] www.python.org

Table 1. Comparison of Answers to Q5-7 (anonymous) with Q8-10 (public disclosure)

Compared Questions	Total scores	Percentage change	Individual Changes (decrease, same, increase)
Q5 vs Q8	188 vs 180	-4.2%	7 dec, 15 same, 5 inc
Q6 vs Q9	28 vs 39	+39.3%	9 dec, 12 same, 7 inc
Q7 vs Q10	-211 vs -202	+4.3%	6 dec, 14 same, 8 inc

The first four survey questions are mainly for warm-up and to obtain insights to (the online interaction habits of) the subjects. The answers to the first and third questions indicate that the participants are quite active online both in e-commerce and social networking. They also take into account opinions of others when making decisions online as indicated by the results of question two. The answers to question four indicate that they are somewhat careful when it comes to their privacy online and do some kind of risk management. While most subjects do not hesitate to share information which they do not see as critical (e.g. E-mail or real name), half of them have never shared their address, telephone, or credit card information online.

The questions five to ten are about evaluation of an Internet (e-commerce) company and giving it a trust score between -10 and 10, meaning no trust and full trust, respectively. A zero score means neutral. In questions five, six, and seven, these scores are anonymous. The questions eight, nine, and ten are repetitions of five, six, and seven, respectively, but now the participants are told that their scores are shared publicly using their real names.

Overall, the answers to questions five to ten are consistent with the instructions given. It is observed that roughly half of the participants do not change their answers when the scores are not anonymous, i.e. they do not hesitate to openly announce their opinions. For the other half, anonymity is a consideration when sharing their trust scores with others. The changes in scoring are summarized in Table 1 below. The mixed nature of the results indicate that multiple factors may be concurrently affecting the subject behavior.

4.2 Dynamic Experiment Results

In each part of the dynamic experiment, the participants give a trust score to an online company based on a slightly different scenario. This is repeated for five rounds in each scenario/part. In each round, they see the average trust score of the previous one, and have a chance to change their own score accordingly.

In the first part, the subjects are asked to decide on a trust score for a company they trust initially and update it four more times using the average score (of all participants) in the previous round. First, they are told that their scores are shared publicly with their real name. The second part is a variation of the first but this time the scores are kept anonymous. In both cases, it is observed in Figures 12 and 13 that a significant portion of the subjects *do not change* their trust level based on the average trust level. Others do so only slightly and mostly in the first round.

The last two parts of the dynamic experiment investigate the effect of the opinion of a "good friend" on the decision of a subject. In the third part, the subjects decide on a trust score of a company they trust. At the same time, "a good friend" tells them that he gives a low trust score of -8 to the company, i.e. the company is not trustworthy. In the fourth part, the situation is reversed and they score a company they do not trust but this time the "good friend" tells them otherwise and gives the company a trust score of $+8$. In both of these parts, the subjects still see the average trust score in each round. It is observed in Figures 14 and 15 of parts three and four of the dynamic experiment, respectively, that the suggestion of a "good friend" has a significant effect on the decisions of most subjects.

5   Discussion and Interpretation

Although the scale of this experimental study is modest and participants have uniform backgrounds (Internet-savvy undergraduate and graduate engineering students), it is possible to gain some insights to the nature of digital trust decisions and applicability of the game theoretic model discussed in Section 2.

The survey results indicate that roughly half of the participants share their trust decisions or ratings openly in an e-commerce setting without hesitation. The other half exhibits, on the other hand, mixed reactions to the anonymity versus disclosure in trust decisions. Multiple factors may be in play here. One possibility is that the abstract nature of the experiment may have affected subjects, preventing a conscious evaluation of the long term effects of anonymity versus public disclosure. Another interpretation is that evaluating a company is a rather regular activity in e-commerce today (e.g. providing feedback in Amazon, E-bay, online hotel booking, etc.). Therefore, most subjects seem to have no misgivings about openly disclosing their opinions. It can be speculated that online environments empower customers in general and do not hinder flow of information among customers. This was not possible before the Internet age. It should be noted, however, in settings other than e-commerce where disclosing such trust decisions may have direct repercussions (social networks, evaluation of instructors, peer review) we may expect a different picture and a stronger anonymity trend.

Another motivation behind the last six survey questions is calibration. The goal here is to make users think about their scores before starting the dynamic experiment. The consistency of the observed scores, low for not trusted and high for trusted, show that such a calibration is useful and arguably makes the results in the dynamic segment more reliable [9].

An important result of the dynamic experiment is that users do <u>not</u> seem to adopt an *iterative best or gradient response* strategy when making trust decisions. This is in contrast to the assumed behavior in simulations of the model in [2]. A simulation result is reproduced in Figure 1 for comparison. In dynamic experiments, however, the subjects mostly either stick to their original decisions or update it often only once in the second round after seeing the average opinion of others. This may be explained with inertia and the inherent (mental) cost of making decisions. It is a well-known fact,

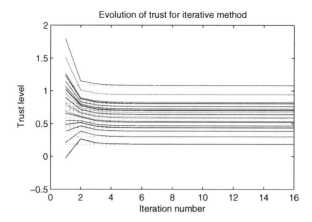

Fig. 1. Evolution of trust scores of 20 **simulated** symmetric users who update their decisions iteratively in parallel according to best response scheme in (3) (figure reproduced from [2])

for example, that customers prefer flat pricing over metered ones in order to decrease mental load.

The game theoretic model in [2] has not taken the inertia factor into account which seems to have a profound effect on actual decision making dynamics. A better model could be one where users express their preferences once and update them only if an external event such as a significant new observation or substantial change in circumstances happens. Such a state and information-based approach may capture the observed subject behavior better than the customary iterative dynamic update models.

This outcome has also consequences for mechanism design. It indicates that a mechanism designer hoping to make use of iterative user updates to infer their preferences has a much more difficult task than expected as inertia decreases variability, and hence hinders the inherent flow of information in a designed mechanism.

As a side note, iterative dynamic update models may still have a place in certain situations. For example, if the users rely on software agents acting on their behalf, then these agents may use best response and gradient algorithms to find the best solution. Even in such cases, the users may update their input to agents or preferences rarely.

Another important result of the dynamic experiment relates to the social influences. The first two parts of the dynamic experiment (Figures 12 and 13) show that participants care relatively little about anonymous and aggregate signals such as the average trust score of all others. They change their decisions only slightly if at all based on such signals. However, the last two parts of the experiment (Figures 14 and 15) show that they take any signal from their friends and trusted third parties seriously. Comparing Figures 12 and 14 clearly shows that such a signal has a profound effect on trust decision of the subjects.

This result verifies that social networks are very important and play a significant role in trust decisions. In the advertisement industry, this phenomenon is well-known and even has its name: *word of mouth advertisement*. An interesting conclusion for game theoretic models in digital decision making is that networking effects cannot be simplified to aggregate signals and have to be explicitly modeled. Again, this is missing in the current model of Section 2.

Finally, despite its shortcomings, the game theoretic approach has captured some of the fundamental aspects of digital trust decisions quite well. The chosen cost function can be easily modified to take into account the mentioned issues. "Inertia" can also be incorporated into the existing model. For example, the approach in [13] can be used, where users stick to their previous decisions with some probability. Alternatively, the probabilistic inertia can be replaced by a rule based variant.

6 Conclusions

An experimental study of the digital trust game in [2] is presented. The study consists of an initial survey followed by a four-part dynamic experiment investigating various aspects of digital trust decisions in online environments. It is observed that while the game theory provides a suitable analytical framework for quantitative analysis of digital trust decisions, the current model has its shortcomings. Firstly, the subjects do not really adopt an iterative best or gradient response strategy. They exhibit significant inertia and mainly respond to new information, e.g. suggestion of a friend, or situation changes such as public disclosure of their scores. Secondly, they take into account signals from their social circle much more than aggregate signals such as average scores. Both of these results have far-reaching implications for future game theoretic modeling efforts.

The next research step is to follow the scientific process and update the models in accordance with experimental findings. Then, they can be studied again with an extended set of experiments. Specific aspects that can be immediately incorporated into the model include networking effects as a factor influencing decisions of users and (mental) inertia in decision updates. It is worth noting that both of these issues are characteristics of digital trust and are much less emphasized in offline settings.

References

1. Alpcan, T., Başar, T.: Network Security: A Decision and Game Theoretic Approach. Cambridge University Press (2011)
2. Alpcan, T., Örencik, C., Levi, A., Savaş, E.: A game theoretic model for digital identity and trust in online communities. In: Proc. of the 5th ACM Symp. on Information, Computer and Communications Security, ASIACCS 2010, pp. 341–344. ACM, New York (2010), http://doi.acm.org/10.1145/1755688.1755735
3. Anderhub, V., Engelmann, D., Güth, W.: An experimental study of the repeated trust game with incomplete information. Journal of Economic Behavior and Organization 48(2), 197–216 (2002), http://www.sciencedirect.com/science/article/pii/S0167268101002165
4. Camerer, C.F.: Behavioral game theory: Experiments in strategic interaction. Princeton University Press (2003), http://press.princeton.edu/titles/7517.html
5. Dellarocas, C.: The digitization of word-of-mouth: Promise and challenges of online feedback mechanisms. Tech. rep., MIT Dspace, United States (2003), http://dspace.mit.edu/dspace-oai/request, http://hdl.handle.net/1721.1/1851

6. Despotovic, Z., Aberer, K.: P2p reputation management: Probabilistic estimation vs. social networks. Computer Networks 50(4), 485–500 (2006), management in Peer-to-Peer Systems, `http://www.sciencedirect.com/science/article/pii/S1389128605002161`

7. Engle-Warnick, J., Slonim, R.L.: The evolution of strategies in a repeated trust game. Journal of Economic Behavior & Organization 55(4), 553–573 (2004), trust and Trustworthiness, `http://www.sciencedirect.com/science/article/pii/S0167268104000721`

8. Engle-Warnick, J., Slonim, R.L.: Inferring repeated-game strategies from actions: evidence from trust game experiments. Economic Theory 28, 603–632 (2006), `http://dx.doi.org/10.1007/s00199-005-0633-6`, doi: 10.1007/s00199-005-0633-6

9. Hubbard, W.D.: How to measure anything finding the value of 'intangibles' in business. John Wiley & Sons, Hoboken (2007)

10. Josang, A., Ismail, R., Boyd, C.: A survey of trust and reputation systems for online service provision. Decision Support Syst. 43(2), 618–644 (2007)

11. King-Casas, B., Tomlin, D., Anen, C., Camerer, C.F., Quartz, S.R., Montague, P.R.: Getting to know you: Reputation and trust in a two-person economic exchange. Science 308(5718), 78–83 (2005), `http://www.sciencemag.org/content/308/5718/78.abstract`

12. Lorenz, J., Rauhut, H., Schweitzer, F., Helbing, D.: How social influence can undermine the wisdom of crowd effect. Proc. of the National Academy of Sciences (2011), `http://www.pnas.org/content/early/2011/05/10/1008636108.abstract`

13. Marden, J., Arslan, G., Shamma, J.: Joint strategy fictitious play with inertia for potential games. In: 44th IEEE Conf. on Decision and Control and 2005 European Control Conf. CDC-ECC 2005, pp. 6692–6697 (December 2005)

14. Mcluhan, M., Fiore, Q.: The Medium is the Massage. Gingko Press, `http://www.amazon.com/exec/obidos/redirect?tag=citeulike07-20&path=ASIN/1584230703`

15. Resnick, P., Zeckhauser, R.: Trust among strangers in Internet transactions: Empirical analysis of eBay's reputation system. In: Baye, M.R. (ed.) The Economics of the Internet and E-Commerce, Advances in Applied Microeconomics, vol. 11, pp. 127–157. Elsevier Science (2002), `http://www.si.umich.edu/~presnick/papers/ebayNBER/index.html`

16. Resnick, P., Zeckhauser, R., Swanson, J., Lockwood, K.: The value of reputation on ebay: A controlled experiment. Experimental Economics 9, 79–101 (2006), `http://dx.doi.org/10.1007/s10683-006-4309-2`, doi: 10.1007/s10683-006-4309-2

17. Yan, Z., Holtmanns, S.: Trust modeling and management: from social trust to digital trust. In: Subramanian, R. (ed.) Computer Security, Privacy, and Politics: Current Issues, Challenges, and Solutions. IRM Press, IGI Global (2008)

Appendix A: Survey and Experiment Results

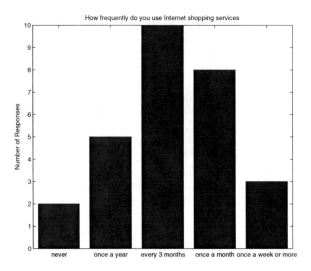

Fig. 2. Answers to Question 1: How frequently do you use Internet shopping services on average (e.g. E-bay, Amazon)?

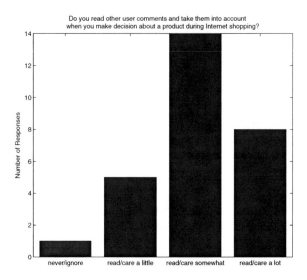

Fig. 3. Answers to Question 2: Do you read other user comments and take them into account when you make decision about a product during Internet shopping (e.g. when you buy something)?

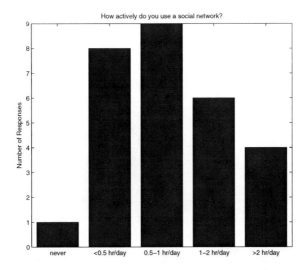

Fig. 4. Answers to Question 3: How actively do you use a social network (e.g. Facebook)?

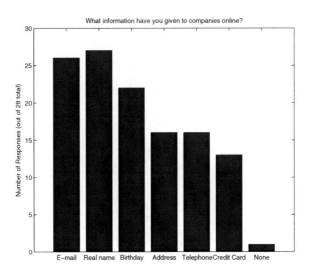

Fig. 5. Answers to Question 4: Which of the informations below have you given to companies online (Amazon, Facebook, etc. Banks who have the information already from offline sources do not count)?

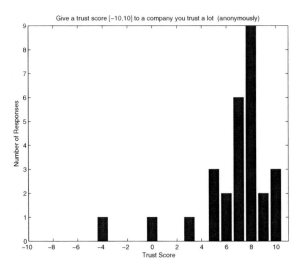

Fig. 6. Answers to Question 5: You evaluate an Internet company by giving it a score **anonymously**. Any positive score indicates your trust in it (maximum +10) and any negative score (-10) indicates mistrust. Zero, 0, means neutral. You do not have to give maximum scores when you trust or mistrust, any score is fine. Now consider a scenario where you give a score to a company **you trust a lot**. What score would you give?

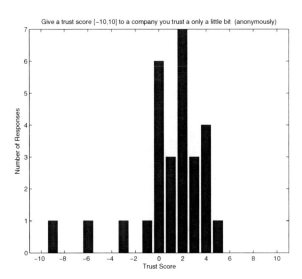

Fig. 7. Answers to Question 6: You evaluate an Internet company by giving it a score **anonymously**. Any positive score indicates your trust in it (maximum +10) and any negative score (-10) indicates mistrust. Zero, 0, means neutral. You do not have to give maximum scores when you trust or mistrust, any score is fine. Now consider a scenario where you give a score to a company **you trust only a little bit**. What score would you give?

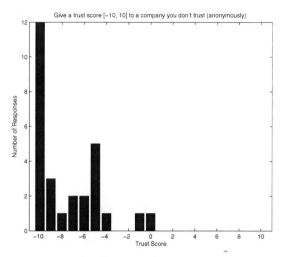

Fig. 8. Answers to Question 7: You evaluate an Internet company by giving it a score **anonymously**. Any positive score indicates your trust in it (maximum +10) and any negative score (-10) indicates mistrust. Zero, 0, means neutral. You do not have to give maximum scores when you trust or mistrust, any score is fine. Now consider a scenario where you give a score to a company **you don't trust**. What score would you give?

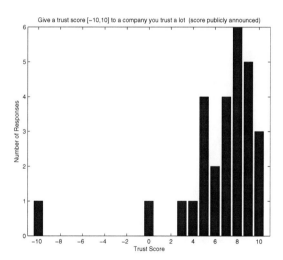

Fig. 9. Answers to Question 8: You evaluate an Internet company by giving it a score. **This time, your scores are shared publicly with your real name on the Internet (e.g. as in amazon.com)**. Any positive score indicates your trust in it (maximum +10) and any negative score (-10) indicates mistrust. Zero means neutral. You do not have to give maximum scores when you trust or mistrust, any score is fine. Now consider a scenario where you give a score to a company **you trust a lot**. What score would you give?

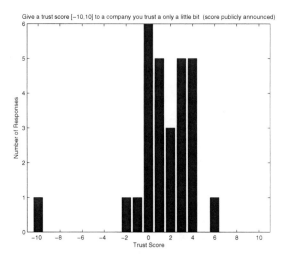

Fig. 10. Answers to Question 9: You evaluate an Internet company by giving it a score. **This time, your scores are shared publicly with your real name on the Internet (e.g. as in amazon.com).** Any positive score indicates your trust in it (maximum +10) and any negative score (-10) indicates mistrust. Zero means neutral. You do not have to give maximum scores when you trust or mistrust, any score is fine. Now consider a scenario where you give a score to a company **you trust only a little bit**. What score would you give?

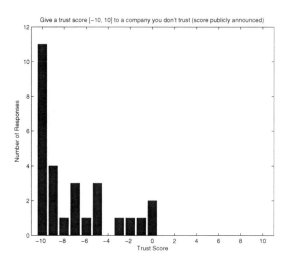

Fig. 11. Answers to Question 10: You evaluate an Internet company by giving it a score. **This time, your scores are shared publicly with your real name on the Internet (e.g. as in amazon.com).** Any positive score indicates your trust in it (maximum +10) and any negative score (-10) indicates mistrust. Zero means neutral. You do not have to give maximum scores when you trust or mistrust, any score is fine. Now consider a scenario where you give a score to a company **you don't trust**. What score would you give?

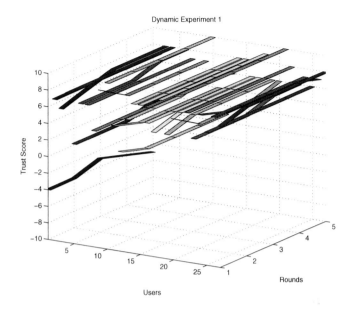

Fig. 12. User scores in dynamic experiment 1. The text of the given scenario is: You evaluate an Internet company **which you mainly trust** by giving it a score. **You have the chance to change your score up to 5 times before finalizing it and you will see the average score the company gets from everyone.** Any positive score indicates your trust in it (maximum +10) and any negative score (-10) indicates mistrust. Zero means neutral. *Your scores are shared publicly with your real name on the Internet.*

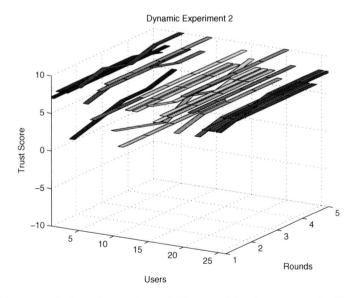

Fig. 13. User scores in dynamic experiment 2. The text of the given scenario is: You evaluate an Internet company **which you mainly trust** by giving it a score. *Your scores are anonymous.*

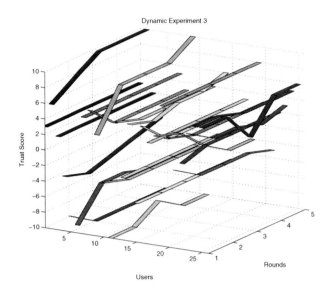

Fig. 14. User scores in dynamic experiment 3. The text of the given scenario is: You evaluate an Internet company **which you mainly trust** by giving it a score. *Your scores are shared publicly with your real name on the Internet.* **A good friend of yours tells you that he does not trust this company and gives the company a score of -8.**

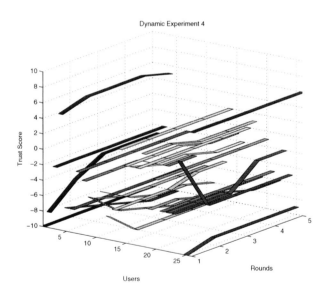

Fig. 15. User scores in dynamic experiment 4. The text of the given scenario is: You evaluate an Internet company **which you do not trust** by giving it a score. *Your scores are shared publicly with your real name on the Internet.* **A good friend of yours tells you that he does trust this company a lot and gives the company a score of +8.**

Appendix B: Web-Based Game Experiment Software

The software specially developed for this experimental study by Tansu Alpcan is a web application. The programming language used on the server side, Python[3], is very suitable for academic usage as it is high-level, easy-to-learn, and has support for scientific computing such as the SciPy[4] library. The software is built upon the python-based Django web framework, which allows fast prototyping and decreases the programming time significantly. In addition, it provides a useful administration interface for accessing the underlying database.

The survey part of the software has a flexible structure that allows entering questions and different types of answers (single/multiple choice, number, text) separately, and hence creation of a question and answer database. Then, surveys can be constructed using question/answer combinations in a flexible manner.

The dynamic experiment section of the software is its unique aspect. Currently, only synchronized iterative update is supported with a hard-coded number of responses needed to move to the next round. It is possible to give subjects a different piece of information at each round as well as aggregate metrics such as the average score in this experiment.

A screenshot of the main screen, a survey question, and a round in a dynamic experiment are shown in the Figures 16, 17, and 18, respectively. It is possible to use a html/css template to increase the visual appeal of the system. Subsequent to some basic improvements, this software is planned to be shared with the community in the future as an open source project.

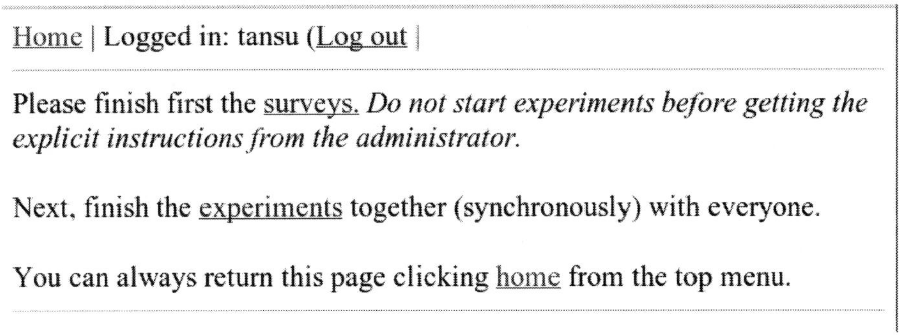

Home | Logged in: tansu (Log out |

Please finish first the surveys. *Do not start experiments before getting the explicit instructions from the administrator.*

Next, finish the experiments together (synchronously) with everyone.

You can always return this page clicking home from the top menu.

Fig. 16. Screenshot of the main screen after login

---

[3] python.org
[4] www.scipy.org

Home | Previous

Question 3:
How actively do you use a social network (e.g. Facebook)?

Answer:

- never/I do not use
- occasionally login and use it up to half an hour per day
- use it everyday half to one hour
- use it everyday one to two hours
- use it everyday more than two hours

Submit

Fig. 17. Screenshot of a survey question

Experiment 1

You evaluate an Internet company **which you mainly trust** by giving it a score. **You have the chance to change your score up to 5 times before finalizing it and you will see the average score the company gets from everyone.** Any positive score indicates your trust in it (maximum +10) and any negative score (-10) indicates mistrust. Zero means neutral. *Your scores are shared publicly with your real name on the Internet.*

Round 2

The average score the company got from all users is: 2.66666666667

Please indicated your (updated) score.

Answer:

Submit

Fig. 18. Screenshot of a dynamic experiment section

Colonel Blotto in the Phishing War

Pern Hui Chia[1] and John Chuang[2]

[1] Centre for Quantifiable Quality of Service (Q2S)[*], NTNU
[2] School of Information, UC Berkeley
chia@q2s.ntnu.no, chuang@ischool.berkeley.edu

Abstract. Phishing exhibits characteristics of asymmetric conflict and guerrilla warfare. Phishing sites, upon detection, are subject to removal by takedown specialists. In response, phishers create large numbers of new phishing attacks to evade detection and stretch the resources of the defenders. We propose the Colonel Blotto Phishing (CBP) game, a two-stage Colonel Blotto game with endogenous dimensionality and detection probability. We find that the optimal number of new phishes to create, from the attacker's perspective, is influenced by the degree of resource asymmetry, the cost of new phishes, and the probability of detection. Counter-intuitively, we find that it is the less resourceful attacker who would create more phishing attacks in equilibrium. And depending on the detection probability, an attacker will vary his strategies to either create even more phishes, or to focus on raising his resources to increase the chance he will extend the lifetime of his phishes. We discuss the implications to anti-phishing strategies and point out that the game is also applicable to web security problems more generally.

Keywords: Phishing, Economics, Colonel Blotto, Web Security.

1   Introduction

Phishing, among other web security issues, has remained a tricky problem today. While it is non-trivial to measure the exact financial losses due to phishing, and that many estimated loss figures appear overstated [9], the damage inflicted by phishing activities is never negligible. Realizing that technical sophistication alone will not be sufficient to fend off phishing activities, over the past few years, researchers have started to look at the ecosystem and modi operandi of phishing activities.

McGrath and Gupta found that phishers misuse free web hosting services and URL-aliasing services, and that phishing domains are hosted across multiple countries with a significant percentage of hosts belonging to residential customers [13]. Moore and Clayton identified different types of phishing attacks

---

[*] Centre for Quantifiable Quality of Service in Communication Systems (Q2S), Centre of Excellence, appointed by the Research Council of Norway, is funded by the Research Council, Norwegian Uni. of Science and Technology (NTNU) and UNINETT.
http://www.q2s.ntnu.no

J.S. Baras, J. Katz, and E. Altman (Eds.): GameSec 2011, LNCS 7037, pp. 201–218, 2011.

according to the way a phishing site is hosted [16]. The most common hosting vectors were found to be compromised web servers and free web-hosting services. While system admins and hosting companies are usually cooperative and quick to take down the phishing pages once notified, noticing them in the first place is challenging [16]. Moreover, victim servers were found to be re-compromised by the attackers to host phishing pages as the vulnerabilities of the servers remain unpatched [17]. Two notorious gangs, known as 'Rock Phish' and 'Avalanche'[1] even showed much technical sophistication in their massive and concerted phishing attacks. Both gangs exploited malware-infested machines and the fast flux method (mapping the domain name to different IP addresses (of different bots) by changing the DNS records in a high frequency) to extend the lifetime of a phishing site. Taking down the phishing pages from a large number of bots is extremely difficult, especially when the ISPs have only limited control and responsibility over malware-infested machines. This forces the defender to take-down the phishing sites by suspending the phishing domain names with the help from registrars and registries.

The above highlights several important challenges in defending against phishing activities. First, it is challenging to detect all phishing attacks out there. Second, taking down phishing attacks that have been identified (e.g., to remove the phishing sites, or to ensure that a vulnerable web server is patched to prevent re-compromise) is also non-trivial. The situation is worsened by a lack of information sharing in the anti-phishing industry [16]. Meanwhile, despite a spike in the count of phishing attacks[2] in 2009 due to the Avalanche gang [2], the number of unique phishing domains found (per six months) has remained steady at around 30,000 over the past few years, except in the second half of 2010 where 43,000 unique phishing domain names were recorded partly due to new data inputs from the China Internet Network Information Center (CNNIC) who operates the .cn registry [3].[3] This suggests that the phishers do factor in the cost consideration when carrying out phishing attacks.

Different from prior studies that have largely taken the empirical approach, we propose in this work a theoretical model to aid researchers and policymakers in better analyzing the different aspects of phishing defense. We build on the Colonel Blotto game, an old but interesting game that has been largely neglected due to its complexity, until the recent work by Roberson [18] which gives a complete characterization to the unique equilibrium payoffs of a two-player asymmetric Colonel Blotto game. The game is particularly suitable to capture the resource allocation problem between a phisher and a defender with asymmetrical resources. In addition to mapping the phishing problem into the Colonel

---

[1] An account of the modi operandi of the Rock Phish and Avalanche gangs can be found in [14] and [2] respectively.

[2] An attack is defined by Anti-Phishing Working Group (APWG) as a unique phishing site targeting a specified brand.

[3] Measurement of unique phishing attacks, uptime of phishing sites and in-depth surveys on the trends and domain name use by phishing sites can be found in a series of reports (e.g., [2,3]) by the APWG on http://www.antiphishing.org.

Blotto game, our model extends the two-stage Colonel Blotto game in [10] to include a detection probability to factor in the consideration of asymmetric information that not all phishes will be known to the defender. We regard the defender in this work as a takedown company (e.g., MarkMonitor[4], BrandProtect[5] and Internet Identity[6]) that has been contracted by its clients (e.g., financial institutions, e-commerce services) to remove phishing sites that masquerade as the clients' legitimate sites. Although the defender is in a disadvantage position for not being able to detect all phishes that have been created, and that the attacker can always exploit the next weakest link whenever a phishing server is taken down, we expect that the defender can garner more resources than the attackers from the contract with its clients, plus the support from the ISPs, service providers, law enforcers, registrars and registries.

In the following, we first give a quick introduction to the Colonel Blotto game and related work in Section 2. We propose the Colonel Blotto Phishing (CBP) game in Section 3 to model phishing attack and defense. We present the results from our analysis based on the CBP model in Section 4. And lastly, we discuss the implications to the anti-phishing strategies in Section 5.

2   Background and Related Work

The *Colonel Blotto* game was first introduced in 1921 by Borel [6] as a two-player constant-sum game, where the players strategically distribute a fixed and *symmetrical* amount of resources over a finite number of n contests (battlefields). The player who expends a higher amount of resources in a contest wins that particular battlefield, similar to an all-pay auction. The objective of the players is to maximize the number of battlefields won. Gross and Wagner [8] in 1950 described the game with *asymmetrical* resources between the two players, but have only solved the case where the number of battlefields $n = 2$.

The complexity for the case when there are $n \geq 3$ battlefields and the lack of pure strategies have arguably led to the Colonel Blotto game being largely neglected by the research community. A resurgence of interests in the Colonel Blotto game (e.g., [4,5,7,11,12,19]) follows the recent work by Roberson [18] which has successfully characterized the unique equilibrium payoffs for all configurations of resource asymmetry, and the equilibrium resource allocation strategies (for most configurations) of a constant-sum Colonel Blotto game with $n \geq 3$ battlefields. Roberson and Kvasov have later studied the non-constant-sum version in [19]. We summarize the main results from Roberson [18] below:

Theorem 1. (case a, b and c correspond to Theorem 2, 3 and 5 in [18])
Let n denote the number of battlefields, while R_w and R_s denote the resources of the weak (w) and strong (s) players respectively such that $R_w \leq R_s$, the Nash equilibrium univariate distribution functions (for allocating resources to individual battlefields strategically), and the unique equilibrium payoffs (measured in

---

[4] http://www.markmonitor.com
[5] http://www.brandprotect.com
[6] http://internetidentity.com

the expected proportion on battlefields won), depending on the $\frac{R_w}{R_s}$ ratio and the number of battlefields n, are given in the following:

case a: $\frac{2}{n} \le \frac{R_w}{R_s} \le 1$

In the unique Nash equilibrium, player w and s allocate x_j resources in each battlefield $j \in \{1, ..., n\}$ based on the following univariate distribution functions:

$$F_{w,j}(x) = (1 - \frac{R_w}{R_s}) + \frac{nx}{2R_s}(\frac{R_w}{R_s}) \qquad\qquad , x \in [0, \frac{2R_s}{n}]$$
$$F_{s,j}(x) = \frac{nx}{2R_s} \qquad\qquad , x \in [0, \frac{2R_s}{n}]$$

The unique equilibrium payoffs (expected proportions of battlefields won) of player w and s are independent of the number of battlefields, given as follows:

$$\pi_w = \frac{R_w}{2R_s}$$
$$\pi_s = 1 - \frac{R_w}{2R_s}$$

case b: $\frac{1}{n-1} \le \frac{R_w}{R_s} < \frac{2}{n}$

In the unique Nash equilibrium, player w and s allocate x_j resources in each battlefield $j \in \{1, ..., n\}$ based on the following univariate distribution functions:

$$F_{w,j}(x) = (1 - \frac{2}{n}) + \frac{x}{R_w}(\frac{2}{n}) \qquad\qquad , x \in [0, R_w]$$
$$F_{s,j}(x) = \begin{cases} (1 - \frac{R_s}{nR_w})(\frac{2x}{R_w}) & , x \in [0, R_w) \\ 1 & , x \ge R_w \end{cases}$$

The expected proportions of battlefields won by player w and s are as follows:

$$\pi_w = \frac{2}{n} - \frac{2R_s}{n^2 R_w}$$
$$\pi_s = 1 - \frac{2}{n} + \frac{2R_s}{n^2 R_w}$$

case c: $\frac{1}{n} < \frac{R_w}{R_s} < \frac{1}{n-1}$

In a Nash equilibrium, player w allocates zero resources to $n-2$ of the battlefields, each randomly chosen with equal probability. On the remaining 2 battlefields, he randomizes the resource allocation over a set of bivariate mass points. On the other hand, player s allocates R_w resources to each of $n - 2$ randomly chosen battlefields. On the remaining 2 battlefields, player s also randomizes the resource allocation over a set of bivariate mass points. Let $m = \lceil \frac{R_w}{R_s - R_w(n-1)} \rceil$ such that $2 \le m < \infty$, the unique expected proportions of battlefields won by player w and s are given as follows:

$$\pi_w = \frac{2m-2}{mn^2}$$
$$\pi_s = 1 - \frac{2m-2}{mn^2}$$

Note that the univariate distribution functions constitute the players' mixed strategies in Nash equilibrium. The allocation of resources across the n battlefields must additionally be contained in the set of all feasible allocations

$\{\mathbf{x} \in \mathbb{R}_+^n \,|\, \sum_{j=1}^n x_{i,j} \le R_i\}$ where $i = w, s$.[7] In general, player s uses a stochastic 'complete coverage' strategy (which expends non-zero resources in all battlefields, and locks down in a random subset of battlefields by allocating R_w resources to them in case b and c), while player w uses a stochastic 'guerrilla warfare' strategy (which optimally abandons a random subset of the battlefields). Despite the resource asymmetry, player w can expect to win a non-zero proportion of the battlefields, except in the case of $R_s \ge nR_w$, where the player s can trivially lock down (win) all battlefields by allocating R_w resources to each of them.

Note that also the proportion of battlefields won by the player w is a function of n in the case b and c of Theorem 1. In a recent work, Kovenock et al. [10] presented a two-stage Colonel Blotto game which endogenizes the dimensionality of the classic Colonel Blotto game, allowing the players to create additional battlefields in the additional 'pre-conflict' stage. They show that with such possibility, player w will optimally increase the number of battlefields in the 'pre-conflict' stage, given a low battlefield creation cost, so to thin the defender's resources and reduce the number of battlefields player s can lock down in the 'conflict' stage. We outline the main results from [10] below:

Theorem 2. (see Theorem 2 in [10])
In the pre-conflict stage of the game with n_0 initial battlefields and resource asymmetry that satisfies $\frac{1}{n_0-1} \le \frac{R_w}{R_s} \le 1$, assuming that the cost to create additional battlefields, c is strictly increasing and strictly convex, the optimal numbers of new battlefields that player w and s will create, n_w^* and n_s^* respectively, in the subgame perfect equilibrium, are given as follows:

<u>*case a:*</u> *If $\frac{R_w}{R_s}$ satisfies $\frac{2}{n_0} \le \frac{R_w}{R_s} \le 1$, then $n_s^* = n_w^* = 0$.*

<u>*case b:*</u> *If $\frac{R_w}{R_s}$ satisfies $\frac{1}{n_0-1} \le \frac{R_w}{R_s} < \frac{2}{n_0}$, then $n_s^* = 0$, and let $n_{wr} \in (0, \frac{2R_s}{R_w} - n_0)$ denotes the real number that solves:*

$$-\frac{2}{(n_0+n_{wr})^2} + \frac{4R_s}{R_w(n_0+n_{wr})^3} - c'_{n_{wr}} = 0$$

then, n_w^* is either $\lceil n_{wr} \rceil$ or $\lfloor n_{wr} \rfloor$ depending on which of it results in a higher utility for player w, given $n_s^* = 0$.

Note that Theorem 2 has not formally treated the case c of Theorem 1. The analysis of case c will be more complicated as the expected proportion of battlefields won by both players have points of discontinuity, but the underlying intuition is the same as case b. [10] Note that also Theorem 2 assumes that the cost of creating additional battlefields is expended separately from the players' resources.

---

[7] We refer interested readers to Roberson [18] for proofs and details on how the equilibrium univariate distribution functions give a n-variate joint distribution function satisfying the constraint that $\sum_{j=1}^n x_{i,j} \le R_i$ where $i = w, s$.

3   Modeling

With an introduction to the classic Colonel Blotto game and the extension with
endogenous dimensionality, we are now ready to model the economics for phishing activities in this section. We will first apply the classic Colonel Blotto game
to phishing attack and defense. Then, we will extend the game to model endogenous dimensionality following the two-stage Colonel Blotto game in [10], and
asymmetric information using an additional detection probability to reflect that
not all phishes will be known to the defender in practice.

3.1   Applying Colonel Blotto to Phishing

We map the basics the Colonel Blotto game in the context of phishing attack
and defense in the following.

Players. Like the classic Colonel Blotto, we consider here a two-player constant-
sum game between a phisher and a defender. We regard the defender here to be
a takedown company such as MarkMonitor, BrandProtect and Internet Identity
as aforementioned. The takedown company is contracted by its clients, including
banks and popular brand owners, to remove phishing sites attacking the clients'
brands. On the other hand, the phisher plays to keep alive the phishing sites, or
to launch new attacks, to victimize as many users he can.

Resources. We assume the phisher to be the weak player (w) and the takedown
company to be the strong player (s). Although this may be debatable, assuming
such resource asymmetry is reasonable if we consider that takedown companies
will usually maintain good contacts with and can thus get assistance from the
ISPs, service providers, law enforcers, registrars and registries in the process
of taking down the phishes. By resources, we thus mean not financial figures
but mainly the *technologies*, *infrastructure* (e.g., access to a botnet), *time* and
manpower.[8] Phisher's profitability is also not as lucrative as it appears in the
news. A number of estimates on the losses due to phishing attacks have been
criticized to be overstated [9]. The resources, R_s and R_w respectively, are finite
with $R_s \geq R_w$. They are of the 'use-it-or-lose-it' nature, meaning that unused
resources will give no value to the players in the end of the game.

Battlefields. We define a battlefield to be a unique phishing site (a fully quali-
fied domain name or IP address, or a site on a shared hosting service) targeting
a specific brand, following the definition of a phishing attack by APWG (see e.g.,
page 4 in [3]). Different URLs directing to the same phishing page, crafted to
evade spam filters or to trick the URL-based anti-phishing toolbars, are considered the same battlefield. Defined this way, creating a battlefield hence involves
some costs ranging from *low* (e.g., to register a subdomain on a shared hosting service, to copy the login page of a brand) to *high* (e.g., to register a new

---

[8] Resource asymmetry should not be confused with asymmetry in coverage where the
defender needs to protect all assets while the attacker can target any of them.

domain name, to compromise a vulnerable web server). In this paper, we use the terminologies 'a phish' and 'a phishing attack' interchangeably.

Objectives & Contests. We model the objective of the phisher and the defender to be maximizing the expected proportion of phishing attacks kept online and taken down, respectively. We consider that either the phisher or the defender can outperform the other party to win a battlefield by allocating more resources to it. And given that we have not factored in the uptime and the number of victims per attack in our model, we loosely define that a specific battlefield (phishing attack) is won by the phisher if the phish has a *long enough uptime*. For example, having the resources of a botnet infrastructure, an attacker can use 'fast-flux' IP addresses and malware-controlled proxies, to make it hard for the defender to take down the phishing server, prolonging the uptime of the phishes, as the defender will have to turn to the responsible registrar or registry to suspend the domain name. We elaborate on other tricks used by phishers, including the two infamous Rock Phish and Avalanche gangs, in Section 3.2.

Given the above configurations, we can already gain a number of useful insights. For example, we can expect that there will be always some phishes that will have long uptime unless that the defender is much more resourceful than the phisher (i.e., $R_s \geq nR_w$). However, the classic colonel blotto game alone does not describe the practical scenario quite yet. Why are there a large number of phishing attacks instead of just a few? Indeed, it is to the phisher's advantage to create an optimal number of additional phishes (battlefields), so to thin the defender's resources in removing each of them. Furthermore, how does the asymmetric information affect the strategies of the phisher? We extend the two-stage Colonel Blotto game in [10] to include an additional parameter, the expected probability of detection P_d, to reflect that not all phishes will be known to the defender – a major challenge in the anti-phishing industry. [16]

Table 1. The flow of the Colonel Blotto Phishing game

	Stage	Phisher (w)	Defender (s)
i)	create – detect	a. create and market n_w^* new phishes b. learn about detection	a. detect new phishes b. publish findings
ii)	resist – takedown	c. expend ε resources to undetected phishes, while allocating $R_w - \varepsilon$ resources to phishes known to the defender to resist removal	c. expend all R_s resources strategically to remove the newly detected and known existing phishes in a promptly manner

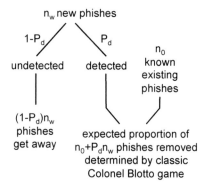

Fig. 1. Expected proportion of phishes in different states

3.2 The Colonel Blotto Phishing Game

We name our model as the Colonel Blotto Phishing (CBP) game. It consists of two stages: (i) create–detect, (ii) resist–takedown, similar to the 'pre-conflict' and 'conflict' stages in [10]. Table 1 summarizes the flow of the CBP game. We detail on the game stages in the following.

Stage 1: Create–Detect. We consider that game starts with the phisher having a number of phishes n_0 that are known to the defender, and both players are allowed to increase the dimensionality of the game by introducing new battlefields in the first stage. Obviously, the defender will not create any phishes. However, it is to the phisher's advantage to create a number of new phishing attacks n_w so to stretch the defender's resources, in hope to increase the expected proportion of phishes that will stay online for more than a certain period of time. Hence, we have the total phishing attacks $n = n_0 + n_w$. We expect the phisher then advertises the newly created phishes through spams and online social networks.[9] We assume a linear cost c for creating and advertising the new phishes; c can be low or high depending on the way the phisher carries out the attack (e.g., through free subdomain services, paying for a newly registered domain, taking the effort to hack a vulnerable web server, and so on).

A new aspect we incorporate into the classic Colonel Blotto game is the situation where some of the newly created phishes might not be detected by the takedown company. We analyze both cases where the expected detection probability P_d is (i) exogenously determined, and (ii) endogenously influenced by the number of new phishing attacks in Section 4. The expected proportions of phishes that trivially get away undetected, or that will possibly stay online long enough depending on the resource allocations of both the phisher and defender in the second stage, are depicted in Figure 1. In practice, takedown companies learn about new phishing attacks through their own infrastructures (e.g., spam

---

[9] McGrath and Gupta [13] observed that most domains created for phishing become active almost immediately upon registration.

filters) in addition to 'raw' feeds bought, negotiated or obtained from the ISPs or phishing clearinghouses, such as the APWG and PhishTank[10].

An assumption we make here is that the phisher will then learn about which of his phishes have been detected before proceeding to the next game stage. This is reasonable, regardless of whether the takedown company shares their detection results[11], as we expect that the phisher can achieve this using public clearinghouses (e.g., phishtank) or through anti-phishing APIs that come with modern browsers (e.g., Google Safe Browsing API[12] for FireFox and Chrome).

Stage 2: Resist–Takedown. Knowing the identity of the detected phishes $\mathbb{J}_d$, the optimal move for the phisher in the second stage is hence to expend all his resources strategically on the detected phishes only, so to resist the takedown process. Here, we assume that the resources (e.g., technologies, infrastructure, manpower) are of the 'use-it-or-lose-it' nature, typical to a constant-sum game. In other words, unused resources will give no value to the players. We further assume that the phisher will optimally allocate $\varepsilon \approx 0$ resources for the undetected phishes $j \notin \mathbb{J}_d$ given that the defender does not know about them. We note that this assumption is reasonable as the resources are finite.

We regard that either the phisher or the takedown company will 'succeed' with respect to a particular phishing attack depending on the amount of resources they put in: the player who expends more resources wins. Specifically, with $x_{i,j}$ and $x_{-i,j}$ denoting the amount of resources player $i \in \{w, s\}$ and his opponent puts into the phish attack j respectively, the success of player i at attack j is given by:

$$\pi_{i,j}(x_{i,j}, x_{-i,j}) = \begin{cases} 0 & \text{if } x_{i,j} < x_{-i,j} \\ 1 & \text{if } x_{i,j} > x_{-i,j} \end{cases}$$

where in the case of $x_{w,j} = x_{s,j}$ (a tie), we assume that defender s will succeed in taking down the attack promptly. As for undetected phishes, i.e., $\forall j \notin \mathbb{J}_d$, we regard that $x_{s,j} = 0$ and the phisher will trivially win the battlefield with $x_{w,j} = \varepsilon$ resources.

Can the phisher still win in an already detected phish in practice? While it may not be intuitive at first, the answer is 'yes' given our definition that a phishing attack is won by the phisher (defender) if the phish has an uptime more (less) than a certain threshold. The longer a phish can resist being removed, the more users could fall victim to it. While a weak phisher may simply abandon his phishes (given that he cannot win) when facing a much more resourceful

---

[10] PhishTank – a community based phishing collator. http://www.phishtank.com

[11] Individual takedown companies often will validate the 'raw' URLs of potential phishes to remove false positives, and they might not voluntarily share their validated feeds for competitive advantages. Moore and Clayton showed how sharing of phishing data could have helped to halve the lifetime of phishes, translating to a potential loss mitigation of $330 million per year, based on data feeds from two takedown companies [15].

[12] http://code.google.com/apis/safebrowsing/

defender (i.e., when $R_s \geq nR_w$), there have been practical examples of how a skilled phisher attempts to extend the lifetime of his phishes via different tricks. For example, a phisher may configure his phishes not to resolve on every access so to misguide the defender, but remain online to trick more users (see e.g., [3], footnote 5). The phisher may also temporarily remove the phishing pages from a compromised web server so to avoid further actions from the defender or admin (e.g., to patch up specific vulnerabilities) and re-plant the phishes at a later time. Indeed, APWG (see e.g., [3], footnote 5) finds that more than 10% of phishes are re-activated after being down for more than an hour. Moore and Clayton also found that 22% of all compromised web servers are re-compromised within 24 weeks to be used as the host for phishing sites [16].

With more resources, a phisher can even increase technical sophistication so to use malware-controlled proxies and fast-flux IP addresses as demonstrated the large-scale attacks by the infamous 'Rock Phish' and 'Avalanche' phishing gangs. The fast-changing nature of IP address that the phishing site resolving to indicate that the attacker has in control of a large number of compromised machines (bots) make it infeasible for the takedown company and the responsible ISPs to take the phishing servers offline promptly. Instead, the defender will have to work towards suspending the domain names in use, which could take a while if the responsible registrars are not responsive or have limited experience in abuse control. The 'Avalanche' gang was found to have exploited this; at the same time as they launched their massive attacks using domains bought from a few registrars (resellers), the gang scouted for other unresponsive registrars for future use (see page 7 of [2]). Meanwhile, in [14] Moore and Clayton found that the fast-flux phishing gang used 57 domain names and 4287 IP addresses for fast-flux phishing. The 1:75 skewed ratio is interesting as it suggests that the fast-flux phishing gang was highly resourceful (having access to a botnet infrastructure). However, we note that these resources are not unlimited. For example, the operations of the 'Avalanche' gang was disrupted as the security community affected a 'temporary' shut-down of the botnet infrastructure in Nov 2009 [2]. Later, although the gang managed to re-establish a new botnet, they were also found to prefer using their resources for a more profitable opportunity to distribute the Zeus malware, which has been designed to automate identity theft and facilitate unauthorized transactions. [3]

Subgame Perfect Equilibrium. We consider the objective of the phisher (the takedown company) is to maximize the proportion of phishes that he succeeds in keeping alive for a certain period (removing promptly), minus the cost for creating new phishing attacks. With $\mathbf{x}_i$ and $\mathbf{x}_{-i}$ denoting the resource allocations across all phishing sites by player $i \in \{w, s\}$ and his opponent respectively, the utility of player i can be written as:

$$U_i(\{\mathbf{x}_i, n_i\}, \{\mathbf{x}_{-i}, n_{-i}\}) = \frac{1}{n}\left(\sum_{j \in \mathbb{J}_d} \pi_{i,j} + \sum_{j \notin \mathbb{J}_d} \pi_{i,j}\right) - cn_i$$

Note that $\mathbf{x}_i$ and $\mathbf{x}_{-i}$ must be contained in the set of all feasible allocations, given by $\{\mathbf{x}_i \in \mathbb{R}_+^n \mid \sum_{j=1}^n x_{i,j} \leq R_i\}$.

The optimal number of new phishes to create n_i^* and the optimal utility U_i^* in subgame perfect equilibrium can be obtained by backward induction. First, we can work out the expected proportion of success of each player in the 'resist–takedown' stage based on Theorem 1 and the fact that a fraction of phishes will get away undetected as given by P_d. Then, returning to the 'create–detect' stage, the optimization problem of the phisher becomes:

$$\max_{n_w} E(U_w|n_w) = \frac{1}{n}E\left(\sum_{j\in\mathbb{J}_d}\pi_{w,j}\right) + \frac{(1-P_d)n_w}{n} - cn_w$$

$$= \frac{n_d}{n}E(\pi_w) + \frac{(1-P_d)n_w}{n} - cn_w$$

where

$$E(\pi_w) = \begin{cases} \frac{R_w}{2R_s} & \text{if } 1 \geq \frac{R_w}{R_s} \geq \frac{2}{n_d} \\ \frac{2}{n_d} - \frac{2R_s}{(n_d)^2 R_w} & \text{if } \frac{2}{n_d} \geq \frac{R_w}{R_s} \geq \frac{1}{n_d-1} \\ 0 & \text{if } \frac{1}{n_d} \geq \frac{R_w}{R_s} \end{cases}$$

$$n_d = P_d n_w + n_0$$
$$n = n_w + n_0$$

As with many real life security problems, the defender in this model is disadvantaged in that he takes only reactive measures against the phisher. Note that also we have omitted the case c of Theorem 1 (i.e., when $\frac{1}{n_d-1} > \frac{R_w}{R_s} > \frac{1}{n_d}$), a relatively small region with points of discontinuity, for simplicity.

4   Analysis

We analyze using the CBP game three different scenarios: (i) the hypothetical case of perfect detection of phishing attacks, i.e., $P_d = 1$, (ii) $P_d < 1$ and is exogenously determined, and (ii) $P_d < 1$ and is endogenously influenced by the number of phishes the attacker creates.

Perfect Phish Detection. Let us start with the hypothetical case where the probability of detection, $P_d = 1$. Figure 2 plots the optimal number of additional phishing attacks n_w^* that the phisher will launch depending on cost c, knowing that all newly created phishes will be detected by the defender. Note that this is exactly the scenario analyzed in [10], and that the dashed and solid lines plot the case a and b of Theorem 2 respectively. When the resource asymmetry is small (with $\frac{2}{n_0} \leq \frac{R_w}{R_s} = \frac{1}{2}$, dashed line), the phisher optimally chooses *not* to create additional phishes. There is no advantage to further stretch the defender as the attacker, given his resources, is expected to win in equilibrium a proportion of battlefields equals $\frac{R_w}{2R_s} = \frac{1}{4}$ as shown in Figure 2(b).

However, when the resource asymmetry is large (with $\frac{2}{n_0} > \frac{R_w}{R_s} = \frac{1}{900}$, solid line), the phisher will create additional phishing attacks to reduce the ability

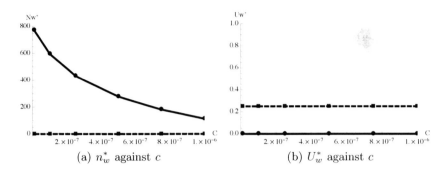

(a) n_w^* against c ⁣ ⁣ ⁣ ⁣ ⁣ (b) U_w^* against c

Fig. 2. The optimal new phishes n_w^* and utility U_w^* given $P_d = 1$. Solid and dashed lines plot the case where $\frac{R_w}{R_s} = \frac{1}{900}$ and $\frac{1}{2}$ respectively, with $n_0 = 1000$.

of the defender in locking down all of them. Especially when cost c (measured in terms of the normalized utility) is negligible, n_w^* approaches $\frac{2R_s}{R_w} - n_0 = 800$ given $\frac{R_w}{R_s} = \frac{1}{900}$ and $n_0 = 1000$. Even so, interestingly, the utility of the phisher is still less than 10^{-3}. Meanwhile, as c increases (see Figure 2(a)), the optimal number of new phishing attacks n_w^* quickly approaches zero.

Imperfect Phish Detection (Exogenous). In practice, we can expect that a significant fraction of phishing attacks will get away undetected by the defender. The problem is exacerbated by non-sharing of data between different security vendors as observed in [15]. Figure 3(a) and 3(b) plot the optimal number of new phishes n_w^* and the corresponding utility of the phisher U_w^* depending on $P_d \in [0, 1]$. We assume that the phisher will be able to estimate P_d based on past experience.

Let us first focus on the game between a resourceful (strong) phisher and the defender, with the resource asymmetry $\frac{2}{n_0} \leq \frac{R_w}{R_s} = \frac{1}{2}$ (as shown by the solid lines). Here, with $Pd < 1$, the phisher will now create additional phishes knowing that the defender will fail to detect some of the attacks, different from the case of perfect detection. The undetected phishes add on to the phisher's utility, which has a lower bound at $\frac{R_w}{2R_s} = \frac{1}{4}$. As for the game between a less resourceful (weak) phisher and the defender given a large resource asymmetry of $\frac{2}{n_0} > \frac{R_w}{R_s} = \frac{1}{900}$ (as depicted by the dashed lines), observe that the optimal numbers of new phishing attacks are now much higher than 800, the upper bound for the case of perfect detection.

Another interesting observation is that the utility gap between a strong and a weak phisher reduces as P_d decreases from 1 to 0. Improving on P_d thus will hurt a weak phisher, but has less impact on a strong phisher as he can leverage on his resources (technologies, infrastructure, manpower, etc.) to resist the takedown of some of his phishes. The trend also suggests that an attacker will optimally vary his strategies to create more phishes when P_d is low, but strive to increase his resources as P_d increases.

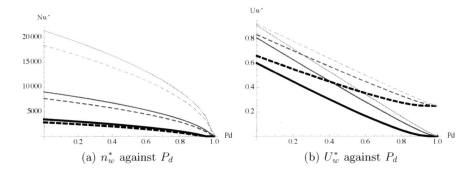

(a) n_w^* against P_d (b) U_w^* against P_d

Fig. 3. Optimal number of new phishes to create n_w^* and the corresponding optimal utility U_w^*. Solid and dashed lines plot the case where $\frac{R_w}{R_s} = \frac{1}{900}$ and $\frac{1}{2}$ respectively, with $n_0 = 1000$. The effect of a decreasing cost c going from 5×10^{-5} to 1×10^{-5} and 2×10^{-6}, measured in terms of the normalized utility, is depicted by the thick-black, normal-black and thin-gray lines, respectively.

Regardless of the extent of resource asymmetry, an increased cost (see the thick-dark lines versus the thin-gray lines) reduces both the optimal number of phishes and the utility of the phisher. But, somewhat counter-intuitively, the lower the detection probability, the more phishes the attacker will want to create. An attacker does not settle with having a fraction of undetected phishes, but will exploit the weakness of the defender in detecting all phishes and create even more phishes to increase his utility.

Another counter-intuitive and interesting finding is that in fact it is optimal for a less resourceful phisher to create more new phishes (than if he is a resourceful phisher) in equilibrium. This can be seen in Figure 3 where the solid lines ($\frac{R_w}{R_s} = \frac{1}{900}$) remain above the dashed lines ($\frac{R_w}{R_s} = \frac{1}{2}$) for all different costs c. This is surprising as large-scale phishing attacks are more often associated with re-sourceful attackers such as the 'Rock Phish' and 'Avalanche' gangs empirically.[13] There could be several reasons to this. First, while the 'Avalanche' phishes can be recognized easily with their distinctive characteristics, we do not know if the bulk of other phishing attacks are not related (carried out by a single organiza-tion) for sure. Secondly, could there be really a 'tragedy of the commons' due to the a large number of phishers (as described in [9]) that has forced the less resourceful attackers out of the phishing endeavor? We note that analyzing the effect of competition between several phishers would be an interesting extension to our current model. Another more likely explanation would be that most of the phishing attacks are in fact detectable by the defender today, forcing the less resourceful attacker to gain too little utility to be profitable (observe that U_w^* for the less resourceful attacker is almost zero as $P_d \rightarrow 1$ in Figure 3(b)). Furthermore, having a large number of phishes can also increase the probability

---

[13] For example, the 'Avalanche' gang was responsible for 84,250 out of 126,697 (66%) phishing attacks recorded by the APWG in the second half of 2009.

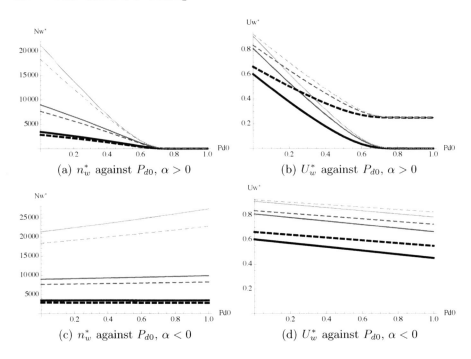

Fig. 4. Optimal n_w^* and U_w^* when the effective probability of detection, $P_d = P_{d0} \times (n_w)^\alpha$. Graphs a and b plot the case where $\alpha = 0.05 > 0$, while graphs c and d plot the case of $\alpha = -0.2 < 0$. Solid and dashed lines plot the case where $\frac{R_w}{R_s} = \frac{1}{900}$ and $\frac{1}{2}$ respectively, with $n_0 = 1000$. The effect of a decreasing cost c going from 5×10^{-5} to 1×10^{-5} and 2×10^{-6} is depicted by the thick-black, normal-black and thin-gray lines, respectively.

of detection by the defender. We analyze the case when P_d depends on the n_w in the next section.

Imperfect Phish Detection (Endogenous). Let us model the effective P_d to depend on the number of phishes an attacker creates with a simple formulation:

$$P_d = P_{d0} \times (n_w)^\alpha$$

where with $\alpha = 0$, we thus have the exogenous case as discussed in the previous section. The interesting analysis here is when $\alpha \neq 0$ as depicted in Figure 4.

There are many examples where increasing the number of phishing attacks (battlefields) can lead to a higher detection rate by the defender. For instance, the way the 'Rock Phish' and 'Avalanche' gangs hosted a number of phishing attacks (i.e., different phishing pages targeting different brands) using the same domain name[14], while reducing cost, increases the chance that all these phishes (battlefields) will be detected and taken down altogether. An attacker

---

[14] A typical 'Avalanche' domain often hosted around 40 phishing attacks at a time [2].

who register multiple domains for phishing purposes may also risk leaving visible patterns in the WHOIS database that is being used by the defender to identify and suspend suspicious domains quickly.[15]

As shown in Figure 4(a) and 4(b), both the n_w^* and U_w^* curves are now steeper than before. The optimal number of additional phishing attacks to create quickly approaches zero as P_{d0} increases. Other than that, the main results from the case of exogenous detection probability (where $\alpha = 0$) remain applicable. First, it is optimal for a weak phisher to create more phishes than a resourceful attacker. The lower the detection probability is the more phishes will an attacker create. Also, improving the baseline detection technologies (P_{d0}) hurts a weaker phisher more than a stronger phisher.

It is harder to think of some practical examples where an increased number of phishes helps to reduce the effective detection rate by the defender (i.e., with $\alpha < 0$). A possible but *unlikely* scenario would be if the phishing attacks that a phisher creates cannot be correlated to each other, and that the larger number of attacks stretch the defender's capability in detecting all of them. We include the plots of optimal n_w^* and U_w^* under such scenario in Figure 4(c) and 4(d) for reference purposes. Notice that the optimal utility of the phisher is now bounded only by the cost of creating new phishes.

5   Discussion: Implications to Anti-phishing Strategies

The success of anti-phishing defense depends on a number of interacting variables. As captured in our model, increasing the cost of creating new phishes c, improving the detection rate of new phishes P_d, as well as, increasing the resource asymmetry between the defender and phisher, $\frac{R_s}{R_w}$ are all crucial factors to be considered.

Increasing the cost for creating new phishes will hurt the attacker especially a weak phisher, who has no resources to resist the prompt removal of his phishes. Raising the cost (both in financial and procedural terms) for registering a domain name can therefore help, but only to a certain extent. Take the decision by CNNIC to make the registration of domain names more restrictive for example, the number of .cn phishing domains dropped, but phishing attacks on Chinese institutions remained high as phishers shifted to use other domain names such as .tk and the co.cc subdomain service (see [3] page 5). Phishers would also usually register new domains using stolen credit cards. Furthermore, studies have found that a larger percentage of phishing attacks (80%) are actually performed using compromised web servers of innocent domain registrants (see e.g., [2,3,17]). To raise the cost c will thus involve patching a large number of vulnerable servers, which is challenging if not impossible without a proper incentive plan.

A more effective alternative is hence to focus on improving the detection rate of new phishes. While automated spam filters help to detect potential phishing

---

[15] APWG reported that attackers often utilize a single or small set of unique names, addresses, phone numbers, or contact email addresses to control their portfolio of fraudulent domain names [1].

URLs, the 'Rock Phish' gang, for example, used GIF image in phishing email to evade detection. The popularity of URL shortening services and wall postings on online social networks add up to the challenge of detecting all phishing advertisements. Calls to share the phishing data in the anti-phishing industry have been made before (e.g., in [15]), but sharing can also create concerns as takedown companies leverage on their phishing data for competitive edges. Here, we see a room to employ and better coordinate the crowds to help improving the detection probability. Collecting user reports against potential phishes (or potentially harmful sites), without necessarily demanding from them higher skilled tasks such as evaluating if a phish is valid (or that a site is secure), can already be helpful.

Naturally, the value of data sharing and crowd-based phish-reporting will depend on the state of information asymmetry (i.e., the detection probability P_d). As can be seen in Figure 3, an 'intelligent' phisher will leverage on a large number of phishes for optimal utility when P_d is low. Meanwhile, as $P_d \to 1$, a phisher will improve his utility by increasing his resources to match the defender's. This includes, for example, to gain access to a botnet infrastructure so to prolong the uptime of his phishes. Should a good estimate of P_d is available, the defender can thus decide whether to prioritize on increasing the cost of creating new attacks (to reduce the number of phishes the attacker can create), or to prioritize on disrupting the channels a phisher can increase his resources (e.g., access to a botnet infrastructure, malicious tools, the underground market to monetize stolen credentials, or domain resellers with shady practices), accordingly.

6   Conclusions

We have proposed the Colonel Blotto Phishing (CBP) game to help better understanding the dynamics of the two-step detect-and-takedown defense against phishing attacks. We gained several interesting insights, including the counterintuitive result that it is optimal for the less resourceful attacker to create even more phishing attacks than the resourceful counterpart in equilibrium, and that the attacker will optimally vary his strategies to either increase the number of phishes or to focus on raising his resources depending on the detection probability. We then discussed the implications to the anti-phishing industry.

Capturing the conflicts between an attacker and a defender with asymmetric resources and information, it is our hope that the CBP game can be eventually used to analyze other interesting problems, including measuring the effects of competition between multiple phishers, and the benefits of cooperation between multiple takedown companies. We also see the suitability of the CBP game to be applied to web security problems in general. Indeed, various web security problems, including malicious sites, illegal pharmacies, mule-recruitment and so fourth, are currently mitigated through a detect-and-takedown process similar to in the anti-phishing industry.

Future Work. Like other stylized models, the CBP game can be extended in several directions. A potential extension is to include the time dimension into the game, for example, using repeated games to model the uptime of a phish, which is often used to measure the damage caused by phishing activities. Using the variants of the classic Colonel Blotto game, such as the non-constant sum version [19] in which players might optimally choose not to expend all their resources, may also yield interesting results. We note that it may be interesting also to test our CBP model through experimental studies. Existing studies as conducted in [4,5,7,12] have largely found that subjects were able to play the equilibrium strategies of the classic Colonel Blotto game, with the weak and strong players adopting the 'guerrilla warfare' and 'stochastic complete coverage' strategies respectively. Testing how the subjects will play our two-stage CBP game can be an interesting future work.

Acknowledgment. This research was supported in part by the National Science Foundation under award CCF-0424422 (TRUST).

References

1. Anti-Phishing Working Group (APWG). Advisory on utilization of whois data for phishing site take down, http://www.antiphishing.org/reports/apwg-ipc_Advisory_WhoisDataForPhishingSiteTakeDown200803.pdf
2. APWG. Global phishing survey: Trends and domain name use in 2H2009, http://www.antiphishing.org/reports/APWG_GlobalPhishingSurvey_2H2009.pdf
3. APWG. Global phishing survey: Trends and domain name use in 2H2010, http://www.antiphishing.org/reports/APWG_GlobalPhishingSurvey_2H2010.pdf
4. Arad, A., Rubinstein, A.: Colonel blotto's top secret files. In: Levine, D.K. (ed.) Levine's Working Paper Archive 814577000000000432 (January 2010)
5. Avrahami, J., Kareev, Y.: Do the Weak Stand a Chance? Distribution of Resources in a Competitive Environment. Cognitive Science, 940–950 (2009)
6. Borel, E.: La théorie du jeu les équations intégrales à noyau symétrique. Comptes Rendus de l'Académie des Sciences 173, 1304–1308 (1921); English translation by Savage, L.: The theory of play and integral equations with skew symmetric kernels. Econometrica 21, 97–100 (1953)
7. Chowdhury, S.M., Dan Kovenock, J., Sheremeta, R.M.: An experimental investigation of colonel blotto games. CESifo Working Paper Series 2688, CESifo Group Munich (2009)
8. Gross, O.A., Wagner, R.A.: A continuous colonel blotto game. RAND Corporation RM–408 (1950)
9. Herley, C., Florêncio, D.: A profitless endeavor: phishing as tragedy of the commons. In: Proceedings of the Workshop on New Security Paradigms (NSPW), pp. 59–70. ACM (2008)
10. Kovenock, D., Mauboussin, M.J., Roberson, B.: Asymmetric conflicts with endogenous dimensionality. Purdue University Economics Working Papers 1259, Purdue University, Department of Economics (December 2010)
11. Kovenock, D., Roberson, B.: Conflicts with multiple battlefields. Purdue University Economics Working Papers 1246. Purdue University, Department of Economics (August 2010)
12. Kovenock, D., Roberson, B., Sheremeta, R.M.: The attack and defense of weakest-link networks. Working Papers 10-14. Chapman University, Economic Science Institute (September 2010)

13. McGrath, D.K., Gupta, M.: Behind phishing: an examination of phisher modi operandi. In: Proceedings of the 1st Usenix Workshop on Large-Scale Exploits and Emergent Threats, pp. 4:1–4:8. USENIX Association (2008)
14. Moore, T., Clayton, R.: Examining the impact of website take-down on phishing. In: Proceedings of the 2nd APWG eCrime Researchers Summit, pp. 1–13 (2007)
15. Moore, T., Clayton, R.: The consequence of noncooperation in the fight against phishing. In: Proceedings of the 3rd APWG eCrime Researchers Summit, pp. 1–14 (2008)
16. Moore, T., Clayton, R.: The impact of incentives on notice and takedown. In: Johnson, M. (ed.) Managing Information Risk and the Economics of Security (2008)
17. Moore, T., Clayton, R.: Evil Searching: Compromise and Recompromise of Internet Hosts for Phishing. In: Dingledine, R., Golle, P. (eds.) FC 2009. LNCS, vol. 5628, pp. 256–272. Springer, Heidelberg (2009)
18. Roberson, B.: The colonel blotto game. Economic Theory 29(1), 1–24 (2006)
19. Roberson, B., Kvasov, D.: The non-constant-sum colonel blotto game. CESifo Working Paper Series 2378. CESifo Group Munich (2008)

Investment in Privacy-Preserving Technologies under Uncertainty

Murat Kantarcioglu[1], Alain Bensoussan[2], and SingRu(Celine) Hoe[3]

[1] University of Texas at Dallas, USA
[2] University of Texas at Dallas, USA, The Hong Kong Polytechnic University, HK, and Ajou University, Korea
[3] Texas A&M University-Commerce, USA

muratk@utdallas.edu, alain.bensoussan@utdallas.edu, hoceline02@yahoo.com

Abstract. Entrepreneurs face investment decisions on privacy-preserving technology (PPT) adoption as privacy-concerned consumers may decide whether to use firms' services based on the extent of privacy that firms are able to provide. Kantarcioglu et.al. (2010)[9] contributes to guidelines for entrepreneurs' adoption decisions through a novel framework, which combines copula functions and a Stackelberg leader-follower game with consumers taking the role of the follower (referred as static-copula-game model hereafter). The valuation requires a clearly defined bivariate distribution function of two random variables, the consumer's valuation of private information and the consumer's profitability to a firm. Copula functions are used to construct the bivariate distribution function from arbitrarily univariate marginals with various dependence structures fitting into different market/industry segments. This study extends the static-copula-game model to include project value uncertainty, simultaneously considering different market competition structures and the regulatory promise of random arrival of government mandatory adoption. The project value from the static-copula-game model is used as an estimate of the initial (current) project value for the stochastic evolution. By doing so, we retain the advantages of applying copulas and preserve the established valuation property exclusively applicable to the valuation of PPT adoption. The extension model makes several improvements including: (1) Reduce concerns about myopic PPT adoption decisions that may result when static valuation is employed. (2) Overcome the potential biased PPT adoption decision that may arise due to negligence of market competition impact. (3) Understand the regulatory influence of government mandatory adoption with uncertainty. We find that: (1) If one can link univariate marginals and dependence structures to industry groups, one can determine for which industries project value uncertainty has no impact on the entrepreneur's immediate PPT adoption decision. For these industries, there is no need for government intervention/regulation to accelerate/induce PPT adoption even though the project value is uncertain. (2) Under project value uncertainty, competition may suggest either a later or an earlier PPT adoption compared with the monopoly case. (3) The promise of government mandatory adoption has the potential to accelerate PPT adoption. The PPT adoption guidelines considering competition and regulatory promises of government mandatory

J.S. Baras, J. Katz, and E. Altman (Eds.): GameSec 2011, LNCS 7037, pp. 219–238, 2011.
© Springer-Verlag Berlin Heidelberg 2011

adoption when the project value is uncertain bring useful recommendations to both entrepreneurs and policymakers.

Keywords: Privacy-Preserving Technology, Stackelberg Game, Government Intervention, Random Competition, Uncertainty, Copulas.

1   Introduction

A recent survey by California Health Foundation[12] shows that one of the barriers to consumer acceptance of digitized personal health records is the fear that their private information may not be adequately protected. Since privacy-concerned consumers may consider privacy issues when making their service usage decisions, entrepreneurs encounter investment assessments on the implementation of the privacy-preserving technology (PPT). Perceiving entrepreneurs' investment decision-making problems, Kantarcioglu et.al. (2010)[9] propose a valuation model for PPT adoption (referred as static-copula-game model hereafter).

The value that an individual puts on his private information incentivizes his demand for privacy protection when using a firm's services. Knowing how individual appraises his private information is essential for entrepreneurs facing the PPT adoption decision. The reason is that entrepreneurs can approximate the demand for PPT adoption based on such knowledge. Literature related to quantifying privacy value often invokes the specification of an individual's utility function with postulated risk preferences. However, by means of experimental auctions, [5] is able to quantify individual privacy value without claiming a particular form of utility function with certain risk preferences. In [5], an individual's valuation of his private information is directly affected by his trait's desirability in relation to the categorized group. It shows that an individual in a group would demand a higher value for private information if his trait deviates from that of average population segment in the group, and the further his trait is away from that of average population segment, the higher the value of private information is demanded. For example, overweight people tend to valuate their private weight information with higher prices. By extending the assertion of [5] that an individual valuates privacy based on his trait's desirability, Kantarcioglu et.al. (2010)[9] propose a unique measure for estimating the consumer base that a firm can exploit given PPT adoption. This unique measure links a consumer's decision threshold for service usage considering privacy protection to his valuation of private information.

In the static-copula-game model, a firm's consumer profiles are characterized by two distributions which are the consumer's valuation of private information and the consumer's profitability to a firm. The expected project value is obtained by applying a Stackelberg leader-follower game and copula functions. By applying copulas, Kantarcioglu et.al. (2010)[9] are able to explore how the two underlying distributions and their dependence structures impact an entrepreneur's project valuation on the proposed static-copula-game model. The results from the static-copula-game model provide useful guidelines for entrepreneurs' investment decisions, and give implications about the necessity of government intervention to encourage firms' PPT adoption. The focus on investment assessments and government policy inferences as well as the integration of copulas and the uniquely defined privacy related decision measure with a Stackelberg leader-follower game distinguishes our earlier work from other privacy related literature.

Two major reasons call for improvements on the static-copula-game model. First, there may be a concern that the the project value may deviate from the static value obtained under the static-copula-game model, for example, the variation arising from some consumers' nuisance decision-making. Second, competition among firms is not considered in the static-copula-game model. The current static-copula-game extension model is established to overcome the above-mentioned shortcomings by incorporating project value uncertainty and competition into the valuation. In addition, to gain policy implications about the regulatory effect of legally mandatory adaption under project value uncertainty, the extension model also studies the random arrival of government mandatory adoption.

To model the project value uncertainty considered in the extension model, we use the project value from the static-copula-game model as the estimate of initial project value (i.e., initial state value), which then evolves stochastically to capture the potential variation. By doing so, we preserve the advantages and properties of the static-copula-game model exclusively applied in valuating PPT adoption. In the extension model, given the uncertain project value evolution, the entrepreneur solves his optimal adoption time (i.e., optimal stopping time in the terminology of control theory) that maximizes his expected discounted project value from undertaking PPT adoption. This methodology is in essence the application of option pricing theory and has been termed "real options" by financial economists (see for example Dixit et.al. (1994)[4]).

Various market structures are studied in the extension model. The monopoly situation enables us to directly compare valuation outcomes under project value uncertainty with those from the static-copula-game model. We consider two specific competition structures.[1] One is the Stackelberg leader-follower competition and the other is stochastic arrival of competition. The social network market segment prompts our interest in studying the Stackelberg leader-follower competition. In this market segment, Diaspora[2] is the leader for implementing PPT, and Facebook, the leader of the social network in terms of the market share, may follow the PPT adoption optimally later. On the other hand, we may also expect to observe market segments where entrepreneurs may join the PPT adoption at their discretions randomly. The random arrival of competitors' entries

---

[1] In our other companion work, we study another market competition structure defined by preemption game.

[2] Amid complaints of Facebook's erosion of personal privacy, a team of students at NYU's Courant Institute of Mathematical Sciences is developing a social network built on privacy[13].

depict such a market structure. We summarize the main results and state the contribution in the following two sections, Sect. 1.1 and Sect. 1.2, respectively.

1.1   Summary of Main Results

The following summarizes main observations of this study:

- The project value uncertainty may or may not alter the immediate PPT adoption decision suggested by the static-copula-game model. The outcome depends on the underlying univariate distributions and dependence structures. Therefore, for those market segments in which the PPT adoption brings significant profits, the optimal policy generated from the static-copula-game model will not be myopic even though the project value is uncertain. For such industry segments, the entrepreneur shall be concerned about project value uncertainty only when the uncertainty is above an identified threshold level.
- Depending on the competition structure of the existing market segment, competition may either delay or accelerate PPT adoption compared with the monopoly case. Under the Stackelberg game framework, although the leader's minimum triggering project value for PPT adoption coincides with the monopolist's, in some states of the world, the leader may indeed defer PPT adoption compared with the monopoly case. This is due to the fact that the leader's optimal PPT adoption rule is characterized by a **two-interval strategy**.[3] When the project value from PPT adoption reaches leader's minimum triggering project value for PPT adoption(coincides with the monopolist's PPT adoption trigger), the leader will adopt the PPT immediately as if he were a monopolist. However, unlike the monopolist, the leader will not adopt the PPT once the the project value reaches the range containing a level close to the follower's triggering project value for PPT adoption. In this range, the leader will postpone PPT adoption until the project value reaches his largest triggering project value for PPT adoption. The reason is that, in the range of project values where the follower will undertake PPT adoption soon, the period at which the leader enjoys the monopoly rent is too short to justify PPT adoption under project value uncertainty. In addition, for the industry segments for which the initial project value suggested by the static-copula-game model is at least as large as the leader's largest triggering project value for PPT adoption, we may observe that the leader and the follower make immediate PPT adoption almost simultaneously.

Next, when the market structure presents random arrival of competitors' PPT adoption, the entrepreneur may adopt PPT earlier than if he were a monopolist. This suggests that if the entrepreneur anticipates uncertain multiple entries of competitors, the fear of obtaining smaller project value by

---

[3] A two-interval strategy means that the entrepreneur adopts the technology within two identified intervals, and does not adopt the technology otherwise. In the current case, the intervals are characterized by 3 distinct project values.

adopting late may induce an earlier PPT adoption. In this case, the deferral of PPT adoption due to project value uncertainty may be less significant compared with the industry where only one firm monopolizes the market.

– The promise of random arrival of legally mandatory adoption may induce an earlier PPT adoption. The regulatory promise of mandatory adoption lowers the triggering project value required by the entrepreneur for PPT adoption. The entrepreneur may even undertake PPT adoption despite a negative net payoff if he expects that government mandatory adoption is forthcoming. The prediction of entrepreneur's acceleration on PPT adoption due to regulatory promise may be supported by the FTC's (Federal Trade Commission) recent pushes towards legislation for pushing industry towards self-regulation regarding privacy issues. For example, at the Senate hearing on March 16, 2011, the FTC recommended imposing more stringent measures to protect Internet users against unauthorized tracking, including a universal Do Not Track browser setting. Knowing business entities have little incentive to adopt this stringent privacy protection measure, FTC called for such a universal mechanism to be accomplished by legislation or potentially through robust, enforceable self-regulation[14,15].

1.2   Contribution of This Study

– Kantarcioglu et.al. (2010)[9] contributes to guidelines for entrepreneurs' PPT adoption decisions through integrating copula functions with a Stackelberg leader-follower game. The current study extends the contribution to incorporate the impact of project value uncertainty. We are able to identify industry segments for which there is no need for government intervention/regulation to accelerate/induce PPT adoption even under project value uncertainty unless the uncertainty reaches a high level.
– The second contribution of this study is to provide optimal PPT adoption policies with different market competition structures under project value uncertainty. We find that the effect of competition on PPT adoption under project value uncertainty is inconclusive compared with the monopoly case. The impact depends on the nature of market competition structures.
– The third contribution is that we shed light on the the regulatory effect of legally mandatory adoption under project value uncertainty. The important message to policymakers is that a credible promise of government mandatory adoption may be a powerful tool to induce PPT adoption.

We omit proofs to keep the presentation concise; all proofs are available on http://www.utdallas.edu/~mxk055100/publications/pptoptions.pdf. The remainder of the paper is organized as follows. In Section 2, we give a short review of relevant literature. In Section 3, we briefly review the static-copula-game valuation model. The project value from the static-copula-game model serves as an input parameter for implementing the extension model. In Section 4, we introduce various models under project value uncertainty and give the corresponding optimal PPT adoption rules. In Section 5, we make qualitative discussions and comparisons of the optimal PPT adoption rules for each scenario. We present concluding remarks in Section 6.

2   Review of Relevant Literature

Privacy is a central concern in the information age, and has attracted much research attention. One major research area on privacy related works focuses on issues related to clarifying the privacy tradeoffs that individuals will make to gain access to specific services or quantifying individuals' privacy values. [5] seeks to quantify how general privacy attitudes impact the price participants set for revealing private information. Through experimental auctions, they show that an individual's trait's desirability in relation to the group plays a key role in the amount people demand to publicize private information. Working on clarifying the tradeoffs between gaining service access and relinquishing private information, [7] concludes that the information-seeking organization has to offer financial incentives and convenience (i.e., privacy mitigation strategies) in exchange for individuals to relinquish personal information.

Researchers have extended the study of clarifying tradeoffs between preserving privacy and gaining specific service access to encompass broader issues. For example, studies devote to finding optimal online service strategies for firms competing against personal information, optimal online privacy protection regimes for firms, consumers and society,...etc. Modelling consumer's utility function based on the privacy-benefit tradeoffs, [2] analyzes the equilibrium strategy of online personalization service offerings under duopolistic markets, and gives policy implications and managerial recommendations. [11] studies the optimal online privacy protection regimes for consumers, retailers and society. The regimes studied include self-regulation, mandatory standards and caveat emptor. They show that the optimal regime depends on the number of individuals facing a loss from privacy violations and the size of loss they face.

Observing that research [3,7,10] has uncovered a dichotomy between stated attitudes and actual behavior of individuals facing decisions on privacy and personal information security, [1] provides an analysis of the dichotomy, outlining an experimental design to test their hypotheses about the observed inconsistency. More recently, multidisciplinary fields of human-computer interaction (HCI) have emerged with a raft of work on privacy in computing. [6] updates the development of HCI from psychology aspects. They give explicit attention to the emergence of computer-supported cooperative work and point out that having both "useful and usable" computing systems are of paramount importance. In accordance with the "useful and usable" criterion postulated by [6], [8] proposes a privacy expectations and security assurance offer system.The proposed system bears the benefits of enhancing consumer privacy choices, creating a market for privacy preferences, and providing direct incentives for privacy offering organizations to care about the security of personal information.

[9] is the closest related paper to the current work; both studies focus on the analysis of entrepreneurs' investment decision on PPT adoption. In [9], two

important factors, a customer's valuation of his private information and a customer's profitability to the firm, affect entrepreneurs' valuation. The valuation is viewed as a consequence of a Stackelberg leader-follower game under complete information with consumers taking the role of the follower. Their uniquely defined measure of consumers' service usage decision is inspired by [5]. Copula functions are applied to obtain the expected project value. They formulate guidelines for entrepreneurs' optimal adoption decisions and identify several cases where the government intervention may be required to have firms invest in privacy-preserving technologies requiring significant costs. The current study extends the valuation model of [9] by retaining the advantages of applying copulas and keeping the established valuation property exclusively applied to appraising PPT adoption. The extension model improves the static model of [9] by integrating project value uncertainty and competition structures into the valuation model. In addition, to gain policy implications from government promises of mandatory adoption, the current study extends the static framework of [9] to model random arrival of legally mandatory adoption.

3   Static-Copula-Game Model - Integrating Copula Functions with a Stackelberg Game

We first briefly review the static-copula-game model and the resulting PPT adoption rule proposed in our earlier work, Kantarcioglu et.al. (2010)[9].

3.1   Basic Information for Model Setup

We consider a firm facing a PPT investment problem. The adoption of this privacy-preserving technology P, will pose a fixed investment cost K,[4] which measures both the fixed adoption cost as well as the opportunity cost. The opportunity cost measures the potential loss caused by preventing firms from using certain private information as a consequence of PPT adoption(eg. facebook's new privacy controls). For a firm's potential consumer group, I, each individual is assigned a consumer profile characterized by his valuation of private information and his profitability to the firm. For example, for customer $i \in I$, his profile revealed to the firm is (x_i, y_i), where x_i represents consumer i's valuation of private information and y_i represents consumer i's profitability to the firm. We assume

---

[4] It does not necessarily mean that this analysis work is only appropriate for valuating a single PPT adoption. Rather, we can consider the privacy-preserving technology P as any possible combination of available technologies resulting in different services and costs. That is, P can be considered as an element of the power set $\mathcal{P}(S)$, $S = \{s_1, s_2, ..., s_n\}$, $n \geq 1$ where s_i, $i = 1, 2...n$ represents different technologies. Thus, there would be at most 2^n possible combinations as well as associated costs. When $S = \emptyset$, it indicates that no PPT is evaluated; hence $K = 0$ in this scenario. When $n = 1$, it resorts to the valuation of single PPT adoption. When $n > 1$, we can valuate all potential combinations and rank them in order to make the best investment decision.

complete information, that is, x_i and y_i are publicly available information.[5] The random variables associated with consumers' valuation of private information, X, and consumers' profitability to a firm, Y, can be best described by their descriptive probability distributions, which can be characterized by corresponding cumulative distribution functions, $F_X(x)$ and $F_Y(y)$ respectively.

To obtain the project value of adopting the privacy-preserving technology P with the privacy-protection level α_P, we view the firm's investment valuation as a Stackelberg leader-follower game under complete information with consumers taking the role of the follower.

3.2   Customer's (Follower's) Decision Function

We solve the problem with backward induction. Given firm's adoption of the privacy-preserving technology P, an individual consumer chooses his utility maximization strategy. We assume homogenous individuals. For each individual consumer, we define his utility function as:

$$U(x, D) = (2D - 1)\left(\alpha_P - \left(a \times \frac{|x - \mu_X|}{\sigma_X} + b\right)\right)$$

where $D = \{0, 1\}$, μ_X and σ_X represent the mean and the standard deviation of X respectively, both $a > 0$ (a given weight in the model) and $b > 0$ (a basic level of privacy related to values of private information common to the population; exogenously determined in the model) are constants, and $\alpha_P > b$.

The term $a \times \frac{|x - \mu_X|}{\sigma_X} + b$ measures individual consumer's fair level of privacy. This specification is motivated by [5] in which the authors show that an individual in a group would demand a higher value for private information if his trait deviates from that of the average population segment in the group. Moreover, the further an individual's trait is away from that of average population segment, the higher the value for private information he demands. It is clear that rational consumers would choose $D = 1$ if $\alpha_P \geq (a \times \frac{|x - \mu_X|}{\sigma_X} + b)$ and $D = 0$ otherwise. That is, a consumer's optimal strategy in response to his utility maximization solution relies solely on his value of private information and privacy protection that the firm's PPT can provide. We define such a rule as a consumer's decision function given:

$$D(x) = \mathbb{1}_{a \times \frac{|x - \mu_X|}{\sigma_X} + b \leq \alpha_P}.$$

3.3   Firm's (Leader's) Project Value and Investment Decision

Once consumers' optimal decisions have been solved, the firm integrates consumers' decisions into their valuation. The expected project value from PPT

---

[5] In practice, we may identify these information through marketing research for example.

adoption is the expected profits that a firm can receive from consumers using services:

$$v_0 = \mathrm{E}[YD(X)] = \int_{X \times Y} yD(x)dF_{X,Y}(x,y). \tag{1}$$

where $F_{X,Y}(x,y)$ is the joint distribution function. From (1), a clearly defined joint distribution function is required for valuation. We propose to employ copula functions for the bivariate distribution function. This allows us to study project values through a richer class of joint distribution functions fitting into different market segments and to investigate the impact of dependence structure on the project value.[6]

In return for the project value obtained by (1), the firm incurs an adoption cost K. Clearly, the market mechanism makes it the case that the firm will make the investment if and only if

$$v_0 \geq K. \tag{2}$$

4   Static-Copula-Game Extension Model

The project value obtained from (1) is static. However, there may be uncertainty embedded in this project value estimate. For example, the consumer's actual service usage decision may deviate from the estimated formula given by $D(x)$ since consumers may simply decide not to use the service and vice versa. A concern about myopic adoption decisions may thus arise due to the potential project value deviation from the static estimate given by (1).

To model project value uncertainty, we propose to use v_0 obtained from (1) in Sect. 3.3 as the initial state value, which then evolves according to a stochastic process to capture the potential variation of the project value. By doing so, we achieve two major desired properties. First, the valuation is exclusively applied to PPT adoption since v_0 is obtained under the valuation model developed solely for PPT adoption. Second, we retain the major advantage of the static-copula-game model brought from employing copula functions to obtain bivariate joint distribution functions required for the valuation, that is, the ability to exploit rich classes of distributions as well as the flexibility in describing dependence structures fitting into different market/industry segments.

Therefore, the project value initiates from varieties of univariate marginals and copulas based on the characteristics of industry/market segments. And it then evolves stochastically. We describe the uncertain project value evolution by a geometric Brownian motion process. We choose this process for two main reasons. First, the project value defined by (1) is the expected profits that a firm can receive from consumers using the service, thus it shall be nonnegative. The geometric Brownian motion process satisfies this condition. Second, Browian

---

[6] In practice, for generating empirically validated or theoretically supported cumulative distribution functions, we may first identify individual marginals, and then use existing copula families or construct empirical copulas for dependence structures.

motion used to describe random movements of particles (Particle Theory) captures the randomness (uncertain movements) of the project value. We define the project value evolving according to:

$$dV(t) = \alpha V(t)dt + \varsigma V(t)dW(t), \ V(0) = v_0 \tag{3}$$

where $W(t)$ is a standard Wiener process(i.e. Brownian motion), and both α and ς are constants, representing the drift rate and the volatility of the project value respectively. As we are mainly interested in the uncertainty of the project value, we would take $\alpha = 0$ in our study. Note that in such a specification, given zero uncertainty, i.e. $\varsigma = 0$, the entrepreneur will undertake the investment as soon as $V(0) = v_0 \geq K$, recovering our static valuation rule proposed in (2), Sect. 3.3.

After setting up the stochastic evolution of the project value due to potential nuisances, we proceed to study optimal PPT adoption rules under different scenarios, which consists of various market competition structures as well as the random arrival of legally mandatory adoption.

4.1   The Case of Monopoly

Given the evolution of the project value defined by (3), the entrepreneur's problem is to find the optimal time to adopt the privacy-preserving technology P (i.e., an optimal stopping time) by maximizing the expected discounted project value:

$$M(v) = \sup_{\tau \geq 0} \mathrm{E}\left[e^{-\mu\tau}\left(V_{v_0}(\tau) - K\right)\mathbf{1}_{\tau < \infty}\right], \tag{4}$$

where μ is the required rate of return required by the entrepreneur.

Assuming that the function $M(v)$ is sufficiently smooth, $M(v)$ solves the following variational inequality (V.I.) as a consequence of Dynamic Programming:

$$\begin{cases} \frac{1}{2}M''(v)v^2\varsigma^2 - \mu M(v) \leq 0 \\ M(v) \geq v - K \\ [M(v) - (v - K)]\left[\frac{1}{2}M''(v)v^2\varsigma^2 - \mu M(v)\right] = 0 \\ M(0) = 0; \ M(v) \geq 0; \ M(v) \ \ \text{has linear growth at infinity.} \end{cases} \tag{5}$$

Theorem 1.

$$M(v) = \begin{cases} \dfrac{K}{\beta - 1}\left(\dfrac{v}{v^*}\right)^\beta & v \leq v^* \\ v - K & v \geq v^* \end{cases}, \tag{6}$$

where $\beta = \frac{1}{2} + \sqrt{\frac{1}{4} + \frac{2\mu}{\varsigma^2}} > 1$, *and* $v^* = \frac{\beta K}{\beta - 1}$.

The optimal stopping rule (i.e., the optimal time to adopt the PPT) which achieves the supremum in (4) is: $\tau^*(v) = \inf\{t|V_{v_0}(t) \geq v^*\}$. That is, the entrepreneur will adopt the PPT as soon as the project value reaches v^* from below.

4.2   The Case of a Stackelberg Leader-Follower Game

We next consider a duopoly market setting with a Stackelberg leader-follower type. This is a market segment where a clear market leader of PPT adoption exists, and the other firm (the follower) will undertake PPT adoption only after the leader's adoption. The social network market segment may be a good example of such market segments, where Diaspora is the leader for PPT adoption and Facebook may be considered to be the follower for PPT adoption.

Under the Stackelberg leader-follower competition, the leader can enjoy the whole project value from PPT adoption at the onset, but he must surrender a portion of project value to the follower upon the follower's optimal PPT adoption. We assume that the leader retains πV portion of the project value upon the follower's optimal entry with $\pi \in (0,1)$, leaving $(1-\pi)V$ to the follower. We note that to assume that the leader and the follower obtain a portion of project value once they both undertake PPT adoption has the same effect as modelling the market demand function faced by each firm. For the case of symmetric firms, we take $\pi = \frac{1}{2}$.

A. Optimal Adoption Time for the Follower. The follower solves the same optimal PPT adoption problem as the monopolist after the leader has adopted the PPT. Thus, the follower's optimal PPT adoption strategy is identical to that described in the monopolist's case. However, since the follower can only enjoy $(1-\pi)V$ portion of the project value, the value function and the triggering investment threshold are slightly different from those of the monopoly case, which are given:

$$F(v) = \begin{cases} \frac{K}{\beta-1}\left(\frac{v}{\hat{v}}\right)^{\beta} & v \leq \hat{v} \\ (1-\pi)v - K & v \geq \hat{v} \end{cases}, \tag{7}$$

where $\beta = \frac{1}{2} + \sqrt{\frac{1}{4} + \frac{2\mu}{\varsigma^2}} > 1$, and $\hat{v} = \frac{\beta K}{(1-\pi)(\beta-1)}$.

The follower's optimal time to adopt the privacy-preserving technology P which maximizes the expected discounted project value is: $\hat{\tau}(v) = \inf\{t|V_{v_0}(t) \geq \hat{v}\}$.

We note that the adopting time $\hat{\tau}(v)$ is the optimal adoption timing if the follower could adopt PPT at time zero. Since the follower will adopt PPT only after the leader has already done so (who adopts at time θ), for finite θ, the follower will adopt at time:[7]

$$\hat{\tau}_\theta = \theta + \hat{\tau}\left(V_{v_0}(\theta)\right). \tag{9}$$

---

[7] For any test function $\Psi(x,s)$, we have:

$$E[\Psi(V_{v_0}(\hat{\tau}_\theta), \hat{\tau}_\theta)|\mathcal{F}_\theta] = \Psi(V_{v_0}(\theta), \theta)\mathbb{1}_{V_{v_0}(\theta)\geq \hat{v}} + \mathbb{1}_{V_{v_0}(\theta)<\hat{v}}E[\Psi(\hat{v}, t+\hat{\tau}(v))]|_{v=V_{v_0}(\theta), t=\theta} \tag{8}$$

B. Optimal Adoption Time for the Leader. We now proceed to the case of the leader. Recall in such a market segment, the follower will only adopt the PPT following the leader's adoption. Thus, when the leader adopts the PPT at time $\theta < \infty$, by paying cost K, he anticipates to receive $V_{v_0}(\theta)$; however, he also anticipates that a rational follower will enter at $\hat{\tau}_\theta$, at which time, he surrenders $(1 - \pi)V_{v_0}(\hat{\tau}_\theta)$. So at time θ, if $\theta < \infty$, the leader receives:

$$V_{v_0}(\theta) - K - (1 - \pi)\mathrm{E}\left[e^{-\mu(\hat{\tau}_\theta - \theta)}V_{v_0}(\hat{\tau}_\theta)\mathbb{1}_{\hat{\tau}_\theta < \infty}|\mathcal{F}_\theta\right]$$
$$= \pi V_{v_0}(\theta)\mathbb{1}_{V_{v_0}(\theta) \geq \hat{v}} + (V_{v_0}(\theta) - \beta F(V_{v_0}(\theta)))\mathbb{1}_{V_{v_0}(\theta) < \hat{v}} - K,$$

where we use the fact that $\dfrac{(1 - \pi)\hat{v}}{(1 - \pi)\hat{v} - K} = \beta$ and $F(v)$ is defined in (7). The leader's problem can be expressed as:

$$L(v) = \sup_{\theta \geq 0} \mathrm{E}\left[e^{-\mu\theta}\Psi(V_{v_0}(\theta))\mathbb{1}_{\theta < \infty}\right], \tag{10}$$

where

$$\Psi(v) = \pi v\mathbb{1}_{v \geq \hat{v}} + (v - \beta F(v))\mathbb{1}_{v < \hat{v}} - K. \tag{11}$$

and $L(v)$ must satisfy:

$$L(v) \geq 0; \quad L(v) \geq \Psi(v). \tag{12}$$

The obstacle, $\Psi(v)$, is only continuous, not $C^1(0, \infty)$ with the only point of non-differentiability, $\hat{v}$. Setting $U(v) = L(v) - \pi v + K$, we have:[8]

$$U(V) = \sup_{\theta \geq 0} \mathrm{E}\left\{e^{-\mu\theta}\left(\Psi(V_{v_0}(\theta)) - \pi V_{v_0}(\theta) + K\right)\mathbb{1}_{\theta < \infty}\right.$$
$$\left. + \int_0^\theta e^{-\mu s}\left(-\pi\mu V_{v_0}(s) + \mu K\right)ds\right\}$$
$$= \sup_{\theta \geq 0} \mathrm{E}\left\{e^{-\mu\theta}\chi(V_{v_0}(\theta))\mathbb{1}_{\theta < \infty} + \int_0^\theta e^{-\mu s}f(V_{v_0}(s))ds\right\}. \tag{13}$$

This formulation leads to an optimal stopping time problem with an obstacle $\chi(v) = \Psi(v) - \pi v + K$ and a running profit $f(v) = -\mu\pi v + \mu K$.
We have:

$$0 \leq \chi(v) \leq K, \quad \text{and} \quad 0 \leq U(v) \leq K. \tag{14}$$

Theorem 2. *There exist three points $0 < v_1 < v_2 < \hat{v} < v_3$ such that:*

$$\begin{cases} -\dfrac{1}{2}U''(v)\varsigma^2 v^2 + \mu U(v) = -\pi\mu v + \mu K \\ \qquad\qquad 0 < v < v_1 \text{ and } v_2 < v < v_3 \\ U(v) = \chi(v) \ \text{for } v_1 \leq v \leq v_2 \\ U(v) = 0 \ \text{for } v \geq v_3 \end{cases} \tag{15}$$

with matching conditions: $U'(v_1) = \chi'(v_1)$, $U'(v_2) = \chi'(v_2)$, and $U'(v_3) = 0$.

---

[8] It is straightforward that the uniqueness and existence of solution, $U(v)$, guarantees the uniqueness and existence of solution, $L(v) = U(v) + \pi v - K$.

We can define the leader's optimal stopping rule as:

$$
\hat{\theta}(v) = \begin{cases} \inf\{t|V_{v_0}(t) \geq v_1\} & \text{if } 0 \leq v < v_1 \\ 0 & \text{if } v_1 \leq v \leq v_2 \\ \inf\{t|V_{v_0}(t) \leq v_2 \text{ or } V_{v_0}(t) \geq v_3\} & \text{if } v_2 < v < v_3 \\ 0 & \text{if } v \geq v_3 \end{cases} \tag{16}
$$

C. Summary of Optimal Rules for the Stackelberg Leader and Follwer.

We summarize the leader's and the follower's optimal PPT adoption rules as follows:

1. If $v < v_1$, the leader waits to adopt the PPT until $v \geq v_1$, and the follower undertake the adoption when $v \geq \hat{v}$.
2. If $v_1 < v < v_2$, the leader adopts the PPT immediately, and the follower undertake the adoption when $v \geq \hat{v}$.
3. If $v_2 < v < v_3$, the leader waits to adopt the PPT until v moves outside the interval bounded by v_2 and v_3. The follower will only undertake PPT adoption when $v \geq \hat{v}$ after the leader has already implemented PPT. The reason is that in this market segment, the follower will undertake PPT adoption only after the leader has already done so.
4. If $v \geq v_3$, the leader adopts PPT immediately, and the follower then makes his move. When undertaking PPT adoption yields big profits, we anticipate to observe almost simultaneous PPT adoption.

The reason that the leader will not adopt the PPT in the interval (v_2, v_3) is because it contains a level close to $\hat{v}$, the follower's triggering project value for PPT adoption. It implies that the follower would adopt the PPT pretty soon after the leader's adoption. As a consequence, the time that the leader enjoys monopoly rents is too short to justify his PPT adoption. Therefore, the rational leader would wait until V moves outside this interval, which either grants him a longer period of enjoying monopoly rents or compensates him with a sufficiently large project value for the loss of monopoly rents. The PPT adoption pattern observed in the social network market segment may support our theoretical prediction.

4.3   Random Arrival of Competitors' Entries

We now consider another market competition structure. Under this scenario, the entrepreneur is aware that more entrepreneurs may adopt the PPT subsequently at some random time in the future, causing unpredictable but sizable drops in the project value. Given this market structure, the project value, V, follows a mixed Brownian motion/jump process:

$$
\begin{cases} dV(t) = \alpha V(t)dt + \varsigma V(t)dW(t) - V(t)dq(t), \ V(0) = v_0, \\ dq(t)_{(\varsigma,\phi)} = \begin{cases} \phi, & \text{with probability } \zeta dt, \\ 0, & \text{with probability } 1 - \zeta dt, \end{cases} \end{cases} \tag{17}
$$

where $W(t)$, α and ς are defined the same as in (3), and we assume $E[dq(t)dW(t)] = 0$. Equation (17) indicates that V will fluctuate as a geometric Brownian motion, which captures the continuous nuisance, but there is a probability ζdt that the value will drop to $(1-\phi)V$ over each time interval dt due to more entrepreneurs' PPT adoption to share the market. As in the geometric Brownian motion case, we take $\alpha = 0$ for we are only interested in the impact of uncertainty. Thus, when $\varsigma = 0$ and $\zeta = 0$, it resorts to the static valuation case and the adoption rule $v_0 \geq K$ applies.

Optimal Adoption Time. The entrepreneur's problem is to find the optimal time to adopt privacy-preserving technology P by maximizing the expected discounted project value, as described in (4), with V subject to the evolution described by (17). We denote the corresponding value function by $M_J(v)$ to distinguish the different underlying process assumption.

Assuming that the function $M_J(v)$ is sufficiently smooth, $M_J(v)$ solves the following V.I. as a consequence of Dynamic Programming:

$$\begin{cases} \frac{1}{2}M_J''(v)v^2\varsigma^2 + \zeta M_J[(1-\phi)v] - (\mu+\zeta)M_J(v) \leq 0 \\ M_J(v) \geq v - K \\ [M_J(v)-(v-K)]\left[\frac{1}{2}M_J''(v)v^2\varsigma^2 + \zeta M_J[(1-\phi)v] - (\mu+\zeta)M_J(v)\right] = 0 \\ M_J(0)=0; \ M_J(v) \geq 0; \ M_J(v) \ \text{has linear growth at infinity.} \end{cases}$$

$$(18)$$

Theorem 3

$$M_J(v) = \begin{cases} \frac{K}{\beta-1}\left(\frac{v}{\hat{v}_J}\right)^\beta & v \leq \hat{v}_J \\ v - K & v \geq \hat{v}_J \end{cases}, \qquad (19)$$

where β is the positive root of $\frac{1}{2}\varsigma^2\beta^2 - \frac{1}{2}\varsigma^2\beta + \zeta(1-\phi)^\beta - (\mu+\zeta) = 0$ and $\hat{v}_J = \frac{\beta}{\beta-1}K$.

The optimal stopping rule is: $\hat{\tau}_J(v) = \inf\{t|V_{v_0}(t) \geq \hat{v}_J\}$. That is the entrepreneur will adopt the PPT as soon as the project value reaches $\hat{v}_J$ from below.

4.4   Random Arrival of Legally Mandatory Adoption

The project value still follows the geometric Brownian motion process given by (3). However, to gain potential implications of regulatory promises, we additionally consider the random arrival of government mandatory adoption. Once the government regulation for mandatory adoption arrives, the entrepreneurs have no choice but to adopt the PPT. We consider the random arrival time of government mandatory adoption, T, as a Poisson process with the mean arrival rate $1/\lambda$, independent of Weiner process $W(t)$ defined in (3). To focus on the regulatory effect of government mandatory adoption, we do not consider competitions under this setting. At the time of the mandatory adoption enacted, the project value will jump down. The drop size of the project value depends on the number of entrepreneurs required to obey the law. We assume that upon the arrival of this government regulation, the project value will jump down to $(1 - \psi)V$ with $\psi \in (0,1]$. The parameter ψ captures the effect of project value dilution due to sharing the market with other participants resulting from regulatory enforcement.

Optimal Adoption Time. The entrepreneur's objective function is:

$$J_{v_0}(\tau) = \mathrm{E}\left[e^{-\mu\tau\wedge T}\left[\mathbb{1}_{\tau<T}\left(V_{v_0}(\tau) - K\right) + \mathbb{1}_{\tau\geq T}\left((1 - \psi)(V_{v_0}(T) - K)\right)\right]\right]$$

$$= \mathrm{E}\left[e^{(-\mu+\lambda)\tau}\left(V_{v_0}(\tau) - K\right)\mathbb{1}_{\tau<\infty}\right]$$

$$+ \lambda\mathrm{E}\left[\int_0^\tau e^{-(\lambda+\mu)s}\left((1 - \psi)V_{v_0}(s) - K\right)ds\right] \tag{20}$$

and the associated value function is:

$$M_R(v) = \sup_\tau J_{v_0}(\tau). \tag{21}$$

Assuming that the function $M_R(v)$ is sufficiently smooth, $M_R(v)$ solves the following V.I. as a consequence of Dynamic Programming:

$$\begin{cases} \frac{1}{2}M_R''(v)v^2\varsigma^2 - (\mu + \lambda)M_R(v) + \lambda\left((1 - \psi)V - K\right) \leq 0 \\ M_R(v) \geq v - K \\ \left[M_R(v) - (v - K)\right]\left[\frac{1}{2}M_R''(v)v^2\varsigma^2 - (\mu + \lambda)M_R(v) + \lambda\left((1 - \psi)V - K\right)\right] = 0 \\ M_R(v) \geq \frac{(1-\psi)\lambda}{\lambda+\mu}v - \frac{K\lambda}{\lambda+\mu} \\ M_R(v) \quad \text{has linear growth at infinity.} \end{cases} \tag{22}$$

Theorem 4

$$M_R(v) = \begin{cases} \frac{K}{\beta - 1}\frac{\mu}{\mu + \psi\lambda}\left(\frac{v}{\hat{v}_R}\right)^\beta + \frac{(1 - \psi)\lambda}{\lambda + \mu}v - \frac{K\lambda}{\lambda + \mu} & v \leq \hat{v}_R \\ v - K & v \geq \hat{v}_R \end{cases} \tag{23}$$

where $\hat{v}_R = \frac{\beta}{\beta-1}\frac{\mu}{\psi\lambda+\mu}K$ and $\beta = \frac{1}{2} + \sqrt{\frac{1}{4} + \frac{2(\mu+\lambda)}{\varsigma^2}}$.

The optimal stopping rule is: $\hat{\tau}_R(v) = \inf\{t|V_{v_0}(t) \geq \hat{v}_R\}$. That is the entrepreneur will adopt the PPT as soon as the project value reaches $\hat{v}_R$ from below.

Remark 1. When $\lambda = 0$, i.e., no promise of potential government mandatory adoption, the solution recovers the monopoly case in Section 4.1.

Table 1. Summary of Adoption Rules under Various Scenarios

Scenarios	Adoption Trigger	Adoption Rules
Monopoly	$v^* = \frac{\beta}{\beta-1}K$, $\beta = \frac{1}{2} + \sqrt{\frac{1}{4} + \frac{2\mu}{\varsigma^2}}$	$v \geq v^*$
Stackelberg Leader	v_1, v_2, v_3; $v_1 > v_2 > v_3$; $v_1 = \frac{\beta}{\beta-1}K$	$v_1 \leq v \leq v_2$, $v \geq v_3$
	$\beta = \frac{1}{2} + \sqrt{\frac{1}{4} + \frac{2\mu}{\varsigma^2}}$	
Stackelberg Follower	$\hat{v} = \frac{\beta}{(1-a)(\beta-1)}K$, $\beta = \frac{1}{2} + \sqrt{\frac{1}{4} + \frac{2\mu}{\varsigma^2}}$	$v \geq \hat{v}$
Random Arrivals of	$\hat{v}_J = \frac{\beta}{\beta-1}K$, β is the positive root of	$v \geq \hat{v}_J$
Competitor's Entries	$\frac{1}{2}\varsigma^2\beta^2 - \frac{1}{2}\varsigma^2\beta + \zeta(1-\phi)^\beta - (\mu + \zeta) = 0$.	
Random Arrival of	$\hat{v}_R = \frac{\beta}{\beta-1}\frac{\mu}{\psi\lambda+\mu}K$, $\beta = \frac{1}{2} + \sqrt{\frac{1}{4} + \frac{2(\mu+\lambda)}{\varsigma^2}}$	$v \geq \hat{v}_R$
Mandatory Regulation		

5   Discussion

Table 1 summarizes the results of project values triggering PPT adoption and the optimal adoption rules under various scenarios presented in Sect. 4.

From Table 1, except for the case of random arrival of government mandatory adoption, the project value triggering PPT adoption is greater than the adoption cost K since $\frac{\beta}{\beta-1} > 1$. It thus indicates that once the project value uncertainty is considered, the project value triggering PPT adoption is higher than the one without uncertainty in which the triggering project value is K (see (2)). Thus, we arrive at the following proposition.

Proposition 1. *In general, the introduction of project value uncertainty would require a higher triggering project value for justifying PPT adoption.*

Proposition 2. *For the industry segments for which the two underlying variables, customers' valuation of private information and customers' profitability to a firm, are either independent or binormally distributed, the concern about project value uncertainty would further impede the entrepreneur's PPT adoption.*

Proposition 2 is a direct result from the static-copula-game model by [9]. Form Proposition 1 and 2 of the static-copula-game model by [9], it identifies that for the industry for which customers' valuation of private information and customers' profitability to a firm are either independent or binormally distributed, the entrepreneur is unlikely to undertake PPT adoption requiring significant investment costs. Therefore, for such industries, the introduction of uncertainty would further discourage PPT adoption since the entrepreneur asks for a higher project value to justify the investment under uncertainty.

Proposition 3. *Under different univariate marginals and copula functions, we obtain different v_0 by (1). For any $v_0 \geq \frac{\beta}{\beta-1}K$, holding other else being constant, the relation holds for $\varsigma \leq \bar{\varsigma}$ where $\bar{\varsigma}$ is determined by v_0.*

Proposition 3 suggests that, for the industry segment with $v_0 \geq \frac{\beta}{\beta-1}K$, the immediate PPT adoption rule suggested by the static valuation will not be myopic due to the introduction of project value uncertainty unless the uncertainty is above the identified level, $\bar{\varsigma}$. In other words, the impact of project value uncertainty is industry specific if one can associate univariate distributions and dependence structures with various industries. With inputs from industrial users, an industry specific table, which identifies the tolerable level of project value uncertainty, may be produced for assisting managerial decision-making.

In our extension model, we introduce competition structures to study the impact of competition on PPT adoption. In the following, we first compare the optimal adoption rules of the Stackelberg leader and follower with that of the monopolist.

Proposition 4. *1. $v_1 = \frac{\beta K}{\beta-1} = v^*$, and $\hat{v} = \frac{\beta K}{(1-\pi)(\beta-1)} > \frac{\beta K}{\beta-1} = v^*$.*
2. For any v_0 from (1) with $v_0 \geq v_3$, holding other else being constant, the relation holds for $\varsigma \leq \bar{\varsigma}$ where $\bar{\varsigma}$ is determined by v_0.

Proposition 4 indicates that the leader's triggering project value for PPT adoption is at least as large as his triggering project value as a monopolist. The necessity of sharing the project value upon the follower's optimal PPT adoption does not affect the entrepreneur's minimum triggering project value for PPT adoption compared with the monopoly case. The reason is that below this triggering project value, the project value is too low to justify PPT adoption under uncertainty. On the other hand, the follower's triggering project value for PPT adoption is larger than a monopolist's. In addition, from Table 1, we observe that the leader's optimal adoption rule is a ***two-interval strategy***. For $v_2 < v < v_3$, the leader will not adopt the PPT due to the promise of losing monopoly rents soon since the follower is likely to undertake PPT adoption soon. Combining the fact that the follower would delay PPT adoption and the fact that the leader will not adopt the PPT for $v_2 < v < v_3$, the Stackelberg leader-follower competition structure may delay PPT adoption compared with the case when only one firm monopolizes the market.

In addition, for the industry segment with $v_0 \geq v_3$, we will observe almost simultaneously immediate PPT adoption. As proposed in Proposition 3, this immediate PPT adoption decision, which also coincides with the rule generated by the static valuation, will hold for $\varsigma \leq \bar{\varsigma}$ (i.e., the project value uncertainty level is below the threshold level, $\bar{\varsigma}$), which is industry specific. Figure 1 shows the relation of the triggering project value for the monopolist, the Stackelberg leader and the Stackelberg follower.

We continue to compare the optimal PPT adoption rule for the entrepreneur facing random arrival of competitors' entries with that for a monopolist.

Proposition 5. *$\hat{v}_J < v^*$. $\hat{v}_J$ is inversely related to ζ and ϕ.*

Proposition 5 suggests that when the potential market competition structure is introduced in the form of random arrival of competitors' entries, the anticipated decreases in the project value due to potential market competition will induce the entrepreneur to adopt the PPT earlier compared with the monopoly

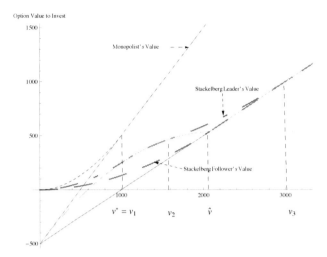

Fig. 1. Optimal Stopping Rules for the Monopolist, the Stackelberg Leader and the Stackelberg Follower

case. The greater the probability that competitors enter into the market soon to share the project value, the lower the triggering project value that the entrepreneur requires to adopt the PPT. Furthermore, the larger the decrease in the project value due to competitors' entries, the lower the triggering project value is required for PPT adoption.

Combining Proposition 4 and 5, we find that market competition structure may present different impacts on PPT adoption. For a Stackelberg leader-follower type framework, PPT adoption may occur later compared with the monopoly case. If the entrepreneur expects to encounter random arrival of competitors' entries, the entrepreneur may be inclined to adopt the PPT earlier than he would as a monopolist or the Stackelberg leader.

Finally, the study of random arrival of legally mandatory adoption arrives at the following two propositions.

Proposition 6. $\hat{v}_R < v^*$. $\hat{v}_R$ *is inversely related to* λ.

Proposition 6 implies that the expected random arrival of government enforcement regulation may accelerate PPT adoption. The higher the possibility that the enforcement regulation will enact soon, the lower the triggering project value that the entrepreneur would require to adopt the PPT.

Proposition 7. $\hat{v}_R \leq K$ *if* $\frac{\beta}{\beta-1}\frac{\mu}{\mu+\psi\lambda} \leq 1$.

Proposition 7 signifies that, due to the anticipation of government mandatory adoption, the entrepreneur may indeed adopt the PPT with the triggering project value smaller than the adoption cost. Combining Proposition 6 and 7, it may suggest that the credible promise of government mandatory enforcement could bring

Table 2. Adoption Triggering Project Value v.s. the Mean Arrival Rate of Government Enforcement Regulation($\frac{1}{\lambda}$); Values of Parameters: $\varsigma = 0.4$, $\mu = 0.05$, $\psi = 1$.

λ	$\frac{\beta}{\beta-1}\frac{\mu}{\mu+\psi\lambda}$	$\hat{v}_R$ versus K
0.01	2.5	$\hat{v}_R > K$
0.03	1.64	$\hat{v}_R > K$
0.05	1.19	$\hat{v}_R > K$
0.07	0.92	$\hat{v}_R < K$

PPT adoption to society sooner. Table (5) illustrates the relations among the mean arrival rate of government enforcement regulation, the triggering project value for PPT adoption, and the adoption cost.

6   Conclusion

This study extends the static-copula-game model by [9] to include project value uncertainty and different market competition structures. The promise of random arrival of government mandatory adoption is also studied for policy implications. The extension model reduces the concern about myopic PPT adoption decisions that may result when static valuation is employed, overcomes the potential biased PPT adoption decision that may arise due to overlooking the impact of competition, and enables us to study the impact of government enforcement regulation under uncertainty.

For the industry segment with $v_0 \geq \frac{\beta}{\beta-1}K$, the immediate PPT adoption rule suggested by the static valuation will not be myopic even under the project value uncertainty unless the uncertainty is above an identified threshold level. If one can associate univariate distributions and dependence structures with various industries, then an industry specific table, which identifies the tolerable level of project value uncertainty, may be produced for assisting entrepreneurial decision-making. Depending on the competition structure of the existing market segment, competition may delay or accelerate PPT adoption compared with the monopoly case. A Stackelberg leader-follower type competition may indeed impede PPT adoption compared with the monopoly case. On the other hand, if the entrepreneur expects to encounter random arrival of competitors' entries, the entrepreneur may be inclined to adopt the PPT earlier than he would as a monopolist or the Stackelberg leader. The promise of random arrival of government mandatory adoption will have the potential to accelerate PPT adoption. Such a government promise may indeed induce the entrepreneur to adopt the PPT with the project value smaller than the adoption cost. The implication for policymakers is that it may be better to implement legislature enforcement regulation than to legislate an investment tax credit. The latter is known to have a detrimental effect on the adoption decision since entrepreneurs would require a higher triggering project value for PPT adoption if the enactment of an investment tax credit is expected in the future.

References

1. Acquisiti, A., Grossklags, J.: Losses, Gains, and Hyperbolic Discounting: An Experimental Approach to Information Security Attitudes and Behavior. In: Proc. 2nd Intl. Workshop Economics and Info. Security (2003)
2. Chellappa, R.K., Shivendu, S.: Managing Piracy: Pricing and Sampling Strategies for Digital Experience Goods in Vertically Segmented Markets. Information Systems Research 16(4), 400–417 (2005)
3. Chellappa, R.K., Sin, R.: Personalization Versus Privacy: An Empirical Examination of the Online Consumers' Dilemma. Information Technology and Management 6(2-3) (2005)
4. Dixit, A., Pindyck, R.S.: Investment Under Uncertainty. Princeton University Press (1994)
5. Huberman, B.A., Adar, E., Fine, L.R.: Valuating Privacy. IEEE Security & Privacy 3(5), 22–25 (2005)
6. Olson, G., Olson, J.: Human-computer Interaction: Psychological Aspects of the Human Use of Computing. Annual Review of Psychology 54, 491 (2003)
7. Hann, I.H., Hui, K.L., Lee, S.-Y.T., Png, I.P.L.: Overcoming Online Information Privacy Concerns: An Information-Processing Theory Approach. Journal of Management Information Systems 24(2), 13–42 (2007)
8. Hunker, J.: A Privacy Expectations and Security Assurance Offer System. In: NSPW 2007, North Conway, NH, USA, September 18-21 (2007)
9. Kantarcioglu, M., Bensoussan, A., Hoe, S(C.): When Do Firms Invest in Privacy-Preserving Technologies? In: Alpcan, T., Buttyán, L., Baras, J. (eds.) GameSec 2010. LNCS, vol. 6442, pp. 72–86. Springer, Heidelberg (2010)
10. Spiekermann, S., Grossklags, J., Berendt, B.: E-Privacy in Second Generation E-Commerce: Privacy Preferences versus Actual Behavior. In: Proc. ACM Conf. Electronic Commerce, pp. 38–47. ACM Press (2001)
11. Tang, Z., Hu, Y., Smith, M.D.: Protecting Online Privacy: Self-Regulation, Mandatory Standards, or Caveat Emptor. In: Proc. 4rth Annual Workshop on Economics and Information Security (2005)
12. Consumers and Health Information Technology: A National Survey. California Health Foundation (April 2010)
13. Diaspora: NYU Students Develop Privacy-Based Social Network,
http://www.huffingtonpost.com/
2010/05/11/diaspora-nyu-students-dev_n_571632.html
14. FTC Regulation of Behavioral Advertising,
http://en.wikipedia.org/wiki/FTC_Regulation_of_Behavioral_Advertising
15. FTC: Privacy Self-Regulation Not Enough, Do Not Track Needed,
http://gigaom.com/2010/12/01/ftcprivacydonottrack/

Modeling Internet Security Investments
Tackling Topological Information Uncertainty

Ranjan Pal[1] and Pan Hui[2]

[1] University of Southern California, USA
rpal@usc.edu
[2] Deutsch Telekom Laboratories, Berlin, Germany
pan.hui@telekom.de

Abstract. Modern distributed communication networks like the Internet are characterized by nodes (Internet users) interconnected with one another via communication links. In this regard, the security of individual nodes depend not only on their own efforts, but also on the efforts and underlying connectivity structure of neighboring network nodes. By the term 'effort', we imply the amount of investments made by a user in security mechanisms like antivirus softwares, firewalls, etc., to improve his security. However, often due to the large magnitude of such networks, it is not always possible for nodes to have complete effort and connectivity structure information about all their neighbor nodes. Added to this is the fact that in many applications, the Internet users are selfish and are not willing to co-operate with other users on sharing effort information.

In this paper, we adopt a non-cooperative game-theoretic approach to analyze individual user security in a communication network by accounting for both, the partial information that a network node possess about its underlying neighborhood connectivity structure and security investment of its neighbors, as well as the presence of positive externalities arising from efforts exerted by neighboring nodes. We analyze the strategic interactions between Internet users on their security investments in order to investigate the equilibrium behavior of nodes and show (i) the existence of monotonic symmetric Bayesian Nash equilibria of efforts and (ii) better connected Internet users choose lower efforts to exert but earn higher utilities than less connected peers with respect to security improvement when user utility functions exhibit strategic substitutes, i.e, are submodular. Our results extend previous work with respect to tackling topological information uncertainty, and provide useful insights to Internet users on appropriately (from improving payoffs perspective) investing in security mechanisms under realistic environments of effort and topological information uncertainty, in order to improve system security and welfare. We also discuss the implications of our results on the parameters of risk management techniques like *cyber-insurance*, and compare the user investment behavior in the incomplete information case with the case when users have increased topological information of their network.

Keywords: security, externality, uncertainty, Bayesian Nash equilibria.

J.S. Baras, J. Katz, and E. Altman (Eds.): GameSec 2011, LNCS 7037, pp. 239–257, 2011.
© Springer-Verlag Berlin Heidelberg 2011

1 Introduction

The Internet has become a fundamental and integral part of our daily lives. Billions of people are using the Internet for various types of applications that demand different levels of security. For example, commercial and government organizations run applications that require a high level of security, since security breaches would lead to large financial damage and loss of public reputation. Another example of a high security application in the Internet is maintaining user anonymity through a censorship-resistant network. On the other hand, an ordinary individual for instance generally uses a computing device for purposes that do not demand strict security requirements. However, all these applications are running on a network, that was built under assumptions, some of which are no longer valid for today's applications, e.g., that all users on the Internet can be trusted and that the computing devices connected to the Internet are static objects. Today, the Internet comprises of both good and malicious users. The malicious users perform illegal activities, are able to aspect many users in a short time period, and at the same time reduce their chances of being discovered. To overcome security related issues, Internet users invest in security mechanisms such as anti-virus solutions and firewalls.

It is commonsense information that due to Internet connectivity, the security strength of an Internet user[1] is dependent on the security strength of other users, especially neighboring users. Thus, from an individual user perspective, two important pieces of information are (i) the amount of security investments of his neighbors in the network and (ii) the knowledge of the underlying connectivity structure of his neighbors. Information on both of these drive optimal user investments. Unfortunately, due to the large magnitude of the Internet, it is not feasible or practical to have exact information about the security investments and connectivity structure of all neighboring Internet users. In addition, most Internet users are selfish in nature and would not be inclined to share investment information with other Internet users. However, users do need to invest properly in security/defense mechanisms to protect themselves as much as possible, in turn improving system security. In this paper, we address the problem of optimal security investments when an individual user is uncertain about both, the underlying network connectivity structure of his neighbors as well as their security investment amounts, and accounts for the network externalities[2] posed by the neighbors when they invest in security mechanisms. We emphasize here that the Internet has a static topology and it is not impossible for users to know the whole topology. However, the size of the Internet is so large that naive users do not care to give efforts to know the topology, and thus a virtual uncertainty arises in their mind regarding complete Internet topology information.

---

[1] An Internet user could be a single individual or an individual organization.

[2] An externality is a positive(negative) effect caused to a user not directly involved in an economic transaction, by other users involved in the transaction. For example, an Internet user investing in security mechanisms benefits all the nodes connected to him and thus creates a positive externality for his neighbors.

In the presence of positive network externalities, we consider models related to two general security scenarios as mentioned in [1]. These scenarios are (i) where the security strength of an individual user depends upon the sum security strength of itself and neighboring individual nodes in the network under operation and (ii) where the security strength of an individual user depends on the strength of the strongest node/s amongst its neighbors. An example of scenario 1 is a peer-to-peer network where an attacker might want to slow down the transfer of a given piece of information, whose transfer speed might depend on the aggregate effort of all relevant nodes concerned. An example of scenario 2 is a censorship-resistant network, where a piece of information will remain available to a public domain as long as atleast one node serving that piece of information is unharmed. Another example of scenario 2 is the flow of traffic between two backbone nodes in the Internet. Modeling each path between two backbone nodes as a node, traffic will flow securely between the backbone as long as there is atleast one node that is unharmed by an attacker, i.e., there exists atleast one path between the backbone nodes. Likewise, there are other examples of applications on the Internet that fit scenarios 1 and 2. We emphasize here that there is another practical scenario as mentioned in [1], viz., one where the security strength of an individual user depends on the strength of the weakest neighboring node. This scenario is mainly an intra-organization scenario, where once a node in an organization is compromised due to a weak password or a security policy, it is easy for an attacker to hack the whole system. However, the information of neighborhood topology structure within an organization may be known to the network users in certainty, whereas in this paper we focus on the case when users have uncertain information about the neighborhood topology structure of the network under operation.

We make the following research contributions in this paper.

1. We present a general model for analyzing individual user security investments in a non co-operative Internet environment. In this regard, we study security investment games where 1) Internet users have incomplete information about the underlying neighboring network connectivity structure as well as on neighborhood security investment amounts and 2) Internet users account for the positive externalities posed by the investments of neighboring Internet users. The novelty of our model over existing security investment models lies in the fact that Internet users in our work account for neighborhood topological information uncertainty in order to decide on their optimal security investments. We discuss the implications of optimal user investments on risk management techniques such as *cyber-insurance*. Our model is based on the work in [27](See Section 3.)

2. We formulate our user security investment problems as Bayesian games of incomplete information and show the existence of a *monotonic symmetric Nash equilibrium* of user investments in these games. The results related to equilibrium show that under incomplete neighboring network topology information, better connected users choose lower efforts to exert and earn

higher payoffs than lesser connected peers when user utility functions ex-hibit strategic substitutes, i.e., are submodular. We also show the existence of monotonic symmetric equilibria in games of increased topological infor-mation and compare user investment behaviors in such games with those in which there is uncertainty regarding complete topological information. We discuss the implications of equilibria on the 'free-riding' behavior of Internet users. (See Section 4.)

2   Related Work

There have been few works related to security investments in the Internet. In this section, we give a brief overview of related work on Internet security investments. We divide the related work into the following three subdivisions:

2.1   Joint Investments in Cyber-Insurance and Self-Protection

The authors in [2][3] have analyzed self-protection[3] investments in Internet se-curity under the presence of cyber-insurance, which is a form of a third-party risk transfer. These papers are based on the inter-dependent risk model in [9]. Under the assumption of users having complete network topology information of neighbors, the works show that (i) cyber-insurance incentivizes users to invest in self-protection, (ii) cyber-insurance entails optimal user investments both in in-surance and in self-protection, and (iii) co-operation amongst Internet users result in higher user self-protection investments when compared to the case when users do not co-operate. However, attractive though the concept may seem, cyber-insurance may not be a market reality due to factors such as inter-dependent security, correlated risks, and information asymmetry between the insurer and the insured [4][5]. In addition, it is also infeasible for Internet users to have complete network topology information of their neighbors.

2.2   Investments In Self-Protection and/or Self-Insurance

For non cyber-insurance environments, in a recent series of works [7][6], the au-thors show that Internet users invest sub-optimally in self-protection measures under selfish environments when compared to the case when user co-operation is allowed. They account for positive network externalities posed by the secu-rity investments of Internet users but base their results by assuming users hav-ing complete network topology information of neighbors. However, as we have discussed previously, in a large network such as the Internet, having complete network topology information is infeasible. In addition, all the mentioned re-lated works do not model the well-known security games mentioned in [1], that are in general played between attackers and defenders (non malicious Internet users) when externalities are present in a network. In this regard, the works in

---

[3] Protection using anti-virus and antispam softwares, firewalls, etc.

[11][12][8][13] tackle the problem of optimal security investments (self-protection and self-insurance) and model the cited security games mentioned in [1], but do not account for any uncertainty of information that a user has regarding the underlying neighboring network topology, or regarding the security investments of other users. In a different type of investment work, the authors in [14] derive optimal liability schemes for increasing software security, where liability schemes are different types of investments by a vendor of a security software to prevent zero-day attacks. However, their work has no relation with the topological elements of a network, i.e., they do not model the network topology in evaluating the probability of zero-day attacks.

2.3   Tackling Information Uncertainty

The authors in [15][16][17][18][19] address certain challenges posed by information uncertainty related to security threats, response mechanisms, and associated expected losses and costs. As a set of contributions, the latter set of papers (i) derive bounds for the ratio of Internet user utilities with and without perfect information on risk parameters, (ii) model uncertainty in risk parameters like user security investments (self-protection and self-insurance), probability of attack, probability of risk propagation, as probability distributions, and (iii) propose Bayesian games of incomplete information to address the strategic interaction amongst Internet users under uncertain environments of risk information and analyze Nash equilibria in the games with their practical applications. However, the works do not consider network topology to be a parameter when users make a decision on their security investments.

In this paper, we advance previous research in security investments by jointly modeling (i) externalities due to user security investments (only self-protection), (ii) the fact that users have uncertain information regarding the connectivity structure of their neighboring nodes, and (iii) user uncertainty about security investments of their neighbors, to arrive at optimal user security investments. Thus, the novelty of our paper over existing security investment models lies in the fact that Internet users in our work account for neighborhood topological information uncertainty in order to decide on their optimal security investments.

3   Modeling Network Security Investment Games

In this section, we propose a general model for analyzing user network security investments using a game-theoretic approach when topological information needs to be accounted for. First, we model the user interaction network in the Internet. Second, we describe the utility/payoff function of the Internet users as a function of their strategies/actions, which are nothing but the security investments of users. Finally, we explain the information structure of Internet users with respect to the underlying connectivity structure of their neighbors and their security investments, and highlight the games of investments that result from the information structure.

3.1   Network Structure

We consider a set $N = \{1, ..., n\}$ of n Internet users, where the connections between them form a graph $G = (V, E)$, where $v_{ij} = 1$ (edge weight between nodes (users) i and j) if the utility of user i is affected by the security investment of user j, i being not equal to j, and 0 otherwise. Let $N_i(v) = \{j | v_{ij} = 1\}$ denote the set of all the one hop neighbors of i, where $v \epsilon \{0, 1\}^{n \times n}$ is a matrix of connections amongst nodes. We also consider the k-hop neighbors of node i and denote the set by $N_i^k(v)$. This set consists of all the nodes that are within k-hops of node i, where $k \geq 1$. Inductively, we have the following relationships between $N_i^k(v)$ and $N_i(v)$:

$$N_i^1(v) = N_i(v). \tag{1}$$

$$N_i^k(v) = N_i^{k-1}(v) \cup (\cup_{j \epsilon N_i^{k-1}(v)} N_j(v)). \tag{2}$$

We represent the degree of a node i by d_i, where d_i equals $|N_i(v)|$. In this paper, we assume that each user has perfect knowledge about his own degree but does not have complete information about the degrees of his neighbors. (More on degree information structure in Section 3.3.)

3.2   User Strategies and Payoffs

In this paper we consider two types of non co-operative security investment games concerning the case when Internet users have incomplete information on the topology of their neighbors and their security investments: (1) *sum of efforts game* - the users are selfish and invest to maximize their own utilities, with the security strength of an individual user depending on the sum of security investments of himself and his neighboring individual nodes and 2) *best-shot game* - the users are selfish and invest to maximize their own utilities, with the security strength of an individual user network depending on the security investments of the most robust node/s amongst his neighbors. In both these types of games, each Internet user is a player and his strategy is the amount of security investment he makes. We assume here that the strategy/action of each user i is x_i and lies in the *compact*[4] set $[0, 1]$. We model the utility/payoff to each user i as U_i, which is a function of the security investments made by himself and his one hop neighbors. Thus, $U_i = U_i(x_i, \overrightarrow{x}_{N_i(v)})$, where $\overrightarrow{x}_{N_i(v)}$ is the vector of security investments of the one hop neighbors of user i. From the structure of user utility functions, we observe that two players having the same degree will have the same utility function. We also model the concept of a positive externality as it forms an integral part of game analyses. A positive externality to a user from its one hop neighbors results when the latter invest in security, thereby improving the individual security strength of the user. We represent the concept mathematically in the following manner: we say that a payoff function exhibits positive externalities if for each U_i and for all $\overrightarrow{x} \geq \overrightarrow{x}'$, $U_i(x_i, \overrightarrow{x}) \geq U_i(x_i, \overrightarrow{x}')$,

---

[4] In mathematical analysis, a compact set is one that is closed and bounded.

where $\overrightarrow{x}$ and $\overrightarrow{x}'$ are the vectors of security investments of one hop neighbors of user i.

In scenarios where the security strength of a user i depends on the sum of investments of himself and other neighboring users, i.e., as in a *sum-of-efforts* game, we mathematically formulate i's utility/payoff function as follows:

$$U_i(x_1, ..., x_{d_i}) = f\left(x_i + \lambda \sum_{j=1}^{d_i} x_j\right) - c(x_i), \qquad (3)$$

where $f(\cdot)$ is a non-decreasing function of $\overrightarrow{x}$, $c(x_i)$ is the cost incurred by user i for putting in effort x_i to make his system more robust, and λ is a real scalar quantity which determines the magnitude of the positive externality experienced by user i due to the security investments made by his one-hop neighbors.

The situation where the security strength of a user depends on the investments made by the strongest neighbor/s, i.e., as in a *best-shot game*, can be modeled as a *special case* of the situation where user security strength depends on the sum of the security investments of his neighbors. We first note that from user i's perspective, the strongest-neighbor situation implies that as long as there is a neighboring node/s that is secure, user i is safe. In Section 1 we have already cited censorship resistant networks and Internet backbone networks to be examples of networks where the former situation might arise leading to a best-shot game. We had also given an example of how the best-shot scenarios arising in these networks can be modeled as a graph to reflect the 'user-neighbor' concept. Once we have modeled a best-shot scenario as a graph, we fix the strategy space of individual users to $\{0,1\}$ and make $f(0) = 0$ and $f(y) = 1$ for all $y \geq 1$. A binary strategy space of $\{0,1\}$ implies that each user decides either to invest or not to invest. If a user or any of his neighbors invest, the former is safe, else he is not. We observe that the 'sum of investments' game gets converted to a best-shot game. In this case user i's payoff follows the following equation:

$$U_i(x_i, (\overrightarrow{x}, 0)) = U_i(x_i, \overrightarrow{x}), \ \forall (x_i, \overrightarrow{x}) \ \epsilon \ [0, 1]^{d_i+1}. \qquad (4)$$

Equation (4) implies that adding a link to a neighbor who invests zero amount in security mechanisms is equivalent to not having the neighbor. This fact captures the intuition of a best-shot game.

In this paper we assume the utility functions of players in both the game types to be of the *strategic substitute* type exhibiting *positive externalities*. We say that a utility/payoff function exhibits strategic substitutes or is *submodular* if it exhibits the property of decreasing differences, i.e., $U_i(x_i, \overrightarrow{x}) - U_i(x_i', \overrightarrow{x}) \leq U_i(x_i, \overrightarrow{x}') - U_i(x_i', \overrightarrow{x}')$. The practical interpretation of a strategic substitute as applicable to this paper is that an increase in the security investments of a user's neighbors reduces the marginal utility of the user, thus de-incentivizing him from investing. This happens due to the positive externality a neighbor exerts on the user through his own investments.

3.3 Information Structure

In this paper we assume that each Internet user (player) knows his own degree[5] but does not have perfect information regarding the degree of his neighbors. It has already been shown by Newman in [20] that nodes (Internet users) in an Internet like network exhibit degree correlations[6]. In this regard, we account for the degree correlations between the neighboring nodes of a user i in our model, i.e., when a user decides on his strategy, he accounts for the amount of information he has on the degree of his neighbors. Information on degree correlations is important as it guides a user to making better security investments when compared to the situation when he has no information about the correlations. For example, a user having the knowledge that his neighbors are connected to a high number of nodes would invest differently than he would if he knows that his neighbors are connected to few nodes.

Let the degrees of the neighbors of user i be the vector $\vec{d}_{N_i(v)}$, whose dimension is d_i. We assume that user i does not know the vector $\vec{d}_{N_i(v)}$ but has information regarding its probability distribution, i.e., he knows the value of $P(\vec{d}_{N_i(v)}|d_i)$. We assume that each player in the network under consideration begins with *ex-ante symmetrical beliefs* and *common priors* regarding the degree of his neighbors. The players may end up with different positions in a network and conditional beliefs, but these beliefs are only updated based on their realized position (their own degree) and not on their identities. Thus, arises a family of conditional distributions, $\mathbf{C} \equiv \{[P(\vec{d}|d)]_{\vec{d} \in \mathbb{N}^d}\}_{d \in \mathbb{N}}$, where $\vec{d}$ is a vector of degrees of the neighbors of a node and d is the degree of a given node.

We model the strategic interactions between the players of the network as a *Bayesian game of incomplete information*. The type space of the Bayesian game is the user knowledge on the potential degrees of his neighboring players. The strategy for each player is his security investment conditioned on the knowledge of the degrees of his neighbors, and the payoff function for each player is as defined in Section 3.2, which depends on the game being a sum of investments game or a best-shot game. Assuming that S is the set of possible investments a user could make, the strategy for player i is a mapping $\gamma_i : \{0, 1, ..., n-1\} \rightarrow \Omega(S)$, where $\Omega(S)$ is the set of distribution functions on S.

We already noted that for a player, his conditional distributions concerning the neighbors' degrees can vary with his own degree. According to our model, players may have different number of neighbors, and the degrees of the neighbors are correlated with each other due to the well-known result in [20]. Thus,

---

[5] We restrict ourselves to having perfect knowledge *only* about a node's own degree because (i) no user has zero knowledge about the Internet topology, which is static, and thus we decide to model partial knowledge of a user, and (ii) for simplicity of analysis we just assume one level of complete information with regard to the neighbors of a node.

[6] Newman show through empirical studies that technological and Internet networks exhibit negative degree correlation whereas social networks exhibit positive degree correlation.

the dimension of the vector of degrees of its neighbors may vary from player to player. In order to address correlation amongst vectors of different dimensions, we adopt the technique of 'affiliation' from the domain of statistics [21]. Affiliation is used to track the correlation patterns of groups of random variables, given the complicated interdependencies that might be present between them. A positive affiliation indicates that higher levels of one variable (in this case a player's degree) implies higher levels of all other variables (in this case a player's neighbors' degrees). On the other hand, a negative affiliation indicates that higher levels of one variable implies lower levels of other variables. Next, we mathematically describe affiliation as appropriate to our work.

Mathematical Description of Affiliation: Given a player i with degree d_i, enumerate the degrees of i's neighbors as $\overrightarrow{d}_{N_i(v)} = (d_1, ..., d_{d_i})$. Now consider a function $F : \{0, 1, ..., n-1\}^m \to \mathbb{R}$, where $m \leq d_i$. Let the following relation hold:

$$E_{P(\cdot|d_i)}[F] = \sum_{\overrightarrow{d}_{N_i(v)}} P(\overrightarrow{d}_{N_i(v)}|d_i)F(d_1, ..., d_m). \tag{5}$$

In Equation (5) we fix a subset $m \leq d_i$ of user i's neighbors, and then take the expectation of F operating on their degrees. We say that the family of distributions $\mathbf{C}$ exhibits positive affiliation if, for all $k' > k$, and any non-decreasing $F : \{0, 1, ...,n-1\}^k \to \mathbb{R}$, we have

$$E_{P(\cdot|k')}[F] \geq E_{P(\cdot|k)}[F], \tag{6}$$

and $\mathbf{C}$ exhibits negative affiliation if

$$E_{P(\cdot|k')}[F] < E_{P(\cdot|k)}[F], \tag{7}$$

for all $k' > k$, and any non-decreasing $F : \{0, 1, ...,n-1\}^k \to \mathbb{R}$. The concept of affiliation simply implies that higher degrees for a given player are correlated with higher of lower degree (depending on whether the affiliation is positive or negative) of all her neighbors.

Practical Implications of Optimal Security Investments: As mentioned in [2], cyber-insurance incentivizes Internet users to invest in self-defense investments. However, self-defense investments have a direct impact on insurance premiums as high investments would result in lesser premiums for a user and low investments would lead to a user paying higher premiums. We will discuss more on the relation between premium amounts and user welfare in Section 4.

4   Game Analysis

In this section, we analyze the *symmetric* Bayesian game of incomplete information played between the users of the network under operation. In any symmetric game, the player payoffs for playing a particular strategy depend only on the strategies of other players and not on who is playing the strategies. In our game, symmetric equilibrium implies that players with the same network characteristic, i.e., network degree, choose the same strategy in a Bayesian Nash equilibrium.

The primary reasons why we consider only symmetric equilibria are (i) the network formation mechanism is anonymous and the population (ex., as in the Internet) is very large, and (ii) the payoff function is strictly concave in its own actions. Under these two conditions, all users of any given degree face the same decision problem and due to the nature of their utility functions choose an unique optimal strategy. We investigate the *existence*, *uniqueness*, and *monotonicity* of our game equilibria. In studying monotonicity of equilibria, we investigate the changes in the best response investment magnitude of a user when other users in the network increase/decrease their best response investment amounts. We also investigate the effect of the increase/decrease in user degrees on the equilibria of the game. We initially give a mathematical definition of our Bayesian game and follow it up with the analysis and practical implications of our game equilibria.

4.1 Game Definition

Consider a player (Internet user) i having degree d_i in a *sum-of-efforts game* or a *best-shot game*. Each player chooses a security investment amount from the set S as its strategy, where S is as defined in Section 3.3. Let $d\rho_{-i}(\overrightarrow{\gamma}, d_i)$ be the probability density over $x_{N_i(v)} \in S^{d_i}$ induced by the beliefs $P(\cdot|d_i)$ held by player i over the degrees of his neighbors, combined with the strategies played via $\overrightarrow{\gamma}$, the vector of strategies of other users in the network. Let

$$EU_i(x_i, \overrightarrow{\gamma}, d_i) = \int_{x_{N_i(v)} \in S^{d_i}} U_i(x_i, x_{N_i(v)}) d\rho_{-i}(\overrightarrow{\gamma}, d_i), \qquad (8)$$

where $EU_i(x_i, \overrightarrow{\gamma}, d_i)$ is the expected utility/payoff of player i with degree d_i and investment x_i when other players choose strategy $\overrightarrow{\gamma}$. The *Bayesian Nash equilibrium* of the game is a strategy vector that *maximizes* the expected utility of each player in the network [22][23]. We note here that the above formulation of a Bayesian game is valid only for continuous payoff functions, which can arise for non-discrete strategy sets. The case for discrete sets has been analyzed by [27]. What is important from this paper's point of view is to relate network structure and user utilities to the Nash equilibria results, which in turn requires us to relate user strategies (security investments) to their degrees. In this regard, we next provide some basic definitions related to our problem model. which would be used in the analysis of game equilibria.

Definition 1. A strategy $\overrightarrow{\gamma}$ is monotonically increasing in player degrees if $\overrightarrow{\gamma}(d')$ first-order stochastically dominates[7] $\overrightarrow{\gamma}(d)$ for each $d' > d$. Similarly,

---

[7] Let X and Y be two random variables representing risks. Then X is said to be smaller than Y in first order stochastic dominance, denoted as $X \leq_{ST} Y$ if the inequality $VaR[X; p] \leq VaR[Y; p]$ is satisfied for all $p \in [0, 1]$, where $VaR[X; p]$ is the value at risk and is equal to $F_X^{-1}(p)$. First order stochastic dominance implies dominance of higher orders. We adopt the stochastic dominant approach to comparing risks because a simple comparison between various moments of two distributions may not be enough for a correct prediction about the dominance of one distribution over another.

a strategy $\overrightarrow{\gamma}$ is monotonically decreasing in player degrees if the domination relationship is reversed, for each $d' > d$.

Definition 2. For a given player i, we say that his expected utility function exhibits degree substitutability if

$$EU_i(x_i, \overrightarrow{\gamma}, d_i) - EU_i(x'_i, \overrightarrow{\gamma}, d_i) \leq EU_i(x_i, \overrightarrow{\gamma}, d'_i) - EU_i(x'_i, \overrightarrow{\gamma}, d'_i), \quad (9)$$

where $x_i > x'_i$, $d_i > d'_i$, and $\overrightarrow{\gamma}$ is non-increasing. Similarly for a given player i, we say that his expected utility function exhibits degree complementarity if

$$EU_i(x_i, \overrightarrow{\gamma}, d_i) - EU_i(x'_i, \overrightarrow{\gamma}, d_i) \geq EU_i(x_i, \overrightarrow{\gamma}, d'_i) - EU_i(x'_i, \overrightarrow{\gamma}, d'_i), \quad (10)$$

where $x_i > x'_i$, $d_i > d'_i$, and $\overrightarrow{\gamma}$ is non-decreasing.

We observe that the concepts of degree substitutability and complementarity are in relation to the marginal expected utilities of a player with increase in his degree. Degree substitutability states that if a high strategy (security investment) is less attractive than a low strategy, for a player of some degree, then the same is true for a player of a higher degree, when the strategy being played by other players is non-increasing. Similarly, degree complementarity states that if a high strategy is more attractive than a low strategy, for a player of some degree, then the same is true for a player of a higher degree, when the strategy being played by other players is non-decreasing. In a recent work, [24] have shown as sufficient conditions that when Equation 4 holds, the user utility functions exhibit strategic substitutes, and the neighbor affiliation of **C** is negative, degree substitution arises. However, the authors did not state these conditions as necessary to ensure degree substitutability. In our work, we only assume the sufficient conditions while considering degree substitutability because the payoff functions for the players in the sum-of-efforts and best-shot games exhibit the strategic substitute property. *We emphasize here that it is yet to be proved through theory or experiments that the the topology of the Internet at the user level exhibits degree substitutes. We assume in this paper that there exists a negative degree of neighbor affiliation (like in the case of degree correlations at the router level [20]) for the Internet at the user level. The analysis case for positive affiliation is an important open problem and is left for future work.*

4.2   Game Equilibria Results

In this section we state the results related to equilibria of our proposed Bayesian game of security investments, and analyze various practical implications of our results. As mentioned earlier, given a symmetric environment; i.e., players participate in a symmetric Bayesian game of security investments, we analyze *symmetric equilibria*. Apart from the reasons previously mentioned on why we address only symmetric game equilibria, asymmetric behavior seems relatively unintuitive, and difficult to explain in a one-shot interaction [25].

Lemma 1. *There exists a symmetric equilibrium in our proposed security investment game when user utility functions exhibit strategic susbstitutes, and the equilibrium is non-increasing, i.e., monotone decreasing.*

Proof. In our game the players (Internet users) have identical strategy set S. The utility functions of each player is the same, and each player's beliefs about the degrees of its neighbors are ex-ante symmetric. Given that action set is compact and the utility/payoff function of users are continuous, there exists a mixed strategy Nash equilibrium of the Bayesian game [22][23]. Regarding monotonicity of equilibria, we use the degree substitute property to show that a player would play a monotone best-reply if the rest of the players play monotone strategies. Thus, the monotone strategies form a compact and convex set, and by the results in [28] there exists a monotonic equilibrium.   **Q.E.D.**

Implications of Lemma 1: The degree substitutes property ensures that there is a game equilibrium that is monotonically decreasing. From a user point of view this implies that his investments monotonically decrease with increase in his own degree, which further implies low user investments on being well connected, leading to a free-riding problem. Assuming the existence of cyber-insurance markets, this problem can be tackled to incentivize well-connected users to invest optimally [2]. Under mandatory cyber-insurance, well-connected users would either pay high premiums or would invest more to avoid high premiums. In the case when there are multiple symmetric Nash equilibria (this case does not arise in best-shot games, It has been shown in [27] that best-shot Bayesian games have a *unique* pure strategy symmetric Nash equilibria which is monotone decreasing) that are *non-monotone*, it may prove good for overall network security because well connected users might put in more investment efforts even if it has high degree, in turn paying less insurance premiums. On the other hand, we cannot be sure if low degree users would exert high investment efforts for non-monotone equilibria.

Lemma 2. *Given that (1) $U_i(x_i, (\overrightarrow{x}, 0)) = U_i(x_i, \overrightarrow{x})$, $\forall (x_i, \overrightarrow{x}) \in S^{d_i+1}$, for each player i and (2) degrees of neighboring nodes of users are independent, then strategic substitutes of user utility functions result in every symmetric equilibrium of our proposed Bayesian game of security investments being monotone decreasing.*

Proof. Let $\overrightarrow{\gamma}^*$ be the strategy played in equilibrium. Consider any $d \in \{0, 1, ..., n\}$ and let $x_d = inf[\mathbf{supp}(\gamma_d^*)]$. If $x_d = 1$, then$x_{d'} \leq x_d$ for all $x_{d'} \in \mathbf{supp}(\gamma_{d'}^*)$ for $d' > d$. Now let us assume $x_d < 1$. Then for any $x > x_d$, Equation 4 holding, and user utility functions exhibiting strategic substitutes, we have for player i

$$A \leq B. \tag{11}$$

Here

$$A = U_i(x, x_{dn_1}, ..., x_{dn_d}, x_s) - U_i(x_d, x_{dn_1}, ..., x_{dn_d}, x_s)$$

and

$$B = U_i(x, x_{dn_1,}, ..., x_{dn_d}) - U_i(x_d, x_{dn_1,}, ..., x_{dn_d}),$$

where $x_s \geq 0$. Given the assumption of stochastically independent neighbor degree distributions, we have

$$EU_i(x, \overrightarrow{\gamma}^*, d+1) - EU_i(x_d, \overrightarrow{\gamma}^*, d+1) < EU_i(x, \overrightarrow{\gamma}^*, d) - EU_i(x_d, \overrightarrow{\gamma}^*,). \quad (12)$$

Now we also know that for all x

$$EU_i(x, \overrightarrow{\gamma}^*, d) - EU_i(x_d, \overrightarrow{\gamma}^*, d) \leq 0. \quad (13)$$

Thus, we have for all $x > x_d$

$$EU_i(x, \overrightarrow{\gamma}^*, d+1) - EU_i(x, \overrightarrow{\gamma}^*, d+1) < 0, \quad (14)$$

which implies γ_d^* first order stochastically dominates γ_{d+1}^*. Iterating our argument, we arrive at the conclusion that γ_d^* first order stochastically dominates $\gamma_{d'}^*$ whenever $d' > d$. **Q.E.D.**

Implications of Lemma 2: Lemma 2 states the conditions under which all symmetric equilibria are monotone, and gives an insight on the topology of the network that could result in all symmetric equilibria being monotone. Lemma 1 only guarantees the existence of a single monotone equilibria when the network topology exhibits degree substitutes. Lemma 2 states that under independence of neighbor degree nodes (ex., as in a *Erdos-Renyi* random graph) every symmetric equilibria is monotone decreasing. However, topologies such as the Erdos-Renyi graph do not represent the Internet. Assuming every equilibrium would be monotone decreasing with respect to the Internet topology, it would enable cyber-insurance markets to flourish (provided that markets exist and cyber-insurance is made mandatory for Internet users). Thus, for user-level Internet topologies and for multiple non-monotone symmetric equilibria, the overall network security strength explanation follows as per the explanation in Lemma 1.

Lemma 3. *Suppose $U_i(x_i, (\overrightarrow{x}, 0)) = U_i(x_i, \overrightarrow{x})$, $\forall (x_i, \overrightarrow{x}) \in S^{d_i+1}$, for each player i. If **C** is negatively affiliated and user utility functions exhibit strategic substitutes, then in every monotonically decreasing symmetric equilibrium of security investment of our proposed Bayesian game, the expected utilities of players are non-decreasing in degree.*

Proof. Let $\overrightarrow{\gamma}^*$ be an equilibrium strategy. Suppose that $x_d \in \textbf{supp}(\gamma_k^*)$ and $x_{d+1} \in \textbf{supp}(\gamma_{d+1}^*)$. Equation 4 implies that

$$U_i(x_d, x_{dn_1}, ..., x_{dn_d}, 0) = U_i(x_d, x_{dn_1}, ..., x_{dn_d}), \quad (15)$$

for all $x_{dn_1}, ..., x_{dn_d}$. Now since the user utilities exhibit positive externalities, it is true for all $x > 0$ that

$$U_i(x_d, x_{dn_1}, ..., x_{dn_d}, x) = U_i(x_d, x_{dn_1}, ..., x_{dn_d}). \quad (16)$$

Now for negative neighbor affiliation, we have

$$EU_i(x_d, \overrightarrow{\gamma}^*, d+1) \leq EU_i(x_d, \overrightarrow{\gamma}^*, d). \tag{17}$$

Since, γ_{d+1}^* is a best response in the network game, and that $x_{d+1} \in supp(\gamma_{d+1}^*)$, we have

$$EU_i(x_{d+1}, \overrightarrow{\gamma}^*, d+1) \leq EU_i(x_d, \overrightarrow{\gamma}^*, d+1). \tag{18}$$

Thus, our result is proved. **Q.E.D.**

Implications of Lemma 3. Lemma 3 provides the relation between network degrees of users and their equilibrium payoffs, and identifies the conditions under which payoffs increase/decrease with network degree. Assuming that the Internet at the user level has negative neighbor degree affiliation, the lemma states that players with more neighbors exert lesser investment efforts and earn higher payoffs as compared to their less connected peers. In general, the lemma provides intuitions about user investments in games exhibiting strategic substitutes. Given that there exist markets for cyber-insurance and that insurance is made compulsory for Internet users, the overall network security strength explanation follows as per the explanation in Lemma 1.

4.3   The Case of Increased Topological Information

In this section, we investigate player investment behavior when he has more information regarding the network topology than just knowing his own degree and the conditional distributions of the degrees of his neighbors. Our goal is to compare user behavior regarding security investments between the 'less information' and 'more information' cases. We consider the case where players apart from knowing his own degree also knows the degrees of his neighbors. In the case when a player has *complete information* about the network topology, it has been shown in [26] that multiple pure strategy Nash equilibria may result (not necessarily monotone).

   For the ease of exposition, we consider the simple comparison setting where the degrees of neighbors of a user are stochastically independent. This assumption also implies the independence of the degrees of neighbors of neighbors. Recall from Lemma 2 that under degree independence and the strategic substitute property of user utility functions, *all* symmetric Nash equilibria of the Bayesian game are monotonic decreasing. However, an interesting trend to study is whether all equilibria are monotone when the ''level of topological information' increases. Note that in the case of increased topology information, the type space of each player in the Bayesian game is of the form $(d_i; dn_{i1}, ..., dn_{id_i})$, where d_i is the degree of player i and $\{dn_i\}$'s are the degrees of i's neighbors. We have the following lemma regarding user behavior in the increased topological information scenario, i.e., the scenario where a user in addition to his own degree also knows the degree of his neighbors.

Lemma 4. *Suppose* $U_i(x_i, (\overrightarrow{x}, 0)) = U_i(x_i, \overrightarrow{x})$, $\forall (x_i, \overrightarrow{x}) \in S^{d_i+1}$, *for each player i. When user utility functions exhibit strategic substitutes and neighbor degrees are stochastically independent, our proposed Bayesian game of security investments has at least one symmetric equilibrium that is monotone decreasing.*

Proof. The proof of this lemma follows from the same logic as that in Lemma 1, i,e., the best-response of a player to a monotone decreasing strategy by all other players is monotone decreasing, given that the set of monotone strategies is convex and compact. The latter condition guarantees the existence of equilibrium. The proof details follow a similar method as proposed in Proposition 10 of [24].   **Q.E.D**

Implications of Lemma 4. The lemma states does not guarantee the existence of every symmetric equilibrium being monotone decreasing, when compared to Lemma 2. Thus, with increasing information, the flourishing of cyber-insurance markets and increments in overall network security might follow the same trends as in the case when users had less information.

A Note on Multiplicity of Nash Equilibria. We observe that our games might have multiple symmetric Nash equilibria, and that the chances of having multiple equilibria increases with the increase in the amount of topological information [27]. There are two important practical implications of this behavior: (i) it is difficult for a player to choose the *best* equilibrium as computing a single Nash equilibria is PPAD-complete [29], and (ii) there might be multiple cyber-insurance contracts for the multiple equilibria, and due to the intractability of computing any Nash equilibria, let alone the best equilibria, clients might go for a contract that either 'over-prices' or 'under-prices' them with regard to insurance premiums, thus leading to chances of market failure. Thus we observe a flip side to having more information on the network topology. However, in most practical cases (approximately 95% of the time) Nash equilibria is reached in polynomial time. Added to this is the fact that having more information on a large network is infeasible and therefore more chances that users will be involved in a game having a single or less number of Nash equilibria.

5   Cyber-Insurance – A Brief Note

In this section we give a brief overview of the need for cyber-insurance in Internet security since we draw practical implications of our model results with respect to this risk management technique.

The Internet has become a fundamental and an integral part of our daily lives. Billions of people nowadays are using the Internet for various types of applications. However, all these applications are running on a network, that was built under assumptions, some of which are no longer valid for today's applications, e,g., that all users on the Internet can be trusted and that there are no malicious elements propagating in the Internet. On the contrary, the infrastructure, the users, and the services offered on the Internet today are all subject to a wide variety of risks. These risks include denial of service attacks, intrusions of various kinds, hacking, phishing, worms, viruses, spams, etc. In order to counter the threats posed by the risks, Internet users[8] have traditionally resorted to antivirus and anti-spam softwares, firewalls, and other add-ons to reduce the likelihood of being affected by threats. In practice, a large industry (companies like *Symantec, McAfee,* etc.) as well as considerable research efforts are centered around developing and deploying tools and techniques to detect threats and anomalies in order to protect the Internet infrastructure and its users from the resulting negative impact.

In the past one and half decade, protection techniques from a variety of computer science fields such as cryptography, hardware engineering, and software engineering have continually made improvements. Inspite of such improvements, recent articles by Schneier [30] and Anderson [31][32] have stated that it is impossible to achieve a 100% Internet security protection. The authors attribute this impossibility primarily to four reasons:

- New viruses, worms, spams, and botnets evolve periodically at a rapid pace and as a result it is extremely difficult and expensive to design a security solution that is a panacea for all risks.
- The Internet is a distributed system, where the system users have divergent security interests and incentives, leading to the problem of 'misaligned incentives' amongst users. For example, a rational Internet user might well spend $20 to stop a virus trashing its hard disk, but would hardly have any incentive to invest sufficient amounts in security solutions to prevent a service-denial attack on a wealthy corporation like an Amazon or a Microsoft [33]. Thus, the problem of misaligned incentives can be resolved only if liabilities are assigned to parties (users) that can best manage risk.
- The risks faced by Internet users are often correlated and interdependent. A user taking protective action in an Internet like distributed system creates positive externalities [9] for other networked users that in turn may discourage them from making appropriate security investments, leading to the 'free-riding' problem [8][7][34][6].
- Network externalities affect the adoption of technology. Katz and Shapiro [35] have determined that externalities lead to the classic S-shaped adoption curve, according to which slow early adoption gives way to rapid deployment once the number of users reaches a critical mass. The initial deployment is subject to user benefits exceeding adoption costs, which occurs only if a minimum number of users adopt a technology; so everyone might wait for others to go first, and the technology never gets deployed. For example DNSSEC, and S-BGP are secure protocols that have been developed to better DNS and BGP in terms of security performance. However, the challenge is getting them deployed by providing sufficient internal benefits to adopting entities.

---

[8] The term 'users' may refer to both, individuals and organizations.

In view of the above mentioned inevitable barriers to 100% risk mitigation, the need arises for alternative methods of risk management in the Internet. Anderson and Moore [32] state that microeconomics, game theory, and psychology will play as vital a role in effective risk management in the modern and future Internet, as did the mathematics of cryptography a quarter century ago. In this regard, *cyber-insurance* is a psycho-economic-driven risk-management technique, where risks are transferred to a third party, i.e., an insurance company, in return for a fee, i.e., the *insurance premium*. The concept of cyber-insurance is growing in importance amongst security engineers. The reason for this is three fold: (i) ideally, cyber-insurance increases Internet safety because the insured increases self-defense as a rational response to the reduction in insurance premium [36][37][38][39], a fact that has also been mathematically proven by the authors in [40][2], (ii) in the IT industry, the mindset of 'absolute protection' is slowly changing with the realization that absolute security is impossible and too expensive to even approach while adequate security is good enough to enable normal functions - the rest of the risk that cannot be mitigated can be transferred to a third party [41], and (iii) cyber-insurance will lead to a market solution that will be aligned with economic incentives of cyber-insurers and users (individuals/organizations) - the cyber-insurers will earn profit from appropriately pricing premiums, whereas users will seek to hedge potential losses. In practice, users generally employ a simultaneous combination of retaining, mitigating, and insuring risks [42].

6   Conclusion

In this paper we proposed a security investment model for the Internet in which Internet users account for the positive externality posed to them by other Internet users and make security investments under situations when they do not have complete information about the underlying connecting topology of his neighbors and their security investments. Our model is based on a game-theoretic approach and we showed (i) the existence of symmetric monotone Bayesian Nash equilibria of efforts and (ii) better connected nodes choose lower efforts to exert but earn higher utilities with respect to security improvement when user utility functions exhibit strategic substitutes. Our results provided ways for Internet users to appropriately invest in security mechanisms under realistic environments of information uncertainty. Our results also clarified how the basic strategic features of the game - as manifest in the substitutes property - combine with different patterns of degree association to shape network behavior and user payoffs. We also stated the implications of our results to successfully realizing risk management schemes such as cyber-insurance, in practice. Finally, we compared between user investment behaviors in 'low information' and 'increased information' scenarios. As a part of future work, we plan to investigate security investments under an asymmetric environment, i.e., a game environment in which user payoffs depend not only on the strategy of other users but also on the identity of the users.

References

1. Varian, H.: System Reliability and Free Riding. In: ACM ICEC (2003)
2. Lelarge, M., Bolot, J.: Economic Incentives to Increase Security in the Internet: The Case for Insurance. In: IEEE INFOCOM (2009)
3. Pal, R., Golubchik, L.: Analyzing Self-Defense Investments In The Internet Under Cyberinsurance Coverage. In: IEEE ICDCS (2010)
4. Bohme, R., Schwartz, G.: Modeling Cyberinsurance: Towards A Unifying Framework. In: WEIS (2010)
5. Shetty, N., Schwarz, G., Feleghyazi, M., Walrand, J.: Competitive Cyberinsurance and Internet Security. In: WEIS (2009)
6. Omic, J., Orda, A., Mieghem, V.P.: Protecting Against Network Infections: A Game-Theoretic Perspective. In: IEEE INFOCOM (2009)
7. Jiang, L., Ananthram, V., Walrand, J.: How Bad are Selfish Investments in Network Security. IEEE Transactions On Networking (2010)
8. Grossklags, J., Christin, G., Chuang, J.: Security and Insurance Management in Networks with Heterogenous Agents. In: ACM EC (2008)
9. Kunreuther, H., Heal, G.: Interdependent Security. Journal of Risk and Uncertainty 26 (2002)
10. Varian, H.R.: Microeconomic Analysis. Norton (1992)
11. Fultz, N., Grossklags, J.: Blue versus Red: Towards a Model of Distributed Security Attacks. In: Dingledine, R., Golle, P. (eds.) FC 2009. LNCS, vol. 5628, pp. 167–183. Springer, Heidelberg (2009)
12. Grossklags, J., Christin, N., Chuang, J.: Secure or Insure? A Game-Theoretic Analysis of Information Security Games. In: WWW (2008)
13. Grossklags, J., Christin, N., Chuang, J.: Security Investments(Failures) in Five Economic Environments
14. Terrence, A., Tunca, I.T.: Who Should Be Responsible for Software Security? Management Science 57(5) (2011)
15. Grossklags, J., Johnson, B.: Uncertainty In Weakest-Link Security Game. In: GameNets (2009)
16. Grossklags, J., Johnson, B., Christin, N.: The Price of Uncertainty in Security Games. Economics of Information Security and Privacy (2010)
17. Grossklags, J., Johnson, B., Christin, N.: When Information Improves Information Security. In: Sion, R. (ed.) FC 2010. LNCS, vol. 6052, pp. 416–423. Springer, Heidelberg (2010)
18. Johnson, B., Grossklags, J., Christin, N., Chuang, J.: Are Security Experts Useful? Bayesian Nash Equilibria for Network Security Games with Limited Information. In: Gritzalis, D., Preneel, B., Theoharidou, M. (eds.) ESORICS 2010. LNCS, vol. 6345, pp. 588–606. Springer, Heidelberg (2010)
19. Johnson, B., Grossklags, J., Christin, N., Chuang, J.: Uncertainty in Interdependent Security Games. In: Alpcan, T., Buttyán, L., Baras, J.S. (eds.) GameSec 2010. LNCS, vol. 6442, pp. 234–244. Springer, Heidelberg (2010)
20. Newman, M.E.J.: Assortative Mixing in Networks. Phy. Rev. Lett. 89 (2002)
21. Esary, J.D., Proschan, F., Walkup, W.: Association of Random Variables With Applications. Annals of Mathematical Statistics 38(5) (1967)
22. Fudenberg, D., Tirole, J.: Game Theory. MIT Press (1991)
23. Osborne, M.J., Rubinstein, A.: A Course in Game Theory. MIT Press (1994)
24. Galeotti, A., Goyal, S., Jackson, M.O., Vega-Redondo, F., Yariv, L.: Network Games. Review of Economic Studies 77(1) (2010)

25. Kreps, D.: Game Theory and Economic Modelling. Oxford University Press (1990)
26. Bramoulle, K., Kranton, R.: Strategic Experimentation in Networks. Journal of Economic Theory 135(1) (2007)
27. Galeotti, A., Goyal, S., Jackson, M.O., Vega-Redondo, F., Yariv, L.: Network Games. Technical Report (2006)
28. Milgrom, P., Shannon, C.: Monotone Comparative Statics. Econometrica 62 (1994)
29. Daskalakis, C., Goldberg, P.W., Papadimitrou, C.H.: The Complexity of Computing A Nash Equilibrium. SIAM Journal of Computing 39(1) (2009)
30. Schneier, B.: Secrets and Lies: Digital Security in a Networked World. John Wiley and Sons (2001)
31. Anderson, R.: Why Information Security is Hard - An Economic Perspective. In: Annual Computer Security Applications Conference (2001)
32. Anderson, R., Moore, T.: Information Security Economics and Beyond. Information Security Summit (2008)
33. Varian, H.: Managing Online Security Risks. The New York Times (June 1, 2000)
34. Ko-Miura, A.R., Yolken, B., Bambos, N., Mitchell, J.: Security Investment Games of Interdependent Organizations. Allerton (2008)
35. Katz, M., Shapiro, C.: Network Externalities, Competition, and Compatibility. The American Economic Review 75(3) (1985)
36. Kesan, J., Majuca, R., Yurcik, W.: The Economic Case for Cyber-Insurance: In Securing Privacy in the Internet Age. Stanford University Press (2005)
37. Kesan, J., Majuca, R., Yurcik, W.: Cyberinsurance As A Market-Based Solution To The Problem of Cyber-Security: A Case Study. In: WEIS (2005)
38. Scheier, B.: Its The Economics Stupid. In: WEIS (2002)
39. Yurcik, W., Doss, D.: Cyberinsurance: A Market Solution To The Internet Security Market Failure. In: WEIS (2002)
40. Lelarge, M., Bolot, J.: Cyberinsurance As An Incentive for Internet Security. In: WEIS (2008)
41. Majuca, R.P., Yurcik, W., Kesan, J.P.: The Evolution of Cyberinsurance. Information Systems Frontier (2005)
42. Schneier, B.: Insurance and the Computer Industry. Communications of the ACM 44(3) (2001)
43. Honeyman, P., Schwarz, G.: Interdependence of Reliability and Security. In: WEIS (2007)
44. Neumann, J.V., Morgenstern, O.: Theory of Games and Economic Behavior. Princeton University Press (2009)
45. Mascollel, A., Winston, M.D., Green, J.R.: Microeconomic Theory. Oxford University Press (1985)
46. Hau, A.: When is A Coinsurance-Type Insurance Policy Inferior or Even Giffen. Journal of Risk and Insurance 75(2) (2008)
47. Lelarge, M., Bolot, J.: A Local Mean Field Analysis of Security Investments in Networks. In: ACM NetEcon (2008)
48. Lelarge, M., Bolot, J.: Network Externalities and The Deployment of Security Features and Protocols in the Internet. In: ACM SIGMETRICS (2008)
49. Internet Wikipedia Source. Information Asymmetry
50. Pal, R., Golubchik, L.: Pricing and Investments in Internet Security. Arxiv (2011)

Author Index